Springer Proceedings in Mathematics & Statistics

Volume 439

This book series features volumes composed of selected contributions from workshops and conferences in all areas of current research in mathematics and statistics, including data science, operations research and optimization. In addition to an overall evaluation of the interest, scientific quality, and timeliness of each proposal at the hands of the publisher, individual contributions are all refereed to the high quality standards of leading journals in the field. Thus, this series provides the research community with well-edited, authoritative reports on developments in the most exciting areas of mathematical and statistical research today.

Víctor Gayoso Martínez • Fatih Yilmaz •
Araceli Queiruga-Dios •
Deolinda M. L. D. Rasteiro •
Jesús Martín-Vaquero • Ion Mierluş-Mazilu
Editors

Mathematical Methods for Engineering Applications

ICMASE 2023, Madrid, Spain, July 12–14

Springer

Editors
Víctor Gayoso Martínez
Centro Universitario de Tecnología y Arte Digital (U-tad)
Madrid, Spain

Araceli Queiruga-Dios
Universidad de Salamanca
Salamanca, Spain

Jesús Martín-Vaquero
Universidad de Salamanca
Salamanca, Spain

Fatih Yilmaz
Ankara Hacı Bayram Veli University
Polatli, Ankara, Türkiye

Deolinda M. L. D. Rasteiro
Instituto Superior de Engenharia de Coimbra
Coimbra, Portugal

Ion Mierluş-Mazilu
Technical University of Civil Engineering
Bucharest, Romania

ISSN 2194-1009 ISSN 2194-1017 (electronic)
Springer Proceedings in Mathematics & Statistics
ISBN 978-3-031-49217-4 ISBN 978-3-031-49218-1 (eBook)
https://doi.org/10.1007/978-3-031-49218-1

This Springer imprint is published by the registered company Springer Nature Switzerland AG
The registered company address is: Gewerbestrasse 11, 6330 Cham, Switzerland

Paper in this product is recyclable.

Organizing Committee

Víctor Gayoso Martínez, U-tad, Spain (Conference Chair)
Fatih Yilmaz, Ankara Hacı Bayram Veli University, Turkey (Organizing Chair)
Araceli Queiruga-Dios, Universidad de Salamanca, Spain
Deolinda M. L. D. Rasteiro, Instituto Superior de Engenharia de Coimbra – ISEC, Portugal
Mustafa Özkan, Gazi University, Turkey
Ion Mierluş-Mazilu, Technical University of Civil Engineering Bucharest, Romania
Jesús Martín-Vaquero, Universidad de Salamanca, Spain
Mücahit Akbiyik, Beykent University, Turkey

Local Organizing Committee

María del Mar Angulo Martínez, U-tad, Spain
Beatriz Martínez Pabón, U-tad, Spain
Miguel Ángel Muñoz Mohedano, U-tad, Spain

Scientific Committee

Agustín Martín Muñoz, Spanish National Research Council, Spain
Abdullah Alazemi, Kuwait University, Kuwait
Ángel Martín del Rey, Universidad de Salamanca, Spain
Ayman Badawi, American University of Sharjah, UAE
Carlos Martins da Fonseca, Kuwait College of Science and Technology, Kuwait
Cesar Benavente-Peces, Technical University of Madrid, Madrid, Spain
Cristina R. M. Caridade, Instituto Superior de Engenharia de Coimbra, Portugal
Daniela Richtarikova, Slovak University of Technology in Bratislava, Slovakia

Daniela Velichova, Slovak University of Technology, Slovakia
Dolores Queiruga, Universidad de La Rioja, Spain
Dursun Taçcı, Gazi University, Turkey
Emília Bigotte, Instituto Superior de Engenharia de Coimbra, Portugal
Gheorghe Morosanu, Babes-Bolyai University, Romania
Ji-TengJia, Xidian University, China
Juan José Bullón Pérez, Universidad de Salamanca, Spain
Judit García Ferrero, Universidad de Salamanca, Spain
Lucian NITA, Technical University of Civil Engineering, Bucharest, Romania
Luis Hernández Álvarez, Spanish National Research Council, Spain
Luis Hernández Encinas, Spanish National Research Council, Spain
Marie Demlova, Czech Technical University in Prague, Czech Republic
María José Cáceres García, Universidad de Salamanca, Spain
Melek Sofyalıoğlu, Ankara Hacı Bayram Veli University, Turkey
Michael Carr, Technological University Dublin, Ireland
Milica Andjelic, Kuwait University, Kuwait
Miguel Ángel González la Torre, Spanish National Research Council, Spain
Miguel Ángel Queiruga-Dios, Universidad de Burgos, Spain
Moawwad El-Mikkawy, Mansoura University, Egypt
Nenad P. Cakic, University of Belgrade, Serbia
Samet Arpacı, Ankara Hacı Bayram Veli University, Turkey
Seda Yamaç Akbıyık, Gelişim University, Turkey
Selçuk Ozcan, Karabuk University, Turkey
Serpil Halıcı, Pamukkale University, Turkey
Snezhana Gocheva-Ilieva, University of Plovdiv Paisii Hilendarski, Bulgaria
Tomohiro Sogabe, Nagoya University, Japan
Said Broumi, Regional Centre of Education and Training, Morocco
Hari M. Srivastava, University of Victoria, Canada
Yonggui Kao, Harbin Institute of Technology, China

Preface

This Springer Nature book includes the papers that have been accepted at International Conference on Mathematics and its Applications in Science and Engineering (ICMASE 2023) which was held in Madrid, Spain, in July 2023. This conference was organized by the Centro Universitario de Tecnología y Arte Digital (U-tad) in Spain and the Ankara Hacı Bayram Veli University in Turkey.

The aim of this conference is to exchange ideas, discuss new developments in mathematics, promote collaborations and interact with professionals and researchers from all over the world in the following interesting topics: Functional Analysis, Approximation Theory, Real Analysis, Complex Analysis, Applied Analysis, Numerical Analysis, Geometry, Algebra, Modern Methods in Summability and Approximation, Operator Theory, Fixed Point Theory and Applications, Sequence Spaces and Matrix Transformation, Spectral Theory and Differential Operators, Boundary Value Problems, Ordinary and Partial Differential Equations, Discontinuous Differential Equations, Convex Analysis and Its Applications, Optimization and Its Application, Mathematics Education, Application on Variable Exponent Lebesgue Spaces, Applications on Differential Equations and Partial Differential Equations, Fourier Analysis, Wavelet and Harmonic Analysis Methods in Function Spaces, and Applications on Computer Engineering.

After the peer-review process, the scientific committee selected 25 papers, from authors from 17 different countries.

The invited speakers of the ICMASE 2023 conference were Tomohiro Sogabe, from the Nagoya University, Japan; Ángel Martín del Rey, from the Universidad de Salamanca, Spain; and Miguel Ángel González-Torre, from the Spanish National Research Council (CSIC), Spain.

Thanks to all ICMASE 2023 committee members.

Víctor Gayoso Martínez – Conference Chair
Fatih Yilmaz – Organizing Chair

Madrid, Spain
Polatli, Ankara, Türkiye

Víctor Gayoso Martínez
Fatih Yilmaz

Salamanca, Spain — Araceli Queiruga-Dios
Coimbra, Portugal — Deolinda M. L. D. Rasteiro
Salamanca, Spain — Jesús Martín-Vaquero
Bucharest, Romania — Ion Mierluş-Mazilu

Contents

Modeling of Nitrogen, Phosphorus, and Potassium Concentrations in Lakes Affected by Soil Fertilization

Déthié Dione, Teubé Cyrille Mbainaissem, Bakary Koné, and Paul Python Ndekou

Abstract The excessive use of fertilizer on agricultural land has led to a significant build-up of chemical nutrient stocks in the soil, which increases the risk of diffuse pollution of surface waters. The reservoirs to which these waters drain are also polluted by the chemical components of the fertilizer. This is a major environmental problem. Indeed, the fishery products contained in these water reservoirs are directly contaminated, and this remains a major public health problem. Thus, using ordinary partial derivatives, we established budget equations for the mass concentrations of the major components of fertilizer chemicals such as nitrogen (N), phosphorus (P) and potassium (K) found in each natural water reservoir. Using the Euler discretization method, we linearized these budget equations in order to formulate them numerically with AMPL and solve them. The numerical solutions obtained thus make it possible to predict the mass concentrations of the chemical elements (N, P, K) at the level of each water reservoir in time by taking the measures for a substantial reduction of the chemical content of fresh water.

1 Introduction

In agricultural land, large quantities of chemicals apply to the soil or incorporated into it as fertilizers. Thus, rainwater or irrigation water displaces many of these

---

D. Dione (✉)
University of Gaston Berger, Dakar, Senegal

T. C. Mbainaissem
University of N'Djaména, N'Djaména, Chad

B. Koné
University of Félix Houphouët-Boigny, Abidjan, Ivory Coast
e-mail: bakary.kone@ucad.edu.sn

P. P. Ndekou
University of Cheikh Anta Diop, Dakar, Senegal

V. Gayoso Martínez et al. (eds.), *Mathematical Methods for Engineering Applications*, Springer Proceedings in Mathematics & Statistics 439,
https://doi.org/10.1007/978-3-031-49218-1_1

chemicals, some of which, such as P, N and K, are taken up by plants, while the rest are transported to groundwater or by runoff to adjacent water reservoirs. The leaching of these chemicals into drinking water reservoirs is a major environmental problem. For example, chemicals such as phosphorus cause eutrophication of freshwater reservoirs [1, 2]. The chemical quantity load to the aquatic environment from various controlled non-point sources is becoming increasingly important. In the past, the subsurface or runoff transport of chemicals such as phosphorus, nitrogen and potassium has long been neglected, given the high uptake of crops. However, authors such as [3] have looked at the transport of chemicals in subsurface or surface environments.

Nowadays, much effort has been made to reduce the pollution of chemicals such as phosphorus, nitrogen and potassium by improving the methods of fertilizer use in areas with high agricultural coverage. To achieve this, the connection rate of industrial and domestic effluents was increased by improving the efficiency of wastewater treatment. According to the authors [4], these efforts seem to bear fruit. However, the phenomenon of chemical pollution of certain water bodies is difficult to reverse given the many diffuse or point sources of chemical contamination. Thus, several diffuse chemical pollution flows remain poorly controlled [5], while additional pressures on the catchment areas are being created. These new flows are linked to factors such as climate change. These are at the origin of the increase in precipitation and temperatures, hence the modification of hydrological regimes. Thus, a disruption of the export regimes of nitrogen, phosphorus and potassium flows is noted, potentially affecting water reservoirs. Therefore, the strengthening of control and mitigation measures for diffuse chemical pollution flows is necessary. Thus, it is important to understand the mechanisms of acquisition and transfer of nutrient load, in order to set up reduction strategies adapted to the studied context [6].

Controlling the chemical load of phosphorus, nitrogen and potassium requires strategic interventions upstream at the level of the sources of the basins and of the industrial activities that are developed in their vicinity [7, 8]. Thus, the quantification and characterization of the concentration of chemical products allows the setting up of strategies for the functioning of catchment areas. This will also allow the control of the nitrogen, phosphorus and the potassium transport system [9]. The mathematical modeling of the concentrations of these chemicals at the scale of a reservoir will provide a decision support tool. The fundamental aim of this tool will be to test hypotheses by studying and evaluating an operational system of variation and the evolution of scenarios.

The objectives of this paper are to quantify, partition and model the concentrations of nitrogen, phosphorus and potassium flows in a catchment area. Thus, we will first quantify the water by including the concentrations of chemical elements such as N, P and K. This comprises understanding and analyzing the variations in concentrations and budget equations, the acquisition and export patterns of the flows of these chemical elements from point and diffuse sources. In addition, modeling is used as a tool for validating the estimation of flows, improving knowledge of the fluxes, to improve knowledge of the system studied. Thus, it will make it possible to predict the effects of changes in the system or external forcing, such as the evolution of precipitation regimes or climate change in the simulated flows, which have been previously corroborated by observations.

The remainder of this article is organized as follows. In Sect. 2 we describe the life cycle of N, P, and K in an ecosystem, considering their transport underground and at the surface. In Sect. 3 we describe the budget equations for water, P, N, and K in a reservoir k during month t. Particularly in connection with fishing, exact trophy predictions are necessary, because the calculation of the fishery yield value is based on the trophy forecast or the primary production, which in turn is based on the predicted of the external phosphorus production load that can be estimated. We use MINOS to solve this model and computational results are presented in Sect. 4. Conclusions and final discussions are given in Sect. 5.

2 Description of the Life Cycle of N, P, K in an Ecosystem

The small-scale life cycle shown in Fig. 1 is: the phosphorus in dead organic matter is progressively mineralized by micro-organisms which simplify the phosphorus compounds until they are transformed into soluble phosphates (PO_4^{3-}). The latter is quickly assimilated and integrated into the living organic matter. Some exchanges take place with non-living elements of the ecosystem: sedimentation, adsorption/desorption on colloidal clay suspensions, but above all immobilization (mainly in connection with Ca^{2+}, Fe^{3+} ions) and release on the deregulation of the phosphorus cycle. The phosphorus cycle is currently deeply disturbed in many aquatic environments by massive external inputs of phosphates from fertilizers and detergents. These inputs lead to a rapid and significant increase in plant production (hyper-eutrophication). This phenomenon, first observed in lakes, is spreading too many lowland rivers and is taking on a worrying dimension.

The nitrogen cycle is a biochemical cycle that describes the succession of changes undergone by the different nitrogen (nitrogen, nitrate, nitrite, ammonia,

Fig. 1 Diagram of reservoir

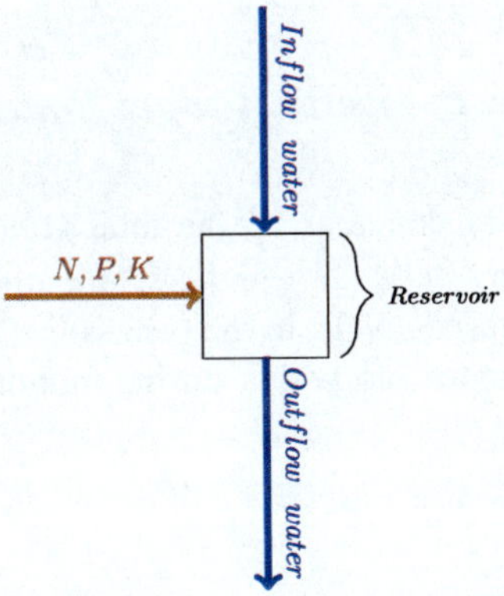

organic nitrogen (proteins)). The atmosphere is the major source of nitrogen, since it contains 78% by volume. Processes are required to convert atmospheric nitrogen into a form that can be taken up by organisms. Atmospheric nitrogen is fixed by bacteria in the soil, such as Azobacter vinelandii, using the enzyme nitrogenase. Some of these bacteria, such as Rhizobium, live in symbiosis with plants, producing ammonia and drawing carbohydrates. Ammonia can also be produced by the decomposition of dead organisms by saprophytic bacteria as ammonium ions NH^{4+}. In well-oxygenated soils, bacteria convert ammonia to nitrite NO^{2-} and then to nitrate NO^{3-} in a process called nitrification. Plants use their roots to absorb nitrate ions NO^{3-} and, to a lesser extent, ammonia from the soil, and incorporate them into amino acids and proteins. Plants are thus the only source of nitrogen that can be assimilated by animals.

Potassium is present only in mineral form. In the soil, potassium is released during the growth phase of the plant from the solid phase of the soil. As with phosphorus, the potassium cycle depends on the physical and chemical characteristics of the soil. It is a fairly mobile element but can be adsorbed onto the cation exchange capacity (CEC) of soils. Potassium in the soil solution is kept by humus or clay; potassium in minerals will only be released slowly. The main pathways of loss of this nutrient are agricultural exports, leaching (more important in sandy soils, poor in organic matter and with a low cation exchange capacity), and finally runoff and erosion. The main pathways are mineral fertilizers and organic inputs. Potassium is taken up by plants from mass flow and diffusion as the K^+ ion. The mobility of K^+ ions in the soil, although higher than that of PO_4^{3-} ions, is not very high (of the order of a centimeter). Depending on its root system, the plant has a greater or lesser capacity to extract the potassium it needs from the soil [10].

3 Budget Equations

The general flow structure around a reservoir is shown in Fig. 2. A survey of the involved parameters is given in Table 1.

3.1 Water Budget Equations

By denoting V_r^t the total volume of water of the reservoir r during month t, and by $Q_r^{i,t}$, $Q_r^{o,t}$, and $Q_r^{e,t}$ the quantities of water in inflow, outflow, and evaporation, respectively, in the reservoir r during month t, we obtain the water budget equation in the reservoir r during month t:

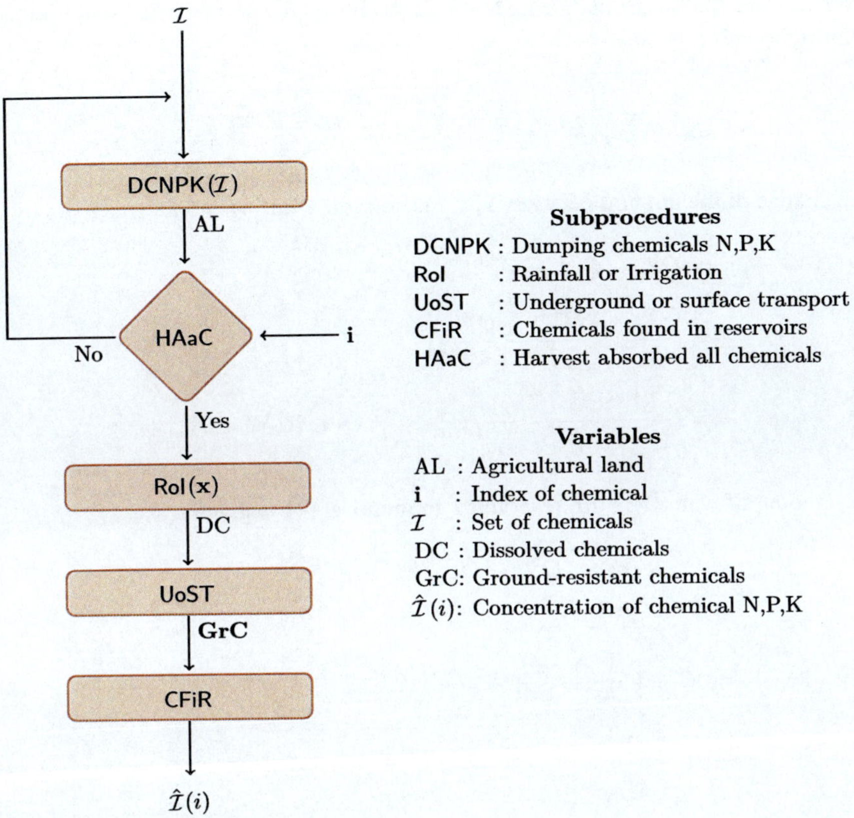

Fig. 2 Scheduling flowchart

Table 1 Parameters of timescale

	Timescale for ...
$\tau_r^{i,t}$	Filling of reservoir r by inflow (precipitation and surface) during month t
$\tau_r^{o,t}$	Outflow of reservoir r water (water retention time) during month t
$\tau_r^{f,t}$	Filling of reservoir r by inflow (inflow time by filling) during month t
$\tau_r^{e,t}$	Evaporation of reservoir r (outflow time by evaporation) during month t

$$\partial(V_r^t) = Q_r^{i,t} - Q_r^{o,t} - Q_r^{e,t} \iff \frac{\partial(V_r^t)}{V_r^t} = \frac{Q_r^{i,t}}{V_r^t} - \frac{Q_r^{o,t}}{V_r^t} - \frac{Q_r^{e,t}}{V_r^t}, \quad \forall r \in \mathcal{R}, \forall t \in \mathcal{T}.$$

However, we have:

$$\frac{V_r^t}{\partial(V_r^t)} = \tau_r^{f,t}, \quad \frac{V_r^t}{Q_r^{i,t}} = \tau_r^{i,t}, \quad \frac{V_r^t}{Q_r^{o,t}} = \tau_r^{o,t}, \quad \frac{V_r^t}{Q_r^{e,t}} = \tau_r^{e,t}, \quad \forall r \in \mathcal{R}, \forall t \in \mathcal{T}.$$

From this, we obtain the budget equation for water inflow and outflow in a reservoir r during month t as

$$\frac{1}{\tau_r^{f,t}} = \frac{1}{\tau_r^{i,t}} - \frac{1}{\tau_r^{o,t}} - \frac{1}{\tau_r^{e,t}}, \quad \forall r \in \mathcal{R}, \forall t \in \mathcal{T}. \tag{3.1}$$

The change of the amount of water $V_r^{c,t}$ in reservoir r during month t is given by

$$\begin{aligned} V_r^{c,t} &= Q_r^{i,t} - Q_r^{o,t} - Q_r^{e,t} \\ &= (\frac{Q_r^{i,t}}{V_r^t} - \frac{Q_r^{o,t}}{V_r^t} - \frac{Q_r^{e,t}}{V_r^t})V_r^t \\ &= (\frac{1}{\tau_r^{i,t}} - \frac{1}{\tau_r^{o,t}} - \frac{1}{\tau_r^{e,t}})V_r^t, \quad \forall r \in \mathcal{R}, \forall t \in \mathcal{T}. \end{aligned} \tag{3.2}$$

The amount of water V_r^{t+1} in reservoir r in month $t+1$ is given by

$$\begin{aligned} V_r^{t+1} &= V_r^t + Q_r^{i,t} - Q_r^{o,t} - Q_r^{e,t} \\ &= V_r^t + V_r^{c,t} \\ &= V_r^t + (\frac{1}{\tau_r^{i,t}} - \frac{1}{\tau_r^{o,t}} - \frac{1}{\tau_r^{e,t}})V_r^t \\ &= (1 + \frac{1}{\tau_r^{i,t}} - \frac{1}{\tau_r^{o,t}} - \frac{1}{\tau_r^{e,t}})V_r^t, \quad \forall r \in \mathcal{R}, \forall t \in \mathcal{T}. \end{aligned} \tag{3.3}$$

3.2 Phosphorus, Nitrogen, and Potassium Budget Equations

The budget equation for the P mass $P_r^t V_r$ in a reservoir r on month t considering sedimentation $\sigma_r^{phi,t}\epsilon^i$ proportional to concentration $P_r^{t,i}$ of the phosphorus in the water inflow and sedimentation $\sigma_r^{ph,t}$ proportional to concentration P_r^t of phosphorus in the reservoir k is written by Will et al. [1, 13]:

$$\frac{\partial(P_r^t V_r^t)}{\partial t} = P_r^{t,i} Q_r^{i,t} - P_r^{t,o} Q_r^{o,t} - \sigma_r^{phi,t}\epsilon^i P_r^{t,i} V_r^t - \sigma_k^{ph,t} P_r^t V_r^t. \tag{3.4}$$

Where a factor $0 \leq \epsilon^i = \frac{Q_r^{i,t}}{V_r^t} \leq 1$ is introduced accounting for the ratio of the reservoir water volume V_r in which $P_r^{t,i}$ can be expected.

Using the rule of the multiplication derivative the change of phosphorus mass in the reservoir on month t can be separated into two components:

$$\frac{\partial(P_r^t V_r^t)}{\partial t} = \frac{\partial(P_r^t)}{\partial t} V_r^t + \frac{\partial(V_r^t)}{\partial t} P_r^t.$$

From this we obtain

$$\begin{aligned}
\frac{\partial(P_r^t)}{\partial t} &= -\frac{\partial(V_r^t)}{\partial t V_r^t} P_r^t + P_r^{t,i} \frac{Q_r^{i,t}}{V_r^t} - P_r^{t,o} \frac{Q_r^{o,t}}{V_r^t} - \sigma_r^{phi,t} \epsilon^i P_r^{t,i} - \sigma_r^{ph,t} P_r^t \\
&= -\frac{1}{\tau_r^f} P_r^t - \frac{1}{\tau_r^{o,t}} P_r^t - \sigma_r^{ph,t} P_r^t + \frac{1}{\tau_r^{i,t}} P_r^{i,t} - \sigma_r^{phi,t} P_r^{i,t} \\
&= (-\frac{1}{\tau_r^{f,t}} - \frac{1}{\tau_r^{o,t}} - \sigma_r^{ph,t}) P_r^t + (\frac{1}{\tau_r^{i,t}} - \sigma_r^{phi,t} \epsilon^i) P_r^{i,t}.
\end{aligned} \tag{3.5}$$

The resulting phosphorus budget equation (3.5) is dependent on the inflow timescale $\tau_r^{i,t}$ and the evaporation precipitation timescale $\tau_r^{e,t}$:

$$\begin{aligned}
\frac{\partial(P_r^t)}{\partial t}.\tau_r^{i,t} &= (-\frac{\tau_r^{i,t}}{\tau_r^{f,t}} - \frac{\tau_r^{i,t}}{\tau_r^{o,t}} - \sigma_r^{ph,t}) P_r^t + (1 - \sigma_r^{phi,t} \tau_r^{i,t} \epsilon^i) P_r^{i,t} \\
&= -(1 - \frac{\tau_r^{i,t}}{\tau_r^{e,t}} - \sigma_r^{ph,t} \tau_r^{i,t}) P_r^t + (1 - \sigma_r^{phi,t} \tau_r^{i,t} \epsilon^i) P_r^{i,t}.
\end{aligned} \tag{3.6}$$

The resulting phosphorus budget equation (3.2) is dependent on the outflow timescale $\tau_r^{o,t}$, the filling timescale $\tau_r^{f,t}$ and the evaporation-precipitation timescale $\tau_r^{e,t}$:

$$\begin{aligned}
\frac{\partial(P_r^t)}{\partial t}.\tau_r^{o,t} &= (-\frac{\tau_r^{o,t}}{\tau_r^{f,t}} - 1 - \sigma_r^{ph,t} \tau_r^{o,t}) P_r^t + (\frac{\tau_r^{o,t}}{\tau_r^{i,t}} - \sigma_r^{phi,t} \tau_r^{o,t} \epsilon^i) P_r^{i,t} \\
&= -(1 + \frac{\tau_r^o}{\tau_r^f} + \sigma_r^{ph,t} \tau_r^{o,t}) P_r^t + (1 + \frac{\tau_r^{o,t}}{\tau_r^{f,t}} + \frac{\tau_r^{o,t}}{\tau_r^{e,t}} - \sigma_r^{phi,t} \tau_r^{o,t} \epsilon^i) P_r^{i,t}.
\end{aligned} \tag{3.7}$$

The N budget equation is:

$$\begin{aligned}
\frac{\partial(N_r^t)}{\partial t} &= -\frac{\partial(V_r^t)}{\partial t V_r^t} N_r^t + N_r^{t,i} \frac{Q_r^{i,t}}{V_r^t} - N_r^{t,o} \frac{Q_r^{o,t}}{V_r^t} - \sigma_r^{ni,t} \epsilon^i N_r^{t,i} - \sigma_r^{n,t} N_r^t \\
&= -\frac{1}{\tau_r^f} N_r^t - \frac{1}{\tau_r^{o,t}} N_r^t - \sigma_r^{n,t} N_r^t + \frac{1}{\tau_r^{i,t}} N_r^{i,t} - \sigma_r^{i,t} N_r^{i,t} \\
&= (-\frac{1}{\tau_r^{f,t}} - \frac{1}{\tau_r^{o,t}} - \sigma_r^{n,t}) N_r^t + (\frac{1}{\tau_r^{i,t}} - \sigma_r^{ni,t} \epsilon^i) N_r^{i,t},
\end{aligned} \tag{3.8}$$

where $\sigma_r^{ni,t}\epsilon^i$ is the sedimentation proportional to concentration $N_r^{t,i}$ of the nitrogen in the water inflow, and $\sigma_r^{n,t}$ is the sedimentation proportional to concentration N_r^t of nitrogen in the reservoir r on month t.

The K budget equation is:

$$\begin{aligned}\frac{\partial(K_r^t)}{\partial t} &= -\frac{\partial(V_r^t)}{\partial t V_r^t}K_r^t + K_r^{t,i}\frac{Q_r^{i,t}}{V_r^t} - K_r^{t,o}\frac{Q_k^{o,t}}{V_r^t} - \sigma_r^{poti,t}\epsilon^i N_r^{t,i} - \sigma_k^{pot,t}K_r^t \\ &= -\frac{1}{\tau_r^f}K_r^t - \frac{1}{\tau_k^{o,t}}K_r^t - \sigma_k^{pot,t}K_r^t + \frac{1}{\tau_r^{i,t}}K_r^{i,t} - \sigma_k^{i,t}K_r^{i,t} \\ &= (-\frac{1}{\tau_r^{f,t}} - \frac{1}{\tau_r^{o,t}} - \sigma_r^{pot,t})K_r^t + (\frac{1}{\tau_r^{i,t}} - \sigma_r^{poti,t}\epsilon^i)K_r^{i,t} \end{aligned} \tag{3.9}$$

where $\sigma_k^{poti,t}\epsilon^i$ is the sedimentation proportional to concentration $K_r^{t,i}$ of the potassium in the water inflow, and $\sigma_r^{pot,t}$ is the sedimentation proportional to concentration K_r^t of potassium in the reservoir r on month t.

Euler's discretization of the budgets equations of phosphorus, nitrogen and potassium (3.9), (3.8) and (3.7), respectively, in each reserve r during month t are computed as:

$$P_r^{t+1} = \left(-\frac{1}{\tau_r^{f,t}} - \frac{1}{\tau_r^{o,t}} - \sigma_r^{ph,t})P_r^t + (\frac{1}{\tau_r^{i,t}} - \sigma_r^{phi,t}\epsilon^i\right)P_r^{i,t}, \quad \forall r \in \mathcal{R}, t \in \mathcal{T}, \tag{3.10}$$

$$N_r^{t+1} = \left(-\frac{1}{\tau_r^{f,t}} - \frac{1}{\tau_r^{o,t}} - \sigma_k^{n,t})N_r^t + (\frac{1}{\tau_r^{i,t}} - \sigma_k^{ni,t}\epsilon^i\right)N_r^{i,t}, \quad \forall r \in \mathcal{R}, t \in \mathcal{T}, \tag{3.11}$$

$$K_r^{t+1} = \left(-\frac{1}{\tau_r^{f,t}} - \frac{1}{\tau_r^{o,t}} - \sigma_k^{pot,t})K_r^t + (\frac{1}{\tau_r^{i,t}} - \sigma_r^{poti,t}\epsilon^i\right)K_r^{i,t}, \quad \forall r \in \mathcal{R}, t \in \mathcal{T}. \tag{3.12}$$

4 Computational Results

We intend to test our numerical results based on three (03) topologically connected water reservoirs denoted respectively by R_1, R_2, and R_3. Indeed, the first reservoir pours into the second which in turn pours into the third as the following model: $(R_1 \longmapsto R_2 \longmapsto R_3)$. The time t of the numerical test will take twelve months where $t \in \{Jan, Feb, Mar, Apr, May, Jun, Jul, Aug, Sep, Oct, Nov, Dec\}$. We linearized the different models of the inflow (Con. Inf) and outflow (Con. Out) concentrations for each reservoir. The concentration in a reservoir is equal to the concentration outflow of that reservoir ($P_r^{o,t} = P_r^t$). Data on the inflow, outflow, and evaporation of water from each tank r are assigned in Table 2. These data are taken from the authors' research work [13]. It should be remembered that these time-scale data are obtained using time measurement tools. However, in this paper, we are not only considering the chemical elements derived from manure but also all the other chemical elements derived from agricultural activity. We have therefore worked with time-scale parameters, taking our three reservoirs as topologically connected lagoons (Tables 3 and 4).

The highest concentrations of Phosphorus, Nitrogen, and Potassium inflow are noted during August and September with average values around 7.3223, 9.95395 and 7.6351 μg/l for P in the reservoirs 1, 2 and 3; 17.881, 24.1739 and 23.3276 μg/l for N in reservoirs 1, 2 and 3; and 7.74307, 10.5227 and 26.4459 μg/l for K in reservoirs 1, 2 and 3. We assume that the remaining amount of these different chemical elements (N, P, K) is used by crops, sedimentation in the soil, etc.

We have water spikes in reservoir 1 in July, August, and September, which also directly implies an increase in manure inflow. Thus, in Reservoir 1, in the same months, we have a direct increase in N, P and K in this reservoir. The same remark is noted in Reservoir 2 and Reservoir 3, as shown in Figs. 3, 4, 5, 6, 7, and 8. We

Table 2 Reservoirs parameters data

	Reservoirs			
Param.	R1	R2	R3	Units
τ_i	2.19916	4.02166	50	[m/month]
τ_e	3.99848	4.06228	10.00001	[m/month]
τ_f	12.21759	402.16666	200.56145	[m/month]
Volume	25,000,000,000	59,400,000	158,000,000	[m^3]
Surface area	560,000,000	22,000,000	21,000,000	[m^2]

Table 3 Results of amount of water in $[\mu \cdot l]$ in reservoir k during month t

	Water inflow [10^6 m^3]			Water outflow [10^6 m^3]			Water in lake [10^6 m^3]		
	R_k								
T_t	R1	R2	R3	R1	R2	R3	R1	R2	R3
Jan	1	1.5	2	0.795310	1.491890	1.800000	1	2.934470	2
Fev	1	1.5	2	0.753413	1.491870	1.780000	1.204690	2.942580	2.200000
Mar	1	1.5	2	0.702939	1.491840	1.758000	1.451280	2.950720	2.420000
Apr	1.2	1.4	1.5	0.842133	1.391820	1.233800	1.748340	2.958870	2.662000
May	1.1	1.9	1.3	0.668882	1.891800	1.007180	2.106210	2.967050	2.928200
Jun	1.4	1.5	1.6	0.880636	1.491770	1.277900	2.537320	2.975260	3.221020
Jul	1.7	1.8	1.9	1.074330	1.791750	1.545690	3.056690	2.983480	3.543120
Aug	2	2	2	1.246260	1.991730	1.610260	3.682360	2.991730	3.897430
Sep	1.5	3	2	0.591976	2.991710	1.571280	4.436100	3	4.287180
Oct	1.7	1.9	1.6	0.606113	1.891680	1.128410	5.344120	3.008290	4.715900
Nov	1.8	1.9	1.3	0.482206	1.891660	0.781252	6.438010	3.016610	5.187480
Dec	1.9	1.7	1.6	0.312467	1.691640	1.029380	7.755810	3.024950	5.706230

Table 4 Results of concentrations inflow and outflow of Phosphorous, Nitrogen and Potassium [µg/l]

		Phosphorous			Nitrogen			Potassium		
		R_k								
	T_t	R1	R2	R3	R1	R2	R3	R1	R2	R3
Con. In.	Jan	4.63712	10.2461	6.41937	11.2616	24.8833	15.5899	4.9021	10.8316	6.78619
	Feb	4.63712	12.396	33.9342	11.2616	30.1047	82.4115	4.9021	13.1044	35.8733
	Mar	4.0527	14.8456	37.7927	9.84228	36.0535	91.7822	4.28429	15.6939	39.9523
	Apr	5.86574	11.5295	11.2113	14.2454	28.0001	27.2274	6.20093	12.1883	11.8519
	May	6.82754	14.2588	13.0445	16.5812	34.6284	31.6795	7.21768	15.0735	13.7899
	Jun	4.4299	11.1215	18.4016	10.7583	27.0094	44.6897	4.68303	11.757	19.4531
	Jul	6.69267	11.4827	25.0164	16.2536	27.8867	60.754	7.07511	12.1389	26.4459
	Aug	7.3223	7.10155	9.60548	17.7827	17.2466	23.3276	7.74072	7.50735	10.1544
	Sep	7.32452	9.95395	7.63551	17.7881	24.1739	18.5434	7.74307	10.5227	8.07182
	Oct	5.16872	6.94079	7.63551	12.5526	16.8562	18.5434	5.46408	7.33741	8.07182
	Nov	5.05465	5.93578	7.62498	12.2756	14.4155	18.5178	5.34348	6.27497	8.06069
	Dec	5.31083	5.93379	7.62369	12.8977	14.4106	18.5147	5.61431	6.27286	8.05933
Con. Out.	Jan	1.514	2.041	0.643	1.4461	4.95671	2.64	0.395	8.69328	7.18341
	Feb	1.514	2.041	1.14123	3.08085	4.95671	2.64	1.56831	0.534	0.413
	Mar	1.514	2.57566	6.30826	2.04736	6.25518	15.3361	1.06121	3.1262	6.76553
	Apr	1.25408	3.052	6.411	2.19723	7.412	15.5676	1.06121	3.1262	6.76553
	May	2.121	2.109	1.348	2.17437	4.99167	3.15047	1.12105	2.25441	1.42647
	Jun	2.153	3.022	2.314	1.14	7.33914	5.61969	0.632438	3.1885	2.44605
	Jul	1.1461	2.015	3.214	2.02	4.89357	7.80543	0.919675	2.13168	3.39768
	Aug	2.506	2.355	4.361	3.06561	5.71929	10.591	1.40492	2.48919	4.6102
	Sep	2.374	1.181	1.293	4.67695	2.86814	3.14014	2.13883	1.24858	1.36689
	Oct	2.374	2.182	1.293	2.88592	5.29914	3.14014	1.82247	2.30666	1.36689
	Nov	1.4152	1.184	1.293	2.34	2.87543	3.14014	1.15073	1.25166	1.36689
	Dec	1.72133	1.182	1.291	2.64144	2.87057	3.13529	1.39827	1.24954	1.36477

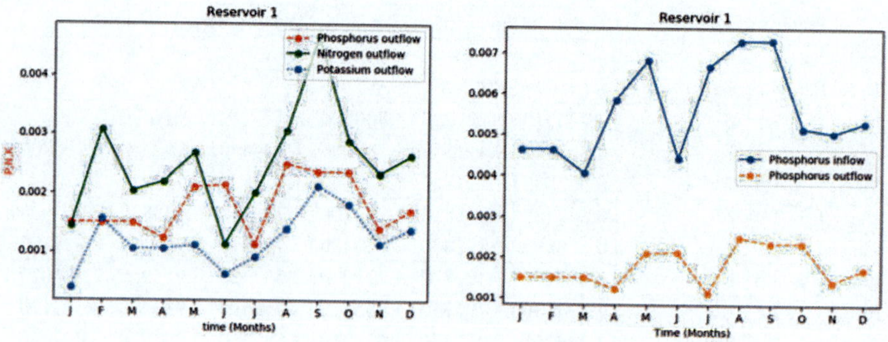

Fig. 3 Temporal variation of nitrogen, phosphorus and potassium inflow and outflow at reservoir 1

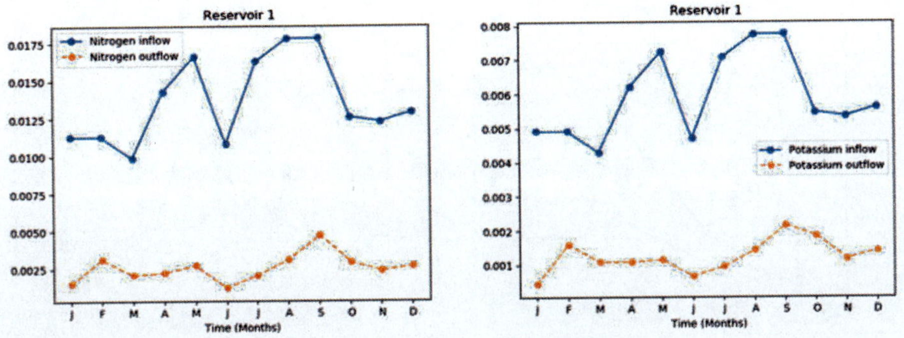

Fig. 4 Comparison of the temporal variation of nitrogen and potassium inflow and outflow at reservoir 1

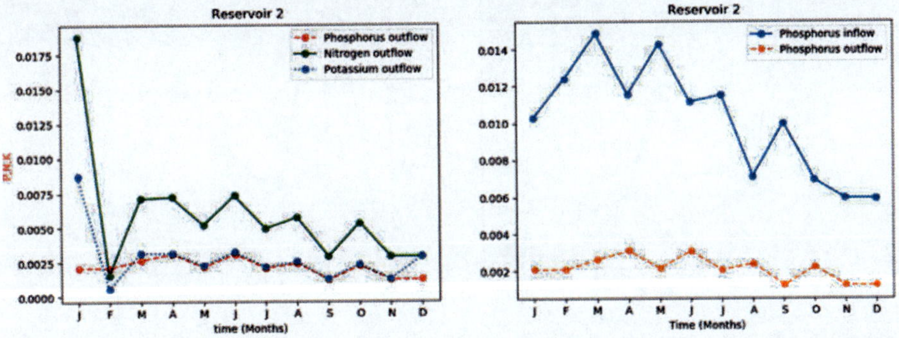

Fig. 5 Temporal variation of Nitrogen, Phosphorus and Potassium inflow and outflow at reservoir 2

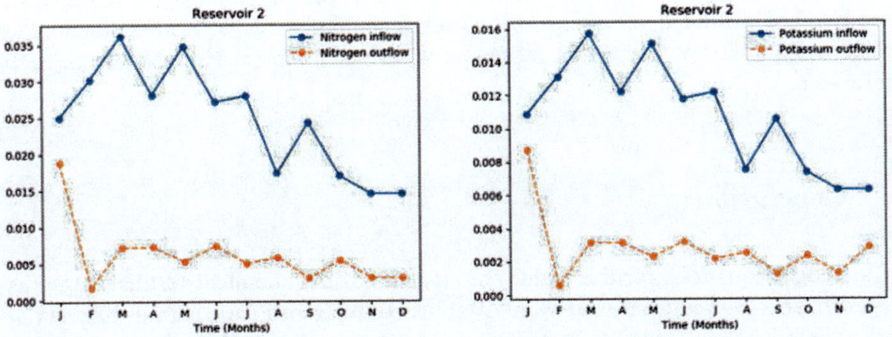

Fig. 6 Comparison of the temporal variation of Nitrogen and Potassium inflow and outflow at reservoir 2

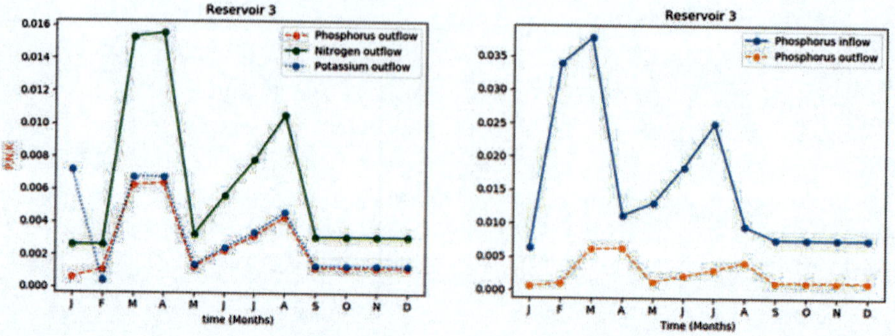

Fig. 7 Temporal variation of Nitrogen, Phosphorus and Potassium inflow and outflow at reservoir 3

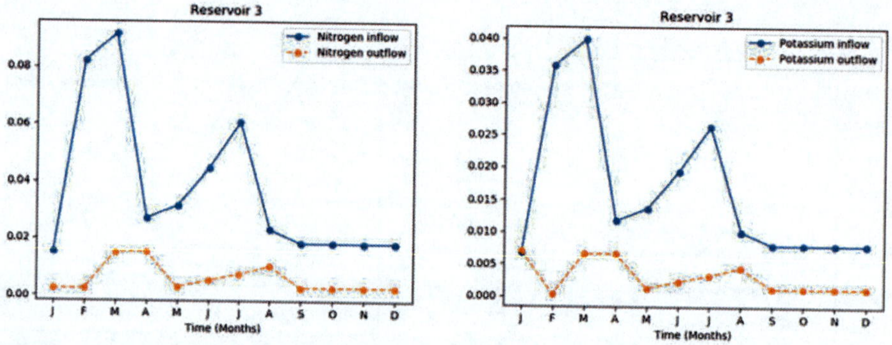

Fig. 8 Comparison of the temporal variation of Nitrogen and Potassium inflow and outflow at reservoir 3

also noticed that the amounts of N, P, K inflow are much higher than the amounts of N, P, K outflow, therefore we have an excess of N, P, K over time in each reservoir, see Fig. 7. The annual variations of the chemical elements N, P, K are compared at Figs. 4, 6 and 8. We find the quantities N, P, and K in an ascending order in each reservoir.

5 Conclusion

The budget equations for the major components of the chemical fertilizer used as a fertilizer in agricultural land were established using ordinary partial differential equations. Knowing that N, P, and K are the major chemical components of the fertilizer, we developed the mathematical model of their different mass concentrations over time for one year. Thus, we previously modeled the budget equation of the amount of water in a reservoir. Using the time scale of the water inflow and outflow a reservoir, we found the budget equations of the concentrations of the chemical elements (N, P, K). We then linearised these budget equations using the Euler discretization criterion. In this way, we could evaluate the discrete values in each month of these different chemical concentrations of the fertilizer components entering each water reservoir, but also of those outflow each reservoir. Thus, in this paper, the fundamental results are the peak concentrations of the chemical elements N, P, and K in the flood months of July, August, and September. For example, the mass of phosphorus is evaluated at 6.69267 Units in July, 7.3223 Units in August, and 7.32452 Units in September. Therefore, the results found allowed us to predict the water quality of each freshwater reservoir in an annual period to take solutions for the use of this water. On another level, chemical elements (N, P, K) are essential nutrients for agriculture and a scarce resource. Their recovery in water reservoirs and their valorization make up a wise solution aiming at transforming a polluting discharge into a useful resource.

As future work, we decided to model the movement of chemical fertilizers in the surfaces and its main components (N, P, K) in the soil and subsoil. We also plan to use other resolution methods based on a linear dynamic system to solve this important environmental problem.

Acknowledgments We are grateful to Prof. Dr. Armin Fügenschuh of Engineering Mathematics and Optimisation Numerics of Brandenburg University of Technology Cottbus—Senftenberg for the realization of these scientific works.

References

1. Vollenweider, R.: Advances in defining critical loading levels for phosphorus in lake eutrophication. Mem. Ist. Ital. Idrobiol **33**, 53–83 (1976)
2. Vollenweider, R., Kerekes, J.: Eutrophication of waters : monitoring, assessment and control. Cooperative Programme for Monitoring of Inland Waters (Eutrophication Control), Paris: 0ECD (1982)
3. Jardine, P.M., Wilson, G.V., Joneset, J.R., Zelazny, L.W., Luxmoore, R.J.: Physical and chemical controls of preferred path flow through a forested hillslope. Geoderma **46**, 139–154 (1990)
4. Rousseau, A.N., Benoy, G., North, R.L., Foulon, E.: A global scan of how the issue of nutrient loading and harmful algal blooms is being addressed by governments, non-governmental organizations, and volunteers. Water Qual. Res. J., 1–22 (2019). https://doi.org/10.2166/wqrj.2019.013
5. Trevisan, D., Dorioz, J.M., Jordan-Meille, L.: Variabilité des mécanismes de transfert du phosphore a l'échelle bassin versant : le cas d'un bassin versant de la region Lémanique et sa portée générale. Océanis **33**, 183–206 (2007)
6. Quetin, P., Prigent-Combaret, C., Trevisan, D., Dorioz, J.M.: Role des facteurs edaphiques et hydrometeorologiques dans la survie et le transfert de bactéries fécales bovines, a l'échelle bassin versant : cas de paturages d'altitude. Etude Gest. des Sols **18**, 217–236 (2011)
7. Wang, D., Poulenard, J., Trevisan, D., Dorioz, J.-M.: The effect of grass buffer strips on phosphorus dynamics - a critical review and synthesis as a basis for application in agricultural landscapes, in France. Agric. Ecosyst. Environ. **117**, 4–21 (2006)

8. Giroux, M., Landry, C., Deslandes, J., Beaudin, I., Lauzier, R., Michaud, A.R.: Prévention des transferts diffus de phosphore en bassins-versants agricoles: perspectives Québécoises et de l'état du Vermont. Océanis **33**, 285–320 (2007)
9. Pezet, F.: Bilans et transferts de phosphore dans le bassin versant du lac du bourget: caractérisation, interprétation et modélisation des flux
10. Unifa, Parlons fertilisation (2022). https://fertilisation-edu.fr/10-cycles-bio-geo-chimiques.htmlstart=2
11. Will, A., Grüneberg, B.: Modelling of phosphor concentration in lakes affected by lignite coal mining (2019)
12. Ahlgren, I., Frisk, T., Kamp-Nielsen, L.: Empirical and theoretical models of phosphorus loading, retention and concentration vs. lake trophic state. Hydrobiologia **170**, 285–303 (1988)
13. Dosso, M., Diouf, S., Kone, B., Dione, D., Ndekou, P.P.: Conservation water environment of a lagoon system by controlling manure distribution. Results Control Optim. **6**, 100086 (2022)

p-Numerical Semigroups of the Triples of the Sequence $(a^n - b^n)/(a - b)$

Takao Komatsu and Ruze Yin

Abstract For a non-negative integer p, we give explicit formulas for the p-Frobenius number and the p-genus of numerical semigroups of (v_n, v_{n+1}, v_{n+2}), where $v_n = (a^n - b^n)/(a - b)$ with $\gcd(a, b) = 1$ and $a > b > 1$. Here, the p-numerical semigroup S_p is the set of integers whose non-negative integral linear combinations of given positive integers are expressed more than p ways. When $p = 0$, S_0 with the 0-Frobenius number and the 0-genus is the original numerical semigroup S_0 with the Frobenius number and the genus. Symmetric properties of numerical semigroups are important to characterize the numerical semigroup. In recent works, some closed formulas of p-Frobenius numbers have been successfully given, but no symmetric property has been found. We give a symmetric property of this numerical semigroup.

1 Introduction

Given the set of positive integers $A := \{a_1, a_2, \ldots, a_k\}$ $(k \geq 2)$, for a non-negative integer p, let S_p be the set of integers whose non-negative integral linear combinations of given positive integers $a_1, a_2, \ldots, a_k$ are expressed more than p ways. For a set of non-negative integers $\mathbb{N}_0$, the set $\mathbb{N}_0 \backslash S_p$ is finite if and only if $\gcd(a_1, a_2, \ldots, a_k) = 1$. Then there exists the largest integer $g_p(A) := g(S_p)$ in $\mathbb{N}_0 \backslash S_p$, which is called the *p-Frobenius number*. The cardinality of $\mathbb{N}_0 \backslash S_p$ is called the *p-genus* and denoted by $n_p(A) := n(S_p)$. The sum of the elements in $\mathbb{N}_0 \backslash S_p$ is denoted by $s_p(A) := s(S_p)$, and is called the *p-Sylvester sum*. This kind of concept is a generalization of the famous Diophantine problem of Frobenius [22, 23] because when $p = 0$, the original Frobenius number $g(A) = g_0(A)$ and the genus $n(A) = n_0(A)$ are recovered. We can call S_p the *p-numerical semigroup*.

T. Komatsu (✉) · R. Yin
Department of Mathematical Sciences, School of Science, Zhejiang Sci-Tech University, Hangzhou, China
e-mail: komatsu@zstu.edu.cn

V. Gayoso Martínez et al. (eds.), *Mathematical Methods for Engineering Applications*, Springer Proceedings in Mathematics & Statistics 439,
https://doi.org/10.1007/978-3-031-49218-1_2

Strictly speaking, the set S_p is not a numerical semigroup but an additive monoid when $p \geq 1$, because the integer 0 has only one representation. Nevertheless, the set $S_p \cup \{0\}$ becomes a numerical semigroup and its maximal ideal is S_p itself. When emphasizing the set A which is generated from, we shall write $S_p(A)$. More detail descriptions of the p-numerical semigroups and their symmetric properties can be found in [12].

The need to care enough is that even in the original case of $p = 0$, it is very difficult to give a closed explicit formula of a general type for three or more variables (see, e.g., [2, 16–18, 20]). Indeed, it is even more difficult when $p > 0$, and finally, we have succeeded in obtaining p-Frobenius number in triangular numbers [7], repunits [9] and Fibonacci and Lucas triplets [11] quite recently.

In this paper, we give explicit closed formulas of the p-Frobenius numbers (Theorem 1) and the p-genus (Theorem 2) for the numerical semigroup generated by the triple

$$\left\{\frac{a^n - b^n}{a - b}, \frac{a^{n+1} - b^{n+1}}{a - b}, \frac{a^{n+2} - b^{n+2}}{a - b}\right\} \quad (n \geq 3),$$

where $\gcd(a, b) = 1$ with $a > b > 1$. The method we follow is the determination of all the elements of the p-Apéry set. Though some situations and techniques seem to be similar to those of Ko22b, it only happened to be successful in this case, and the same thing is not true for the general triplets. In fact, there are triples where the arrangement of the Apéry sets is different (in preparation). Note that the case where $b = 1$ with $p = 0$ is discussed in [19], but not included in our result. The real situation for $b \geq 2$ is not as simple as in the case $b = 1$ at all. It is because each Apéry set is a complete residue system modulo ν_n formed from the sequence $\{b^k \pmod{\nu_n}\}_{k=0}^{\nu_n-1}$.

Symmetric properties of numerical semigroups are important to characterize the numerical semigroup. They can be also applied to Algebraic Geometry, Factorization Theory, Algebraic Combinatorics and several computational theories. In [12], a generalized concept, p-symmetric property, is introduced and studied. In recent works (e.g., [10, 11, 15]), no symmetric property has been found. In this paper, we show that the p-numerical semigroup of the triple $(\nu_n, \nu_{n+1}, \nu_{n+2})$ is symmetric when $p = 0$ and n is even.

2 Preliminaries

We introduce the Apéry set (see [1]) below in order to obtain the formulas for $g_p(A)$, $n_p(A)$ and $s_p(A)$. Without loss of generality, we assume that $a_1 = \min(A)$.

Definition 1 Let p be a non-negative integer. For a set of positive integers $A = \{a_1, a_2, \ldots, a_k\}$ with $\gcd(A) = 1$ and $a_1 = \min(A)$ we denote by

$$\mathrm{Ap}_p(A) = \mathrm{Ap}_p(a_1, a_2, \dots, a_k) = \{m_0^{(p)}, m_1^{(p)}, \dots, m_{a_1-1}^{(p)}\},$$

the p-Apéry set of A, where each $m_i^{(p)}$ $(1 \le i \le a_1)$ satisfied the conditions:

$$\text{(i) } m_i^{(p)} \equiv i \pmod{a_1}, \quad \text{(ii) } m_i^{(p)} \in S_p(A), \quad \text{(iii) } m_i^{(p)} - a_1 \notin S_p(A).$$

Note that $m_0^{(0)}$ is defined to be 0.

It follows that for each p,

$$\mathrm{Ap}_p(A) \equiv \{0, 1, \dots, a_1 - 1\} \pmod{a_1}.$$

When $k \ge 3$, it is hard to find any explicit form of $g_p(A)$ as well as $n_p(A)$ and $s_p(A)$. Nevertheless, the following convenient formulas are known (For a more general case, see [8]). Though finding $m_j^{(p)}$ is enough hard in general, we can obtain it for some special sequences $(a_1, a_2, \dots, a_k)$.

Lemma 1 *Let k, p and μ be integers with $k \ge 2$, $p \ge 0$ and $\mu \ge 1$. Assume that* $\gcd(a_1, a_2, \dots, a_k) = 1$. *We have*

$$g_p(a_1, a_2, \dots, a_k) = \left(\max_{0 \le j \le a_1 - 1} m_j^{(p)}\right) - a_1, \tag{1}$$

$$n_p(a_1, a_2, \dots, a_k) = \frac{1}{a_1} \sum_{j=0}^{a_1-1} m_j^{(p)} - \frac{a_1 - 1}{2}, \tag{2}$$

$$s_p(a_1, a_2, \dots, a_k) = \frac{1}{2a_1} \sum_{j=0}^{a_1-1} \left(m_j^{(p)}\right)^2 - \frac{1}{2} \sum_{j=0}^{a_1-1} m_j^{(p)} + \frac{a_1^2 - 1}{12}. \tag{3}$$

Remark When $p = 0$, the formulas (1), (2) and (3) reduce to the formulas by Brauer and Shockley [3], Selmer [21], and Tripathi [24], respectively:

$$g(a_1, a_2, \dots, a_k) = \left(\max_{1 \le j \le a_1 - 1} m_j\right) - a_1,$$

$$n(a_1, a_2, \dots, a_k) = \frac{1}{a_1} \sum_{j=0}^{a_1-1} m_j - \frac{a_1 - 1}{2},$$

$$s(a_1, a_2, \dots, a_k) = \frac{1}{2a_1} \sum_{j=0}^{a_1-1} (m_j)^2 - \frac{1}{2} \sum_{j=0}^{a_1-1} m_j + \frac{a_1^2 - 1}{12},$$

where $m_j = m_j^{(0)}$ $(1 \le j \le a_1 - 1)$ with $m_0 = m_0^{(0)} = 0$. More general formulas using Bernoulli numbers can be seen in [6, 8]. The formulas for weighted sums can be explicitly given by using Eulerian numbers [13, 14].

3 Main Results

For $n \ge 3$, let $A := \{\nu_n, \nu_{n+1}, \nu_{n+2}\}$, where $\nu_n = \nu_n(a,b) = (a^n - b^n)/(a-b)$ with $\gcd(a,b) = 1$ and $a > b$. Set $q = \lfloor \nu_n/(a+b) \rfloor$ and $r = \nu_n - (a+b)q$. Then, the p-Frobenius number is given explicitly as follows.

Theorem 1 *For* $0 \le p \le \lfloor (a^n - b^n)/(a^2 - b^2) \rfloor$ *we have*

$$g_p(A) = \begin{cases} \dfrac{\big((r-1)a-1\big)a^n - \big((r-1)b-1\big)b^n}{a-b} + \left(p + \left\lfloor \dfrac{a^n - b^n}{a^2 - b^2} \right\rfloor\right) \dfrac{a^{n+2} - b^{n+2}}{a-b} \\ \qquad \text{if } r\nu_{n+1} > ab\nu_n; \\ \dfrac{\big((b-1)a-1\big)a^n - \big((a-1)b-1\big)b^n}{a-b} + \left(p + \left\lfloor \dfrac{a^n - b^n}{a^2 - b^2} \right\rfloor\right) \dfrac{a^{n+2} - b^{n+2}}{a-b} \\ \qquad \text{otherwise}. \end{cases}$$

For simplicity, put $t_{y,z} = \nu_{n+1}y + \nu_{n+2}z$ $(y, z \ge 0)$, where $\nu_n = (a^n - b^n)/(a-b)$. Then all the elements of $\mathrm{Ap}_0(A)$, where $A = \{\nu_n, \nu_{n+1}, \nu_{n+2}\}$, are expressed in the area of Table 1.

Since $\nu_{n+1} \equiv b^n \pmod{\nu_n}$ and $\nu_{n+2} \equiv (a+b) \cdot b^n \pmod{\nu_n}$, the sequence $\{b^k \pmod{\nu_n}\}_{k=0}^{\nu_n - 1}$ is given by

$$\begin{aligned} &t_{0,0}, \dots, t_{a+b-1,0}, t_{0,1}, \dots, t_{a+b-1,1}, \dots, t_{0,q-1}, \dots, t_{a+b-1,q-1} \\ &t_{0,q}, \dots, t_{r-1,q} \end{aligned}$$

in this order. Since $\gcd(\nu_n, b^n) = 1$, the set $\{0, b, b^2, \dots, b^{\nu_n - 1}\}$ is a complete residue system modulo $(a^n - b^n)/(a-b)$.

Table 1 $\mathrm{Ap}_0(\nu_n, \nu_{n+1}, \nu_{n+2})$

$t_{0,0}$	$\cdots$	$\cdots$	$\cdots$	$t_{a+b-1,0}$
$t_{0,1}$	$\cdots$	$\cdots$	$\cdots$	$t_{a+b-1,1}$
$\vdots$				$\vdots$
$t_{0,q-1}$	$\cdots$	$\cdots$	$\cdots$	$t_{a+b-1,q-1}$
$t_{0,q}$	$\cdots$	$t_{r-1,q}$		

From Table 1, it is clear that the largest element of $\mathrm{Ap}_0(A)$ is $t_{r-1,q}$ or $t_{a+b-1,q-1}$. We see that $t_{r-1,q} > t_{a+b-1,q-1}$ if and only if $r v_{n+1} > ab v_n$. In addition,

$$\frac{a^n - b^n}{a-b} \pmod{a+b} = \begin{cases} 0 & \text{if } n \equiv 0 \pmod 2; \\ b^{n-1} & \text{if } n \equiv 1 \pmod 2. \end{cases}$$

Hence, if $n \equiv 0 \pmod 2$, $t_{a+1,q-1}$ is the largest. Thus, by Lemma 1 (1), if $r v_{n+1} > ab v_n$, then we have

$$\begin{aligned} g_0(A) &= t_{r-1,q} - v_n \\ &= \frac{(r-1)(a^{n+1} - b^{n+1})}{a-b} + \left\lfloor \frac{a^n - b^n}{a^2 - b^2} \right\rfloor \frac{a^{n+2} - b^{n+2}}{a-b} - \frac{a^n - b^n}{a-b} \\ &= \frac{\big((r-1)a - 1\big)a^n - \big((r-1)b - 1\big)b^n}{a-b} + \left\lfloor \frac{a^n - b^n}{a^2 - b^2} \right\rfloor \frac{a^{n+2} - b^{n+2}}{a-b}. \end{aligned}$$

Otherwise, we have

$$\begin{aligned} g_0(A) &= t_{a+b-1,q-1} - v_n \\ &= \frac{(a+b-1)(a^{n+1} - b^{n+1})}{a-b} + \left(\left\lfloor \frac{a^n - b^n}{a^2 - b^2} \right\rfloor - 1 \right) \frac{a^{n+2} - b^{n+2}}{a-b} - \frac{a^n - b^n}{a-b} \\ &= \frac{\big((b-1)a - 1\big)a^n - \big((a-1)b - 1\big)b^n}{a-b} + \left\lfloor \frac{a^n - b^n}{a^2 - b^2} \right\rfloor \frac{a^{n+2} - b^{n+2}}{a-b}. \end{aligned}$$

3.1 The Case $p = 1$

All the elements in $\mathrm{Ap}_1(A)$ can be determined from those in $\mathrm{Ap}_0(A)$. Only those elements that have the same residue modulo a as those in the top row of $\mathrm{Ap}_0(A)$ are arranged in order in the form of filling gaps under the same block. Elements that have the same residue modulo a as the other elements of $\mathrm{Ap}_0(A)$ are arranged in a row shift up to the immediately adjacent block. It is shown in Table 2.

This fact is supported by the congruence relations

$$\begin{aligned} t_{y,z} &\equiv t_{y+a+b,z-1} \pmod{v_1} \quad (0 \le y \le a+b-1, 1 \le z \le q-1; \\ &\qquad\qquad\qquad\qquad\qquad\quad 0 \le y \le r-1, z = q), \\ t_{y,0} &\equiv t_{y+r,q} \pmod{v_1} \quad (0 \le y \le a+b-1-r), \\ t_{a+b-r+y,0} &\equiv t_{y,q+1} \pmod{v_1} \quad (0 \le y \le r-1). \end{aligned}$$

In addition, each element of $\mathrm{Ap}_1(A)$ has two representations in terms of v_n, v_{n+1} and v_{n+2}, because

Table 2 $\mathrm{Ap}_1(A)$ from $\mathrm{Ap}_0(A)$

$t_{0,0}$	$t_{1,0}$	$\cdots$	$\cdots$	$t_{a+b-1,0}$	$t_{a+b,0}$	$t_{a+b+1,0}$	$\cdots$	$\cdots$	$t_{2(a+b)-1,0}$
$t_{0,1}$	$t_{1,1}$	$\cdots$	$\cdots$	$t_{a+b-1,1}$	$t_{a+b,1}$	$t_{a+b+1,1}$	$\cdots$	$\cdots$	$t_{2(a+b)-1,1}$
$t_{0,2}$	$t_{1,2}$	$\cdots$	$\cdots$	$t_{a+b-1,2}$	$\vdots$	$\vdots$			$\vdots$
$\vdots$	$\vdots$			$\vdots$	$t_{a+b,q-2}$	$t_{a+b+1,q-2}$	$\cdots$	$\cdots$	$t_{2(a+b)-1,q-2}$
$t_{0,q-1}$	$t_{1,q-1}$	$\cdots$	$\cdots$	$t_{a+b-1,q-1}$	$t_{a+b,q-1}$	$\cdots$	$t_{r+a+b-1,q-1}$		
$t_{0,q}$	$\cdots$	$t_{r-1,q}$	$t_{r,q}$ $\cdots$	$t_{a+b-1,q}$					
$t_{0,q+1}$	$\cdots$	$t_{r-1,q+1}$							

$$
\begin{aligned}
t_{y+a+b,z-1} &= (y+a+b)v_{n+1} + (z-1)v_{n+2} \\
&= (ab)v_n + yv_{n+1} + zv_{n+2}\,, \\
t_{y+r,q} &= (y+r)v_{n+1} + qv_{n+2} \\
&= \big(v_{n+1} - (ab)q\big)v_n + yv_{n+1}\,, \\
t_{y,q+1} &= yv_{n+1} + (q+1)v_{n+2} \\
&= \big(v_{n+1} - (ab)(q+1)\big)v_n + (y+a+b-r)v_{n+1}\,.
\end{aligned}
$$

There are four candidates to take the largest value in $\mathrm{Ap}_1(A)$:

$$
t_{r-1,q+1}, \quad t_{a+b-1,q}, \quad t_{r+a+b-1,q-1}, \quad t_{2(a+b)-1,q-2}\,.
$$

However, it is easy to see that $t_{r-1,q+1} > t_{r+a+b-1,q-1}$ and $t_{a+b-1,q} > t_{2(a+b)-1,q-2}$. In addition, $t_{r-1,q+1} > t_{a+b-1,q}$ if and only if $rv_{n+1} > abv_n$. Hence, if $rv_{n+1} > abv_n$, then by Lemma 1 (1) we have

$$
\begin{aligned}
g_1(A) &= t_{r-1,q+1} - v_n \\
&= \frac{(r-1)(a^{n+1}-b^{n+1})}{a-b} + \left(\left\lfloor \frac{a^n-b^n}{a^2-b^2} \right\rfloor + 1\right)\frac{a^{n+2}-b^{n+2}}{a-b} - \frac{a^n-b^n}{a-b}\,.
\end{aligned}
$$

Otherwise, we have

$$
\begin{aligned}
g_1(A) &= t_{a+b-1,q} - v_n \\
&= \frac{(a+b-1)(a^{n+1}-b^{n+1})}{a-b} + \left\lfloor \frac{a^n-b^n}{a^2-b^2} \right\rfloor \frac{a^{n+2}-b^{n+2}}{a-b} - \frac{a^n-b^n}{a-b}\,.
\end{aligned}
$$

3.2 *The Case $p \geq 2$*

When $p \geq 2$, in a similar manner, each element of $\mathrm{Ap}_p(A)$ is determined by the corresponding element with the same residue modulo v_n in $\mathrm{Ap}_{p-1}(A)$. In each block

Table 3 $\mathrm{Ap}_p(A)$ ($p = 0, 1, 2, 3$) for $q \geq 3$

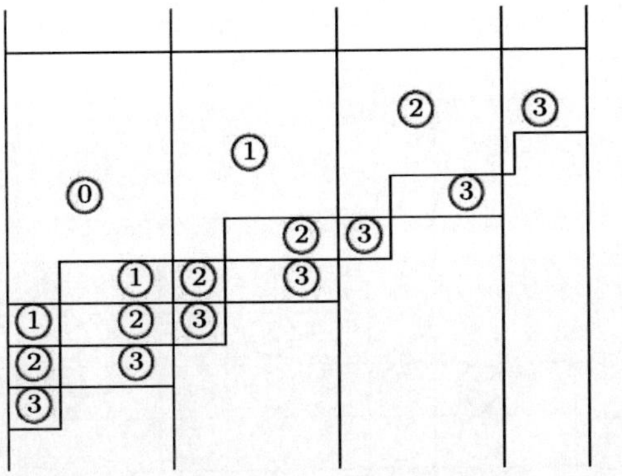

with a lateral length of $a + b$, the elements in the top row in $\mathrm{Ap}_{p-1}(A)$ are arranged in order to fill the gap below the left-most block in $\mathrm{Ap}_p(A)$. The other elements of $\mathrm{Ap}_{p-1}(A)$ are shifted directly to the right block by one in $\mathrm{Ap}_p(A)$.

In Table 3, $\textcircled{\ell}$ denotes the area of elements in $\mathrm{Ap}_\ell(A)$. Here, each $m_j^{(\ell)}$, satisfying $m_j^{(\ell)} \equiv j \pmod{\nu_n}$ $(0 \leq j \leq \nu_n - 1)$, can be expressed in at least $\ell + 1$ ways but $m_j^{(\ell)} - \nu_n$ in at most ℓ ways. Each element of $\mathrm{Ap}_3(A)$ existing in the second block to the fourth block corresponds to each element having the same residue of $\mathrm{Ap}_2(A)$ existing in the block immediately to the left thereof in a form of shifting up one step. The $(a + b)$ elements of $\mathrm{Ap}_3(A)$ existing over two rows (or one row) at the bottom of the first block correspond to the $(a + b)$ elements with the same residue of $\mathrm{Ap}_2(A)$ at the top of the third block. Therefore, since all the elements in $\mathrm{Ap}_2(A)$ form a complete remainder system, so is $\mathrm{Ap}_3(A)$. We omit the details, but it can be seen that all the elements of $\mathrm{Ap}_3(A)$ have exactly four ways of expressing in terms of ν_n, ν_{n+1}, and ν_{n+2}. Within each region of $\mathrm{Ap}_3(A)$, one of the two leftmost (lower left) elements $t_{r-1,q+3}$ and $t_{4,q+2}$ is the largest, so by comparing these elements, the largest element of $\mathrm{Ap}_3(A)$ can be determined. Namely, if $r\nu_{n+1} > ab\nu_n$, $t_{r-1,q+3}$ is the largest. Otherwise, $t_{a+b-1,q+2}$ is the largest.

This structure of $\mathrm{Ap}_p(A)$ continues as long as $p \leq \lfloor \nu_n/(a + b) \rfloor = q$. Eventually, the largest element in $\mathrm{Ap}_p(A)$ is $t_{r-1,q+p}$ or $t_{a+b-1,q+p-1}$. However, when $p = \lfloor \nu_n/(a + b) \rfloor + 1$, this kind of regularity is broken. Therefore, regularity cannot be maintained even for the largest value of $\mathrm{Ap}_p(A)$ too. Therefore, we can conclude as follows. Therefore, for $0 \leq p \leq q = \lfloor (a^n - b^n)/(a^2 - b^2) \rfloor$, if $r\nu_{n+1} > ab\nu_n$, we have

$$
\begin{aligned}
g_p(A) &= t_{r-1,q+p} - \nu_n \\
&= \frac{(r-1)(a^{n+1} - b^{n+1})}{a - b} + \left(p + \left\lfloor \frac{a^n - b^n}{a^2 - b^2} \right\rfloor\right) \frac{a^{n+2} - b^{n+2}}{a - b} - \frac{a^n - b^n}{a - b} \\
&= \frac{((r-1)a - 1)a^n - ((r-1)b - 1)b^n}{a - b} + \left(p + \left\lfloor \frac{a^n - b^n}{a^2 - b^2} \right\rfloor\right) \frac{a^{n+2} - b^{n+2}}{a - b}.
\end{aligned}
$$

Table 4 $\mathrm{Ap}_p(A)$ $(p = \lfloor \nu_n/(a+b) \rfloor)$

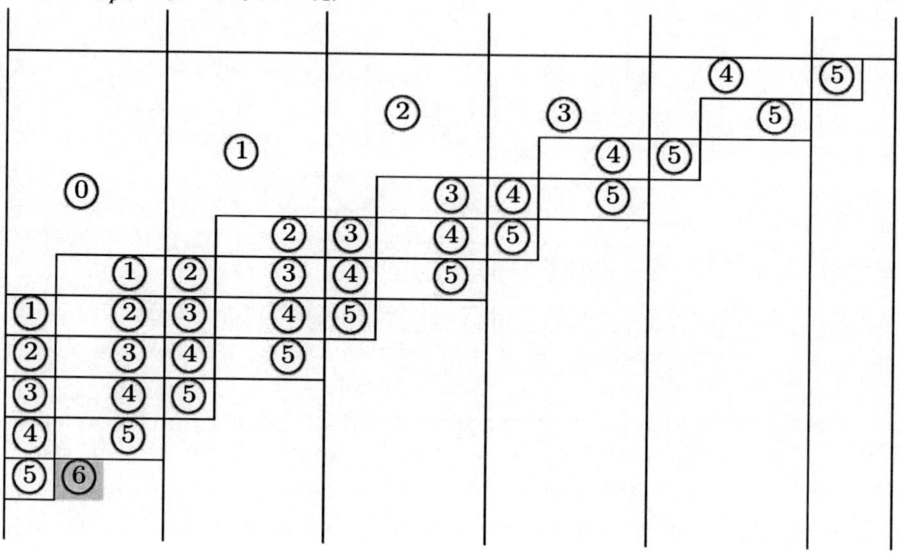

Otherwise, we have

$$
\begin{aligned}
g_p(A) &= t_{a+b-1,q+p-1} - \nu_n \\
&= \frac{(a+b-1)(a^{n+1}-b^{n+1})}{a-b} + \left(p + \left\lfloor \frac{a^n - b^n}{a^2 - b^2} \right\rfloor - 1\right) \frac{a^{n+2}-b^{n+2}}{a-b} - \frac{a^n - b^n}{a-b} \\
&= \frac{\big((b-1)a-1\big)a^n - \big((a-1)b-1\big)b^n}{a-b} + \left(p + \left\lfloor \frac{a^n - b^n}{a^2 - b^2} \right\rfloor\right) \frac{a^{n+2}-b^{n+2}}{a-b} .
\end{aligned}
$$

Table 4 shows the arrangement of the p-Apéry sets ($p = 0, 1, \dots, 5$) when $\lfloor \nu_n/(a+b) \rfloor = 5$. One can see that there will be a deficiency in the arrangement of the lower left for 6-Apéry set.

4 p-Genus

The elements of $\mathrm{Ap}_p(A)$ in the area of the $2p$ staircase parts are

$$
\begin{aligned}
&t_{0,q+p}, \dots, t_{r-1,q+p}, \quad t_{r,q+p-1}, \dots, t_{a+b-1,q+p-1}, \\
&t_{a+b,q+p-2}, \dots, t_{r+a+b-1,q+p-2}, \quad t_{r+a+b,q+p-3}, \dots, t_{2(a+b)-1,q+p-3}, \\
&t_{2(a+b),q+p-4}, \dots, t_{r+2(a+b)-1,q+p-4}, \quad t_{r+2(a+b),q+p-5}, \dots, t_{3(a+b)-1,q+p-5}, \\
&\dots
\end{aligned}
$$

$$t_{(a+b)(p-1),q-p+2}, \dots, t_{(a+b)(p-1)+r-1,q-p+2},$$
$$t_{(a+b)(p-1)+r,q-p+1}, \dots, t_{(a+b)p-1,q-p+1}$$

in order from the lower left, and the elements of $\mathrm{Ap}_p(A)$ in the right-most main area are

$$t_{(a+b)p,0}, \dots, t_{(a+b)p+a+b-1,0},$$
$$t_{(a+b)p,1}, \dots, t_{(a+b)p+a+b-1,1},$$
$$\dots$$
$$t_{(a+b)p,q-p-1}, \dots, t_{(a+b)p+a+b-1,q-p-1},$$
$$t_{(a+b)p,q-p}, \dots, t_{(a+b)p+r-1,q-p}\,.$$

Hence, by $v_n = (a+b)q + r$, we have

$$\begin{aligned}
&\sum_{w\in\mathrm{Ap}_p(A)} w\\
&= \sum_{i=0}^{p-1}\sum_{j=0}^{r-1} t_{(a+b)i+j,q+p-2i} + \sum_{i=0}^{p-1}\sum_{j=0}^{a+b-1-r} t_{(a+b)i+r+j,q+p-2i-1}\\
&\quad + \sum_{i=0}^{a+b-1}\sum_{j=0}^{q-p-1} t_{(a+b)p+i,j} + \sum_{i=0}^{r-1} t_{(a+b)p+i,q-p}\\
&= \frac{v_n}{2}\biggl(v_{n+1}(v_n-1) - ab(2v_n-a-b)q + ab(a+b)q^2\\
&\quad + (a+b)(2v_{n+1}-ab)p - ab(a+b)p^2\biggr)\,. \qquad (4)
\end{aligned}$$

Thus, by Lemma 1 (2), we obtain that

$$\begin{aligned}
&n_p(v_n, v_{n+1}, v_{n+2})\\
&= \frac{1}{v_n}\sum_{w\in\mathrm{Ap}_p(A)} w - \frac{v_n-1}{2}\\
&= \frac{(v_n-1)(v_{n+1}-1)}{2} - \frac{ab(2v_n-a-b)}{2}q + \frac{ab(a+b)}{2}q^2\\
&\quad + \frac{(a+b)(2v_{n+1}-ab)}{2}p - \frac{ab(a+b)}{2}p^2\,.
\end{aligned}$$

Theorem 2 *Let a and b positive integers with* $\gcd(a,b)=1$ *and* $a>b$*. Then, for a positive integer* $n\ge 3$*, and* $0\le p\le q=\lfloor (a^n-b^n)/(a^2-b^2)\rfloor$ *we have*

$$
\begin{aligned}
&n_p\left(\frac{a^n-b^n}{a-b},\frac{a^{n+1}-b^{n+1}}{a-b},\frac{a^{n+2}-b^{n+2}}{a-b}\right)\\
&=\frac{1}{2}\left(\frac{a^n-b^n}{a-b}-1\right)\left(\frac{a^{n+1}-b^{n+1}}{a-b}-1\right)\\
&\quad-\frac{ab}{2}\left(\frac{2(a^n-b^n)}{a-b}-a-b\right)q+\frac{ab(a+b)}{2}q^2\\
&\quad+\frac{a+b}{2}\left(\frac{2(a^{n+1}-b^{n+1})}{a-b}-ab\right)p-\frac{ab(a+b)}{2}p^2\,.
\end{aligned}
$$

5 *p*-Sylvester Sum

The explicit formulas for more general types of sums

$$
\sum_{w\in\mathbb{N}_0\setminus S_p} w^\mu \quad\text{and}\quad \sum_{w\in\mathbb{N}_0\setminus S_p} \lambda^w w^\mu \quad (\lambda\ne 1)
$$

can be given by using the formulas including the elements of the p-Apéry set [6, 8, 13, 14]. The former one can be expressed by using Bernoulli numbers or Apostol-Bernoulli numbers, and the latter one can be expressed by using Eulerian numbers. If $w=0$, the former one is reduced to the p-genus. If $w=1$, the former one is reduced to the p-Sylvester sum. Here we simply give the formula for the p-Sylvester sum.

By $r=\nu_n-(a+b)q$ and $(a+b)\nu_{n+1}-\nu_{n+2}=ab\nu_n$, we have

$$
\begin{aligned}
&\sum_{w\in\mathrm{Ap}_p(A)} w^2\\
&=\frac{\nu_n}{6}(2\nu_n-1)(\nu_n-1)\nu_{n+1}^2\\
&\quad-\frac{\nu_n}{6}qab(a+b)\big((6\nu_n(\nu_n-1)-a-b+3)\nu_{n+1}-ab\nu_n\big)\\
&\quad+\frac{\nu_n}{2}q^2ab\big(2ab\nu_n^2+(a+b)(\nu_{n+2}-\nu_{n+1})\big)\\
&\quad+\frac{\nu_n}{3}q^3ab(a+b)(a^{n+1}+b^{n+1})\\
&\quad+\nu_n p\bigg(\frac{1}{6}(6(a+b)(\nu_n-1)\nu_{n+1}^2-ab(a+b)(2a+2b-3)\nu_{n+1}
\end{aligned}
$$

$$
\begin{aligned}
&- 3ab(2\nu_n - a - b)\nu_{n+2}\big) \\
&- qab(a+b)(2\nu_n - a - b)\nu_{n+1} + q^2ab(a+b)^2\nu_{n+1}\Big) \\
&+ \frac{\nu_n}{2}p^2\big(ab(a+b)((2\nu_n - 2a - 2b + 1)\nu_{n+1} + \nu_{n+2}) + 2\nu_{n+2}^2\big) \\
&- \frac{2\nu_n}{3}p^3ab(a+b)^2 .
\end{aligned}
$$

By Lemma 1 (3), together with (4), we have

$$
\begin{aligned}
s_p(A) &= \frac{1}{2\nu_n}\sum_{w\in\mathbb{N}_0\backslash S_p(A)} w^2 - \frac{1}{2}\sum_{w\in\mathbb{N}_0\backslash S_p(A)} w + \frac{\nu_n^2 - 1}{12} \\
&= \frac{(\nu_n - 1)(\nu_{n+1} - 1)(2\nu_n\nu_{n+1} - \nu_n - \nu_{n+1} - 1)}{12} \\
&\quad + \frac{abq}{12}\Big(3\nu_n(2\nu_n - a - b) \\
&\qquad - (a+b)\big(6\nu_n(\nu_n - 1)\nu_{n+1} - ab\nu_n - (a+b-3)\nu_{n+1}\big)\Big) \\
&\quad + \frac{abq^2}{4}\big(\nu_n(2ab\nu_n - a - b) + (a+b)(\nu_{n+2} - \nu_{n+1})\big) \\
&\quad + \frac{\nu_n}{3}q^3ab(a+b)(a^{n+1} + b^{n+1}) \\
&\quad - \frac{p}{12}\Big(2ab(a+b)(a\nu_{n+2} + b\nu_{n+1}) - 3ab(a+b)(\nu_n + \nu_{n+1} + \nu_{n+2}) \\
&\qquad + 6(a+b)\nu_{n+1}(\nu_{n+1}^2 - \nu_n\nu_{n+1} + \nu_n) + 6ab\nu_n\nu_{n+2} \\
&\qquad - qab(a+b)(2\nu_n - a - b)\nu_{n+1} + q^2ab(a+b)^2\nu_{n+1}\Big) \\
&\quad + \frac{\nu_n}{2}p^2\big(ab(a+b)((2\nu_n - 2a - 2b + 1)\nu_{n+1} + \nu_n + \nu_{n+2}) + 2\nu_{n+2}^2\big) \\
&\quad - \frac{2\nu_n}{3}p^3ab(a+b)^2 .
\end{aligned}
$$

6 Examples

By Theorems 1 and 2 with $a = 3$ and $b = 2$, we can obtain

$$
g_{13}(3^4 - 2^4, 3^5 - 2^5, 3^6 - 2^6) = 17404 ,
$$

$$n_{13}(3^4-2^4,3^5-2^5,3^6-2^6)=15365\,,$$

$$s_{13}(3^4-2^4,3^5-2^5,3^6-2^6)=118607650\,,$$

which match the real situation. But when $p=14>\lfloor(3^4-2^4)/5\rfloor$, the real values cannot be applied by the formulas.

When $a=13$ and $b=4$, Theorems 1 and 2 hold for $0\le p\le q=\lfloor(13^4-4^4)/(13^2-4^2)\rfloor=185$. By $r=0$, when $p=5$, we get

$$g_5\left(\frac{13^4-4^4}{9},\frac{13^5-4^5}{9},\frac{13^6-4^6}{9}\right)=101932084$$

and

$$n_5\left(\frac{13^4-4^4}{9},\frac{13^5-4^5}{9},\frac{13^6-4^6}{9}\right)=53110125\,.$$

By using GAP package numericalsgps [4, 5], the p-Frobenius number and the p-genus can be obtained systematically.

```
gap> s:=NumericalSemigroup((13^4-4^4)/9,(13^5-4^5)/9,(13^6-4^6)/9
);
<Numerical semigroup with 3 generators>
gap> FrobeniusNumber(s);
99252799
gap> Number(Difference([0..FrobeniusNumber(s)+1],SmallElements(s)
));
49626400
gap> FrobeniusNumber(DenumerantIdeal(s,5));
101932084
gap> Number(Difference([0..FrobeniusNumber(DenumerantIdeal(s,5))
+1],
SmallElements(DenumerantIdeal(s,5))));
53110125
```

7 Symmetric Properties

By using the results on p-Frobenius numbers and p-genus, we can find symmetric properties for some cases. The p-numerical semigroup S_p, which is p-generated from A, is called *p-symmetric* if for all $x\in\mathbb{Z}\backslash S_p$, $\ell_0(p)+g_p(A)-x\in S_p$, where $\ell_0(p)$ is the least element of S_p, that is the *p-multiplicity* of S_p if $p\ge 1$; $\ell_0(p)=0$ if $p=0$. When $p=0$, "0-symmetric" is just "symmetric". The following criterion [12] is useful to prove the symmetric property.

Lemma 2 *For a non-negative integer p, S_p, which is p-generated from A, is p-symmetric if and only if*

$$n_p(A) = \frac{g_p(A) + \ell_0(p) + 1}{2}.$$

Since for even n

$$\left\lfloor \frac{a^n - b^n}{a^2 - b^2} \right\rfloor = \frac{a^n - b^n}{a^2 - b^2}$$

and $r = 0$, we have $r v_{n+1} < ab v_n$. Thus,

$$2n_p(A) - g_p(A) = 1 - p(p+1)ab(a+b) + 2pabv_n + pv_{n+2}.$$

Hence, by Lemma 2 with $\ell_0(0) = 0$, if $p = 0$, then S_0 is symmetric. However, as in general,

$$\ell_0(p) = p(a+b)v_{n+1} \neq -p(p+1)ab(a+b) + 2pabv_n + pv_{n+2},$$

for $p \geq 1$, S_p is not necessarily symmetric.

Theorem 3 *If n is even, then $S_0(v_n, v_{n+1}, v_{n+2})$ is symmetric.*

Example In the previous example,

$$S_0\left(\frac{13^4 - 4^4}{9}, \frac{13^5 - 4^5}{9}, \frac{13^6 - 4^6}{9}\right)$$

is symmetric. For a smaller triple,

$$S_0\left(\frac{5^2 - 2^2}{3}, \frac{5^3 - 2^3}{3}, \frac{5^4 - 2^4}{3}\right)$$

is symmetric. Indeed, we can see that

$$\begin{array}{llllllllllllllll} S_0 = & \{\ldots, 226, & 225, & 224, & 223, & 222, & 221, & 219, & \ldots, & 39, & 35, & 28, & 21, & 14, & 7, & 0\ \} \\ \mathbb{Z}\backslash S_0 = & \{\ldots, 1, & 2, & 3, & 4, & 5, & 6, & 8, & \ldots, & 188, & 192, & 199, & 206, & 213, & 220, & 227\ \} \end{array}$$

They satisfy the relation

$$\begin{aligned} &226 + 1 = 225 + 2 = 224 + 3 = 223 + 4 = 222 + 5 = \cdots \\ &\qquad = 28 + 199 = 21 + 206 = 14 + 213 = 7 + 220 = 0 + 227 = 227. \end{aligned}$$

8 Final Comments

In this paper, the appearance of patterns similar to those in [10, 11, 15] should not be mistaken for applying a simple process of repetitive process. It is just a lot of luck, and for many triples we still do not even know how the elements of the Apéry set will be arranged. In fact, it turns out that for some triples of the Pell sequence, the arrangement of the elements of the Apéry set is different, so that the simple and similar process of repetition does not work at all (in preparation).

It would be natural to consider the n-tuple

$$A = \left\{ \frac{a^n - b^n}{a-b}, \frac{a^{n+1} - b^{n+1}}{a-b}, \dots, \frac{a^{2n-1} - b^{2n-1}}{a-b} \right\}.$$

Note that $(a^n - b^n)/(a-b)$ divides $(a^{2n} - b^{2n})/(a-b)$ $(k \ge 0)$. When $b = 1$, such a case has been studied in [19]. For example, it is obtained that for an odd with $a \ge 7$

$$\begin{aligned} & g_0\left(\frac{a^4 - 2^4}{a-2}, \frac{a^5 - 2^5}{a-2}, \frac{a^6 - 2^6}{a-2}, \frac{a^7 - 2^7}{a-2}\right) \\ & = \frac{7(a^5 - 2^5)}{a-2} + \frac{a(a^7 - 2^7)}{a-2} - \frac{a^4 - 2^4}{a-2}. \end{aligned}$$

However, even when $p = 0$ and $b \ge 3$, no explicit formula has been conjectured. When $p > 0$, it is even harder to find explicit closed formulas of $g_p(A)$ and $n_p(A)$ for four or more variables.

References

1. Apéry, R.: Sur les branches superlinéaires des courbes algébriques. C. R. Acad. Sci. Paris **222**, 1198–1200 (1946)
2. Bokaew, R., Yuttanan, B., Mavecha, S.: Formulae of the Frobenius number in relatively prime three Lucas numbers. Songklanakarin J. Sci. Technol. **42**(5), 1077–1083 (2020)
3. Brauer, A., Shockley, B.M.: On a problem of Frobenius. J. Reine. Angew. Math. **211**, 215–220 (1962)
4. Delgado, M., García-Sánchez, P.A., Morais, J.: "numericalsgps": a GAP package on numerical semigroups. http://www.gap-system.org/Packages/numericalsgps.html
5. Delgado, M., García-Sánchez, P.A., Morais, J.: numericalsgps- a package for numerical semigroups. Version 1.3.0 dev, 2022 (Refereed GAP package). https://github.com/numerical-semigroups
6. Komatsu, T.: Sylvester power and weighted sums on the Frobenius set in arithmetic progression. Discrete Appl. Math. **315**, 110–126 (2022)
7. Komatsu, T.: The Frobenius number for sequences of triangular numbers associated with number of solutions. Ann. Comb. **26**, 757–779 (2022)

8. Komatsu, T.: On p-Frobenius and related numbers due to p-Apéry set (2022). arXiv:2111.11021v3
9. Komatsu, T.: The Frobenius number associated with the number of representations for sequences of repunits. C. R. Math., Acad. Sci. Paris **361**, 73–89 (2023). https://doi.org/10.5802/crmath.394
10. Komatsu, T., Pita-Ruiz, C.: The Frobenius number for Jacobsthal triples associated with number of solutions. Axioms **12**(2), Article 98, 18 pp. (2023). https://doi.org/10.3390/axioms12020098
11. Komatsu, T., Ying, H.: The p-Frobenius and p-Sylvester numbers for Fibonacci and Lucas triplets. Math. Biosci. Eng. **20**(2), 3455–3481 (2023). https://doi.org/10.3934/mbe.2023162
12. Komatsu, T., Ying, H.: p-Numerical semigroups with p-symmetric properties. J. Algebra Appl. (online ready). https://doi.org/10.1142/S0219498824502165
13. Komatsu, T., Zhang, Y.: Weighted Sylvester sums on the Frobenius set. Irish Math. Soc. Bull. **87**, 21–29 (2021)
14. Komatsu, T., Zhang, Y.: Weighted Sylvester sums on the Frobenius set in more variables. Kyushu J. Math. **76**, 163–175 (2022)
15. Komatsu, T., Laishram, S., Punyani, P.: p-Numerical semigroups of generalized Fibonacci triples. Symmetry **15**(4), Article 852, 13 pp. (2023). https://doi.org/10.3390/sym15040852
16. Robles-Pérez, A.M., Rosales, J.C.: The Frobenius number for sequences of triangular and tetrahedral numbers. J. Number Theory **186**, 473–492 (2018)
17. Rosales, J.C., Garcia-Sanchez, P.A.: Numerical semigroups with embedding dimension three. Arch. Math. (Basel) **83**(6), 488–496 (2004)
18. Rosales, J.C., Branco, M.B., Torrão, D.: The Frobenius problem for Thabit numerical semigroups. J. Number Theory **155**, 85–99 (2015)
19. Rosales, J.C., Branco, M.B., Torrão, D.: The Frobenius problem for repunit numerical semigroups. Ramanujan J. **40**, 323–334 (2016)
20. Rosales, J.C., Branco, M.B., Torrão, D.: The Frobenius problem for Mersenne numerical semigroups. Math. Z. **286**, 741–749 (2017)
21. Selmer, E.S.: On the linear diophantine problem of Frobenius. J. Reine Angew. Math. **293/294**, 1–17 (1977)
22. Sylvester, J.J.: On subinvariants, i.e. semi-invariants to binary quantics of an unlimited order. Am. J. Math. **5**, 119–136 (1882)
23. Sylvester, J.J.: Mathematical questions with their solutions. Educ. Times **41**, 21 (1884)
24. Tripathi, A.: On sums of positive integers that are not of the form $ax + by$. Amer. Math. Mon. **115**, 363–364 (2008)

Some Identities for Balancing and Lucas-Balancing Numbers in Bidimensional Version

José Chimpanzo, Paula Catarino, and M. Victoria Otero-Espinar

Abstract Several number sequences have been studied for many researchers, most of times in their unidimensional version. Our main goal is to study the bidimensional version of two sequences of known numbers. At the expense of studying the unidimensional version of the numerical balancing sequence and the numerical Lucas-balancing sequence, in particular on the basis of some of their properties, some identities related to bidimensional extensions of these two number sequences involving sums of terms are introduced: with odd index, with even index and with generic index.

1 Introduction

Actually, several sequences of numbers have been the research topic of many researchers. The Fibonacci, Leonardo and Lucas sequences have been studied with great frequency. In addition to these sequences, there are many other sequences that have, in a certain way, contributed to the increase of researchers interested in this area and its varied applications. There is undoubtedly a vast literature related to the study of sequences, among which numerical sequences, polynomials, quaternions, octonions and sedenions stand out. The reader can consult some of these studies for further clarification, the works of Catarino and Campos [9], where the incomplete numbers k-Pell, k-Pell-Lucas and modified k-Pell were introduced; the work of

J. Chimpanzo (✉)
Department of Exact Sciences, Higher Polytechnic Institute of Soyo, Soyo, Angola

P. Catarino
Department of Mathematics, University of Tràs-os-Montes e Alto Douro, Vila Real, Portugal
e-mail: pcatarin@utad.pt

M. V. Otero-Espinar
Department of Statistics, Mathematical Analysis and Optimization, University of Santiago de Compostela, Santiago de Compostela, Spain
e-mail: mvictoria.otero@usc.es

V. Gayoso Martínez et al. (eds.), *Mathematical Methods for Engineering Applications*, Springer Proceedings in Mathematics & Statistics 439,
https://doi.org/10.1007/978-3-031-49218-1_3

Yilmaz et al. [26], where the harmonic complex balancing numbers were defined, a type of symmetric matrices whose inputs are harmonic complex balancing numbers and some of their inequalities were also studied; of Kizilates [15] with the study of another sequence of hybrid numbers related to the Fibonacci and Lucas numbers; of Catarino [7], where the sequence of hybrid numbers k-Pell was introduced; of Alves and Catarino [2], where $h(x)$-polynomials of Jacobsthal was studied; de Catarino [5], where a new sequence of polynomials was introduced, whose terms are quaternions; by Catarino and Borges [8], where the sequence of numbers derived through Leonardo numbers was introduced; by Catarino [6], where the sequence of bicomplex quaternions whose terms are the k-Pell numbers was studied, among many other works. In addition to these works, the papers of Akbiyik and Alo in [1], where an application in the area of cryptography of the Fibonacci bronze numbers defined by a third-order recurrence relation was established; by Ayse Karakas and Murat Karakas [13] with the introduction of the space of sequences by Lucas, where some properties are investigated; by Karataş and Halici [14] with the study of the bicomplex version of the Pell and Pell-Lucas numbers and the work of Qi [22] with the study of the determination of expressions and recurrence relations for generalized Fubini and Eulerian polynomials.

Several properties and identities of sequences result from using the so-called Binet formula for that sequence. This formula allows one to determine any term of a specific sequence without first having to resort to its previous terms. This formula was deduced by Levesque in 1985 [16].

Our purpose in this work is to present and study some identities of two new bidimensional sequences. The number sequences that will serve as the basis for the study that we will present in this article are the numerical sequences called balancing and Lucas-balancing.

The mathematical researchers Behera and Panda in [3] introduced the sequence $\{B_n\}_{n\geq 0}$ of balancing numbers satisfying the recurrence relation

$$B_{n+1} = 6B_n - B_{n-1}, n \geq 1,$$

with the initial conditions $B_0 = 0$ and $B_1 = 1$.

Like any type of number sequence, this sequence has an infinite number of elements, some of them being

$$0, 1, 6, 35, 204, 1189, \ldots,$$

certainly belonging to the sequence A001109 of the OEIS [24] Encyclopedia of numerical sequences.

There is another sequence of numbers derived or coming from the balancing numbers called Lucas-balancing numbers. This number sequence has the same recurrence relation as the balancing numbers, differing only in its initial conditions which, in this case, are $C_0 = 1$ and $C_1 = 3$. There is an infinite number of elements in this sequence, some of them being

$$1, 3, 17, 99, 577, 3363, \ldots,$$

belonging to the sequence A001541 from the OEIS [24].

Panda in [23], shows one of the relations between the terms of order n of the sequences $\{B_n\}_{n\geq 0}$ and $\{C_n\}_{n\geq 0}$, as

$$C_n = \sqrt{8B_n^2 + 1}.$$

(For more information on this, see the articles [4, 10, 17, 18, 21], among others.)

There are several studies related to bidimensional, three-dimensional and n-dimensional versions of different sequences. As an example, in [25] we can see a brief study about the Leonardo sequence, as well as an approach to the bidimensional recurrence relations of this sequence at the expense of its unidimensional version. In the studies carried out in [2], the authors focus on the bidimensional and three-dimensional recurrence relations defined by means of a unidimensional recursive model of the Narayana sequence. In [20], a related study was carried out with the bidimensional and three-dimensional identities for Fibonacci numbers in complex form, and in [19] the Gaussian Fibonacci numbers and their bidimensional recurrence relation were studied and introduced. Thus, it was these and other works that motivated the study of new two-dimensional versions of two numerical sequences, based on their unidimensional model.

In this paper we present the bidimensional extension of these two sequences, namely, bidimensional balancing numbers $B_{(n,m)}, n, m \geq 0$ and bidimensional Lucas-balancing numbers $C_{(n,m)}, n, m \geq 0$. These sequences were introduced by Chimpanzo et al. in [11] and are defined, respectively, as follows:

$$\begin{cases} B_{(n+1,m)} = 6B_{(n,m)} - B_{(n-1,m)}, \\ B_{(n,m+1)} = 6B_{(n,m)} - B_{(n,m-1)}, \end{cases}$$

with the following initial conditions $B_{(0,0)} = 0$, $B_{(1,0)} = 1$, $B_{(0,1)} = i$, $B_{(1,1)} = 1 + i$, where $i^2 = -1$ and n and m are non-negative integers; and

$$\begin{cases} C_{(n+1,m)} = 6C_{(n,m)} - C_{(n-1,m)}, \\ C_{(n,m+1)} = 6C_{(n,m)} - C_{(n,m-1)}, \end{cases}$$

with the following initial conditions $C_{(0,0)} = 1$, $C_{(1,0)} = 3$, $C_{(0,1)} = 1 + i$, $C_{(1,1)} = 3 + i$. (For more information on these bidimensional numerical sequences, we reiterate to see the works described in [11] and in [12].)

About these new sequences, we study in the first two sections some identities satisfied by them involving sums of their terms.

2 Some Identities Involving Bidimensional Balancing Numbers

In this section, we will explore some identities of the bidimensional balancing numbers, using some properties attached to the sequence of balancing numbers. Note that in [11] we study similar sums of terms $B_{(m,n)}$ of this numerical sequence considering, for all non-negative integers m, the cases where n is odd, even and a generic index. Such results are the Propositions 4.1, 4.2 and 4.3 in [11], and they are described respectively as follows:

- The sum of the numbers $B_{(n,p)}$ of odd index p can be described by:

$$\sum_{l=1}^{n} B_{(2l-1,m)} = \left(3\left(B_{n-1}B_n\right) + \frac{1}{2}\left(B_{2n-1}+1\right)\right)\left(B_m - B_{m-1} + B_m i\right) - B_{n-1}B_nB_m i.$$

- The sum of the numbers $B_{(n,p)}$ of even and not null index p can be described by:

$$\sum_{l=1}^{n} B_{(2l,m)} = B_nB_{n+1}\left(B_m - B_{m-1} + B_m i\right) - \left(3B_{n-1}B_n + \frac{1}{2}\left(B_{2n-1}+1\right)\right)B_m i.$$

- The sum of the first p numbers $B_{(n,p)}$, can be described as follows:

$$\sum_{l=1}^{n} B_{(l,m)} = \left(\frac{-1 - B_n + B_{n+1}}{4}\right)\left(B_m - B_{m-1}\right) + B_nB_m i.$$

Now, the next result is new and is related with the case with $B_{(n,p)}$ with p an odd index.

Proposition 1 *For all non-negative integers n, the sum of the numbers $B_{(n,p)}$ of odd index p can be given by:*

$$\sum_{j=1}^{m} B_{(n,2j-1)} = B_n\left(B_m^2 - B_mB_{m+1}\right) + \left(B_n - B_{n-1}\right)B_m^2 i.$$

Proof By Theorem 3.3 in [11], we have

$$\sum_{j=1}^{m} B_{(n,2j-1)} = \sum_{j=1}^{m} \left(B_n \left(B_{2j-1} - B_{2j-2}\right) + \left(B_n - B_{n-1}\right) B_{2j-1} i\right).$$

Thus,

$$\begin{aligned}\sum_{j=1}^{m} B_{(n,2j-1)} &= B_n \sum_{j=1}^{m} \left(B_{2j-1} - B_{2j-2}\right) + \left(B_n - B_{n-1}\right) \sum_{j=1}^{m} B_{2j-1} i \\ &= B_n \left(\sum_{j=1}^{m} B_{2j-1} - \sum_{j=1}^{m} B_{2j-2}\right) + \left(B_n - B_{n-1}\right) \sum_{j=1}^{m} B_{2j-1} i \\ &= \left(B_n + \left(B_n - B_{n-1}\right) i\right) \sum_{j=1}^{m} B_{2j-1} - B_n \sum_{j=1}^{m} B_{2j-2}.\end{aligned}$$

The result follows, using what is described in items (*a*) and (*b*) of Corollary 2.3.6 in [23]. □

Now we present the case where the second index of $B_{(n,p)}$ is even.

Proposition 2 *For all non-negative integers n, the sum of the numbers* $B_{(n,p)}$ *of even index p can be given by:*

$$\sum_{j=1}^{m} B_{(n,2j)} = B_n \left(B_m B_{m+1} - B_m^2\right) + \left(B_n - B_{n-1}\right) B_m B_{m+1} i.$$

Proof Using Theorem 3.3 in [11], we have

$$\sum_{j=1}^{m} B_{(n,2j)} = \sum_{j=1}^{m} \left(B_n \left(B_{2j} - B_{2j-1}\right) + \left(B_n - B_{n-1}\right) B_{2j} i\right).$$

Hence

$$\begin{aligned}\sum_{j=1}^{m} B_{(n,2j)} &= B_n \sum_{j=1}^{m} \left(B_{2j} - B_{2j-1}\right) + \left(B_n - B_{n-1}\right) \sum_{j=1}^{m} B_{2j} i \\ &= B_n \left(\sum_{j=1}^{m} B_{2j} - \sum_{j=1}^{m} B_{2j-1}\right) + \left(B_n - B_{n-1}\right) \sum_{j=1}^{m} B_{2j} i \\ &= \left(B_n + \left(B_n - B_{n-1}\right) i\right) \sum_{j=1}^{m} B_{2j} - B_n \sum_{j=1}^{m} B_{2j-1}.\end{aligned}$$

Applying what is described in Corollary 2.3.6, items (b) and (a) in [23], the result follows. □

Finally, in the next result we consider the case of $B_{(n,p)}$ for a generic index p.

Proposition 3 *For all non-negative integers n, the sum of the numbers $B_{(n,p)}$ can be described by:*

$$\sum_{j=1}^{m} B_{(n,j)} = \frac{1}{4} B_n \left(B_{m-1} + 2B_m + B_{m+1}\right) - \frac{1}{4} \left(B_n - B_{n-1}\right)\left(1 + B_m - B_{m+1}\right) i.$$

Proof Analogously to the proof of the previous results, by Theorem 3.3 in [11], we get:

$$\sum_{j=1}^{m} B_{(n,j)} = \sum_{j=1}^{m} \left(B_n \left(B_j - B_{j-1}\right) + \left(B_n - B_{n-1}\right) B_j i\right).$$

So, we obtain:

$$\begin{aligned}
\sum_{j=1}^{m} B_{(n,j)} &= B_n \sum_{j=1}^{m} \left(B_j - B_{j-1}\right) + \left(B_n - B_{n-1}\right) \sum_{j=1}^{m} B_j i \\
&= B_n \left(\sum_{j=1}^{m} B_j - \sum_{j=1}^{m} B_{j-1}\right) + \left(B_n - B_{n-1}\right) \sum_{j=1}^{m} B_j i \\
&= \left(B_n + \left(B_n - B_{n-1}\right) i\right) \sum_{j=1}^{m} B_j - B_n \left(B_0 + \sum_{j=1}^{m-1} B_j\right).
\end{aligned}$$

Applying what is described in item (6) of Proposition 2.6 in [10] and by the initial conditions of the one-dimensional sequence of balancing numbers $\{B_n\}_{n \geq 0}$, the result follows. □

3 Some Identities Involving Bidimensional Lucas-Balancing Numbers

In this section, we will explore some identities of the bidimensional Lucas-balancing sequence by applying some properties related to the unidimensional version of balancing and Lucas-balancing numbers. Once more, in [11] we study similar sums of terms $C_{(m,n)}$ of this sequence of numbers considering, for all non-negative integers m, the cases where n is odd, even and a generic index. These results are the Propositions 6.1, 6.2 and 6.3 in [11], and they are expressed respectively as follows:

- The sum of the numbers $C_{(n,p)}$ of odd index p can be described by:

$$\sum_{l=1}^{n} C_{(2l-1,m)} = \left(\frac{4 - C_n + C_{n+1} + 2ST_n - 2ST_{n+1}}{4}\right)(B_m - B_{m-1}) + \left(3B_{n-1}B_n + \frac{1}{2}(B_{2n-1} + 1) - B_{n-1}B_n\right)B_m i.$$

- The sum of the numbers $C_{(n,p)}$ of even and not null index p can be described by:

$$\sum_{l=1}^{n} C_{(2l,m)} = \left(\frac{-1 - ST_n + ST_{n+1}}{2}\right)(B_m - B_{m-1}) + \left(B_n B_{n+1} - 3B_{n-1}B_n - \frac{1}{2}(B_{2n-1} + 1)\right)B_m i.$$

- The sum of the first p numbers $C_{(n,p)}$, can be described by:

$$\sum_{l=1}^{n} C_{(l,m)} = \left(\frac{2 - C_n + C_{n+1}}{4}\right)(B_m - B_{m-1}) + B_n B_m i.$$

In what follows, we present the results for three cases, namely, with $C_{(n,p)}$ with odd index p, with even index p and with a generic index p.

Proposition 4 *For all non-negative integers n, the sum of the numbers $C_{(n,p)}$ of odd index p can be given by:*

$$\sum_{j=1}^{m} C_{(n,2j-1)} = C_n\left(B_m^2 - B_m B_{m+1}\right) + (B_n - B_{n-1})B_m^2 i.$$

Proof By Theorem 5.3 in [11], we have

$$\sum_{j=1}^{m} C_{(n,2j-1)} = \sum_{j=1}^{m}\left(C_n\left(B_{2j-1} - B_{2j-2}\right) + (B_n - B_{n-1})B_{2j-1}i\right).$$

Hence

$$\sum_{j=1}^{m} C_{(n,2j-1)} = C_n\sum_{j=1}^{m}\left(B_{2j-1} - B_{2j-2}\right) + (B_n - B_{n-1})\sum_{j=1}^{m} B_{2j-1}i$$

$$= C_n \left(\sum_{j=1}^{m} B_{2j-1} - \sum_{j=1}^{m} B_{2j-2} \right) + (B_n - B_{n-1}) \sum_{j=1}^{m} B_{2j-1} i$$

$$= (C_n + (B_n - B_{n-1})\, i) \sum_{j=1}^{m} B_{2j-1} - C_n \sum_{j=1}^{m} B_{2j-2}.$$

Applying what is described in items (*a*) and (*b*) of Corollary 2.3.6 to [23], the result follows. □

Proposition 5 *For all non-negative integers n, the sum of the numbers* $C_{(n,p)}$ *of even index p can be described by:*

$$\sum_{j=1}^{m} C_{(n,2j)} = C_n \left(B_m B_{m+1} - B_m^2 \right) + (B_n - B_{n-1}) B_m B_{m+1} i.$$

Proof Using Theorem 5.3 in [11], we get:

$$\sum_{j=1}^{m} C_{(n,2j)} = \sum_{j=1}^{m} \left(C_n \left(B_{2j} - B_{2j-1} \right) + (B_n - B_{n-1}) B_{2j} i \right).$$

Thus, we have:

$$\sum_{j=1}^{m} C_{(n,2j)} = C_n \sum_{j=1}^{m} \left(B_{2j} - B_{2j-1} \right) + (B_n - B_{n-1}) \sum_{j=1}^{m} B_{2j} i$$

$$= C_n \left(\sum_{j=1}^{m} B_{2j} - \sum_{j=1}^{m} B_{2j-1} \right) + (B_n - B_{n-1}) \sum_{j=1}^{m} B_{2j} i$$

$$= (C_n + (B_n - B_{n-1})\, i) \sum_{j=1}^{m} B_{2j} - C_n \sum_{j=1}^{m} B_{2j-1}.$$

Using what is described in Corollary 2.3.6, specifically in its items (*b*) and (*a*) in [23], the result follows. □

Proposition 6 *For all non-negative integers n, the sum of the numbers* $C_{(n,p)}$ *can be given by:*

$$\sum_{j=1}^{m} C_{(n,j)} = \frac{1}{4} C_n \left(B_{m-1} - 2B_m + B_{m+1} \right) - \frac{1}{4} (B_n - B_{n-1}) (1 + B_m - B_{m+1})\, i.$$

Proof Similarly to the proof of the previous propositions, by Theorem 5.3 in [11], we get:

$$\sum_{j=1}^{m} C_{(n,j)} = \sum_{j=1}^{m} \left(C_n \left(B_j - B_{j-1} \right) + \left(B_n - B_{n-1} \right) B_j i \right).$$

Thus, we obtain:

$$\begin{aligned}\sum_{j=1}^{m} C_{(n,j)} &= C_n \sum_{j=1}^{m} \left(B_j - B_{j-1} \right) + \left(B_n - B_{n-1} \right) \sum_{j=1}^{m} B_j i \\ &= \left(C_n + \left(B_n - B_{n-1} \right) i \right) \sum_{j=1}^{m} B_j - C_n \left(B_0 + \sum_{j=1}^{m-1} B_j \right).\end{aligned}$$

The result follows immediately applying item (6) of Proposition 2.6 in [10] and by the initial conditions of the unidimensional sequence of balancing numbers $\{B_n\}_{n\geq 0}$. □

4 Conclusion

This research paper presents some very interesting identities about two sequences of numbers in their unidimensional and bidimensional versions. This work is part of an investigation started by the authors around the bidimensional versions of these two number sequences. Some identities involving these number sequences in their bidimensional version are presented, certainly relying on their unidimensional model. We think that this research work will make a great contribution to the field of mathematics and to other areas of science and technology, and will give researchers interested in this topic the opportunity to devote themselves to it.

Acknowledgments The first author gratefully acknowledges Portuguese funding through the Calouste Gulbenkian Foundation and the Angolan institution INAGBE - Instituto Nacional de Gestão de Bolsa de Estudo. The second author is a member of the CMAT-UTAD Research Center and is also a collaborating member of the CIDTFF Research Center. The last author is a member of the Research Group in Mathematics of the University of Santiago de Compostela - USC GI 2136 and also of the Project Homology, Homotopy and Categorical Invariants in Groups and Non-Associative Algebras - PID2020 115155GB I00, under the objective Dynamical Systems, Genetic Algebras and Non-Stochastic Quadratic Operators. This research was partially funded by the Portuguese fund FCT - Fundação para a Ciência e Tecnologia, under the projects UIDB/00013/2020, UIDP/00013/2020 and UIDB/00194/2020.

References

1. Akbiyik, M., Alo, J.: On third-order bronze Fibonacci numbers. Mathematics **9**(20), 2606 (2021). https://doi.org/10.3390/math9202606
2. Alves, F., Catarino, P.: Engenharia Didática de 2ª Geração com o tema: $h(x)$-Polinômios de Jacobsthal. Ensino de Ciâncias e Tecnologia em Revista **8**(3), 28–55 (2018). https://doi.org/10.31512/encitec.v8i3.2357
3. Behera, A., Panda, G.K.: On the square roots of triangular numbers. Fib. Quart. **37**(2), 98–105 (1999). https://www.fq.math.ca/Scanned/37-2/behera.pdf
4. Bouroubi, S.: On the square-triangular and balancing-numbers. Rostock. Math. Kolloq. **72**, 73–80 (2021)
5. Catarino, P.: A note on $h(x)$-Fibonacci quaternion polynomials. Chaos Solutions Fractals **77**, 1–5 (2015). https://doi.org/10.1016/j.chaos.2015.04.017
6. Catarino, P.: Bicomplex k-Pell quaternions. Comput. Methods Funct. Theory **19**(1), 65–76 (2019). https://doi.org/10.1007/s40315-018-0251-5
7. Catarino, P.: On k-Pell hybrid numbers. J. Discret. Math. Sci. Cryptogr. **22**(1), 83–89 (2019). https://doi.org/10.1080/09720529.2019.1569822
8. Catarino, P., Borges, A.: On Leonardo numbers. Acta Math. Univ. Comenian. **89**(1), 75–86 (2019)
9. Catarino, P., Campos, H.: Incomplete k-Pell, k-Pell-Lucas and modified k-Pell numbers. Hacettepe J. Math. Stat. **46**(3), 361–372 (2017). https://dergipark.org.tr/en/download/article-file/518228
10. Catarino, P., Campos, H., Vasco, P.: On some identities for balancing and cobalancing numbers. Ann. Math. Inform. **45**, 11–24 (2015)
11. Chimpanzo, J., Catarino, P., Vasco, P., Borges, A.: Bidimensional extensions of balancing and Lucas-Balancing numbers. J. Discret. Math. Sci. Cryptogr. (in press)
12. Chimpanzo, J., Otero-Espinar, M.V, Borges, A., Vasco, P., Catarino, P.: Bidimensional extensions of cobalancing and Lucas-Cobalancing numbers. https://doi.org/10.2478/amsil-2023-0022
13. Karakas, A.M., Karakas, M.: A new BK-space defined by regular matrix of Lucas numbers. J. Interdiscip. Math. **22**(6), 837–847 (2019). https://doi.org/10.1080/09720502.2019.1668152
14. Karataş, A., Halici, S.: Some properties of bicomplex Pell and Pell-Lucas numbers. J. Inf. Optim. Sci. **42**(3), 701–709 (2021). https://doi.org/10.1080/02522667.2020.1814495
15. Kizilates, C.: A new generalization of Fibonacci hybrid and Lucas hybrid numbers. Chaos Solitons Fractals **130**, 109449 (2020). https://doi.org/10.1016/j.chaos.2019.109449
16. Levesque, C.: On m-th order linear recurrences. Fibonacci Quart. **23**(4), 290–293 (1985). https://www.fq.math.ca/Scanned/23-4/levesque.pdf
17. Liptai, K.: Lucas balancing numbers. Acta Math. Univ. Ostrav. **14**, 43–47 (2006)
18. Olajos, P.: Properties of balancing, cobalancing and generalized balancing numbers. Ann. Math. Inform. **37**, 125–138 (2010)
19. Oliveira, R., Alves, F.: Os números Gaussianos de Fibonacci e relações recorrentes bidimensionais. Revista Thema **16**(4), 745–754 (2019). https://doi.org/10.15536/thema.V16.2019.745-754.1333
20. Oliveira, R., Alves, F., Paiva, B.: Identidades bi e tridimensionais para os números de Fibonacci na forma complexa. C.Q.D-Revista Eletrônica Paulista de Matemática **11**, 91–106 (2017)
21. Panda, G.K., Ray, P.K.: Some links of balancing and cobalancing numbers with Pell and associated Pell numbers. Bul. Inst. Math. Acad. Sin. **6**(1), 41–72 (2011)
22. Qi, F.: Determinantal expressions and recurrence relations for Fubini and Eulerian polynomials. J. Interdiscip. Math. **22**(3), 317–335 (2019). https://doi.org/10.1080/09720502.2019.1624063
23. Ray, P.K.: Balancing and cobalancing numbers. Ph.D. thesis, Department of Mathematics, National Institute of Technology, Rourkela (2009). http://ethesis.nitrkl.ac.in/2750/1/Ph.D._Thesis_of_P.K._Ray.pdf

24. Sloane N.J.A.: The on-line encyclopedia of integer sequences. Available in http://oeis.org/
25. Vieira, R., Alves, F., Catarino, P.: Relações bidimensionais e identidades da sequência de Leonardo. Revista Sergipana de Matemática e Educação Matemática **4**(2), 156–173 (2019). https://doi.org/10.34179/revisem.v4i2.11863
26. Yilmaz, F., Ertaş, A., Jia, J.: On harmonic complex balancing numbers. Mathematics **11**, 210 (2023). https://doi.org/10.3390/math11010210

Sequences of Uncountable Iterated Function Systems: The Convergence of the Sequences of Fractals and Fractal Measures Associated

Ion Mierlus-Mazilu and Lucian Niță

Abstract In this paper, we consider a sequence of uncountable iterated function system (U.I.F.S.). Each term of this sequence is built using an uncountable family of contractions and a linear and continuous operator. For each U.I.F.S. of the sequence we have an associated attractor, a Markov-type operator and a fractal measure.

We study the convergence of the corresponding attractors and fractal measures sequences.

1 Introduction

First, we consider the Markov operator on the Borel normalized measures and thus, we obtain the Hutchinson (scalar) measure (or the fractal (scalar) measure). We prove that the support of the fractal measure is the attractor of the U.I.F.S.

The second step is to consider fractal vector measures. For this aim, we construct the Markov-type operator, using the family of contractions and an uncountable family of linear and continuous operators. Then, we consider a new operator, which adds a fixed measure μ° to the Markov operator. The fixed point of this new operator is called *the fractal vector measure*. The measure μ° can be chosen such that the support of the fractal vector measure is, also, the attractor of the U.I.F.S.

The third step is to consider a sequence of U.I.F.S., using a sequence $(T_n)_n$ of linear and continuous operators and an uncountable family of contractions, $(\omega_\alpha)_\alpha$. Then $[(T_n \circ \omega_\alpha)_\alpha]_n$ is a sequence of U.I.F.S. We suppose that $T_n \overset{\|\cdot\|}{\to} T$, T being, also a linear and continuous operator. We denote: K_n and μ_n^* the attractor and the fractal vector measure associated to $(T_n \circ \omega_\alpha)_\alpha$, and K, respectively μ^* the attractor and the fractal vector measure associated to $(T \circ \omega_\alpha)_\alpha$. The problem that we solve is: it is true that $K_n \to K$ and $\mu_n^* \to \mu$, in some appropriate metrics?

I. Mierlus-Mazilu (✉) · L. Niță
Technical University of Civil Engineering, Bucharest, Romania
e-mail: ion.mierlusmazilu@utcb.ro

V. Gayoso Martínez et al. (eds.), *Mathematical Methods for Engineering Applications*, Springer Proceedings in Mathematics & Statistics 439,
https://doi.org/10.1007/978-3-031-49218-1_4

2 Preliminaries

We recall here some definitions and results that we will need on the entire paper. Most of them were introduced/proved in [1–3] and [4]. For more details about some questions of Functional Analysis and Iterated Function Systems, one can consult [5–7] and [8].

Let (T, d) be a compact metric space, $a, b \in \mathbb{R}, a < b$. For each $\alpha \in [a, b]$, let $\omega_\alpha : T \to T$ be a contraction of ratio $r_\alpha \in [0, 1)$, such that $r = \sup_\alpha r_\alpha < 1$.

Definition 1 The set $(\omega_\alpha)_{\alpha \in [a,b]}$ is called *uncountable iterated function system* (U.I.F.S.).

Theorem 1 *We consider the set* $\mathcal{K} = \{K \in \mathcal{P}(T) | K$ *is compact and nonempty* $\}$ *and the function* $\delta : \mathcal{K} \times \mathcal{K} \to [0, \infty)$, $\delta(K_1, K_2) \stackrel{def}{=} \inf_{r>0}\{K_1 \subset B(K_2, r), K_2 \subset B(K_1, r)\}$. *Then:*

(i) δ *is a metric on* $\mathcal{K}$*;*
(ii) $(\mathcal{K}, \delta)$ *is a compact metric space;*
(iii) The function $S : \mathcal{K} \to \mathcal{K}$, $S(B) = \overline{\bigcup_{\alpha \in [a,b]} \omega_\alpha(B)}$ *is a contraction on* $\mathcal{K}$.

Consequence $\exists! A \in \mathcal{K}$ such that $S(A) = A$ (the set A is called *the attractor associated to the U.I.F.S.*).

Definition 2 Let $\mathcal{B}$ be the family of Borel subsets of T. The measure $\mu : \mathcal{B} \to [0, 1]$ is called *normalized* if $\mu(T) = 1$.

We denote:

$$\tilde{\mathcal{B}} = \{\mu : \mathcal{B} \to [0, 1] | \mu \text{ is normalized measure}\},$$

$$\mathcal{L}_1 = \{f : T \to \mathbb{R} | f \text{ is a Lipschitz function with its Lipschitz constant less or equal to } 1\}.$$

For $\mu, \nu \in \tilde{\mathcal{B}}$, we define

$$d_H(\mu, \nu) = \sup\left\{\left|\int f d\mu - \int f d\nu\right| \Big| f \in \mathcal{L}_1\right\}.$$

Theorem 2

(a) d_H *is a metric on* $\tilde{\mathcal{B}}$ *(called* the Hutchinson metric*);*
(b) $\left(\tilde{\mathcal{B}}, d_H\right)$ *is compact (hence, complete);*

Let $p : [a, b] \to [0, 1]$ be a Lebesgue integrable function, such that

$$\int_{[a,b]} p(x)d\lambda(x) = 1$$

(λ being the Lebesgue measure). For each $\mu \in \tilde{\mathcal{B}}$, we define:

$$M(\mu)(C) = \int_{[a,b]} p(x)\mu\left(\omega_x^{-1}(C)\right) d\lambda(x),$$

provided that the integral exists. For this, we will suppose that the function $x \mapsto p(x)$ and $x \mapsto \mu\left(\omega_x^{-1}(C)\right)$ are continuous, $\forall C \in \mathcal{B}$.

Theorem 3

(a) $\forall \mu \in \tilde{\mathcal{B}}, M(\mu) \in \tilde{\mathcal{B}}$;
(b) *M is a contraction on* $\tilde{\mathcal{B}}$;
(c) $\exists! \mu^* \in \tilde{\mathcal{B}}$ *such that* $M(\mu^*) = \mu^*$ *(*μ^* *is called* the Hutchinson measure *or* the fractal measure *associated to the U.I.F.S. and to the function p).*
(d) $\mathrm{supp}(\mu^*) = A$.

Proof The points (a), (b), (c) were proved in [4].
We will prove here the point (d).
Let $B \in \mathcal{B}$. We have:

$$\mu^*(B) = M(\mu^*)(B) = \int_{[a,b]} p(\alpha) \underbrace{\mu^*\left(\omega_\alpha^{-1}(B)\right)}_{M(\mu^*)\left(\omega_\alpha^{-1}(B)\right)} d\lambda(\alpha) =$$

$$= \int_{[a,b]} p(\alpha) \left[\int_{[a,b]} p(\beta)\mu^*\left(\omega_\beta^{-1}\left(\omega_\alpha^{-1}(B)\right)\right) d\lambda(\beta) \right] d\lambda(\alpha) = \ldots =$$

$$= \int_{[a,b]} p(\alpha_1) \int_{[a,b]} p(\alpha_2) \ldots$$

$$\int_{[a,b]} p(\alpha_n)\mu^*\left(\omega_{\alpha_n}^{-1}\left(\omega_{\alpha_{n-1}}^{-1} \cdots \left(\omega_{\alpha_1}^{-1}(B)\right)\right)\right) d\lambda(\alpha_n) \ldots d\lambda(\alpha_1) \quad (1)$$

We consider $(i_1, i_2, \ldots, i_n) \in [a, b]^n$ such that $p(i_1)p(i_2)\ldots p(i_n) > 0$ and let us take $B = \omega_{i_1} \circ \omega_{i_2} \circ \ldots \circ \omega_{i_n}(A)$. We will use the fact that $\mu^*(A) = 1$ (see, also, [4]). We denote:

$$f(\alpha_1, \dots, \alpha_n) = p(\alpha_1)p(\alpha_2)\dots p(\alpha_n)\mu^*\left(\omega_{\alpha_n}^{-1}\left(\dots\omega_{\alpha_1}^{-1}\left(\omega_{i_1}\circ\omega_{i_2}\circ\dots\circ\omega_{i_n}(A)\right)\dots\right)\right).$$

Then, $f(i_1, \dots, i_n) > 0$ and, f being continuous, we find a neighbourhood V of $(i_1, \dots, i_n)$ such that $V \subset [a, b]^n$ and $f(\alpha_1, \dots, \alpha_n) > 0$ on V. Hence,

$$\mu^*(B) = \int\limits_{[a,b]^n} f(\alpha_1, \dots, \alpha_n)d\lambda(\alpha_n)\dots d\lambda(\alpha_1) > \int\limits_V f(\alpha_1, \dots, \alpha_n)d\lambda(\alpha_n)\dots d\lambda(\alpha_1) > 0. \tag{2}$$

Let now $a \in A$ and $W \subset T$ be an open set, such that $a \in W$. We know that $\text{diam}\left[\left(\omega_{i_1}\circ\dots\circ\omega_{i_n}\right)(A)\right] \le r^n \,\text{diam}(A) \to 0 (n \to \infty)$. We deduce that $\overline{\bigcap\limits_{n=1}^{\infty} \omega_{i_1}\circ\dots\circ\omega_{i_n}(A)}$ has a unique element. On the other hand, $A = \bigcup\limits_{\alpha\in[a,b]} \omega_\alpha(A)$, so $\forall a \in A$, $(\exists)(i_n)_{n\ge1}$ such that $a = \bigcap\limits_{n=1}^{\infty} \omega_{i_1}\left(\omega_{i_2}\dots\left(\omega_{i_n}(A)\right)\right)$. Hence, for n big enough, we have $\omega_{i_1}\left(\omega_{i_2}\dots\left(\omega_{i_n}(A)\right)\dots\right) \subset W \Longrightarrow \mu^*(W) > 0$ (using (2)) $\Longrightarrow a \in \text{supp}(\mu^*) \Longrightarrow A \subset \text{supp}(\mu^*)$.

Conversely, let $a \in \text{supp}(\mu^*)$. Suppose $a \notin A \Longrightarrow a \in T\setminus A \overset{not}{=} V$. But, $\mu^*(V) =$
$= \mu^*(T) - \mu^*(A) = 0$. This fact contradicts the one that $a \in \text{supp}(\mu^*)$. We conclude that $A = \text{supp}(\mu^*)$.

Now, let $(X, \langle\cdot,\cdot\rangle)$ be a real Hilbert space. We denote:

$$\text{cabv}(X) = \{\mu : \mathcal{B} \to X | \mu \text{ is } \sigma\text{–additive with bounded variation}\}$$

and

$$\mathscr{E} = \left\{ f : T \to X,\ f = \sum_{i=1}^{n} \varphi_{A_i} x_i, \text{ where } (A_i)_{1\le i\le n} \subset \mathcal{B},\ x_i \in X \right\}$$

($\mathscr{E}$ is the set of simple functions).

Definition 3 (See [1]) For $f = \sum\limits_{i=1}^{n} \varphi_{A_i} x_i \in \mathscr{E}$ and $\mu \in \text{cabv}(X)$, we define:

$$\int f d\mu = \sum_{i=1}^{n} \langle \mu(A_i), x_i \rangle.$$

We consider the space:

$$TM(X) = \left\{ f : T \to X | \exists (f_n)_n \subset \mathscr{E}, \text{ such that } f_n \xrightarrow{u} f \right\}$$

(the space of *totally measurable functions*).

For $f \in TM(X)$, we define $\int f d\mu = \lim_{n\to\infty} \int f_n d\mu, \forall (f_n)_n \subset \mathscr{E}$ such that $f_n \xrightarrow{u} f$.

If $f : T \to X$ is a Lipschitz function, having its Lipschitz constant equal to L, let us denote $\|f\|_{BL} = \|f\|_L + \|f\|_\infty$ (it can be proved that $\|\cdot\|_{BL}$ is a norm on the set of Lipschitz functions $f : T \to X$).

We denote: $BL_1(X) = \{f : T \to X | f$ is a Lipschitz function and $\|f\|_{BL} \leq 1\}$.

Theorem 4 (See [2]) *For each* $\mu \in \text{cabv}(X)$, *let us define*

$$\|\mu\|_{MK} = \sup \left\{ \left| \int f d\mu \right| \middle| f \in BL_1(X) \right\}.$$

Then, the application $\mu \mapsto \|\mu\|_{MK}$ *is a norm on* cabv(X) *called* the Monge–Kantorovich-type norm.

3 An Integral for Vector Functions with Respect to the Lebesgue Measure. The Markov-Type Operator. The Fractal Vector Measure

Let $(A_i)_{1\leq i\leq m}$ be a partition of the interval $[a, b]$ with Borel sets and

$f = \sum_{i=1}^{m} \varphi_{A_i} x_i (x_i \in X)$ be a simple function, where λ is the Lebesgue measure on $[a, b]$. We define $\int_{[a,b]} f d\lambda = \sum_{i=1}^{m} \lambda(A_i) x_i$.

Definition 4 If $f_n : [a, b] \to \mathbb{R}$ are simple functions and $f_n \xrightarrow{u} f$, then we define

$$\int_{[a,b]} f d\lambda = \lim_{n\to\infty} \int_{[a,b]} f_n d\lambda.$$

For each $\alpha \in [a, b]$, let $\omega_\alpha : T \to T$ be a contraction of ratio $r_\alpha \in [0, 1)$, with $r = \sup_\alpha r_\alpha < 1$ and $R_\alpha : X \to X$ be a linear and continuous operator. We consider $\mu \in \text{cabv}(X)$ and define:

$$H(\mu) = \int_{[a,b]} \left(R_\alpha \circ \mu \circ \omega_\alpha^{-1} \right) d\lambda (\textit{the Markov} - \textit{type operator}).$$

Theorem 5 *Let us consider the normed space* $(\text{cabv}(X), \|\cdot\|_{MK})$. *Then:*

1. $\forall \mu \in \text{cabv}(X)$, $H(\mu) \in \text{cabv}(X)$;
2. $\|H\|_\circ \leq \int\limits_{[a,b]} \|R_\alpha\|_\circ (1 + r_\alpha) d\lambda$;

Let $c > 0$. We denote

$$B_c\left(\mathbb{R}^n\right) = \left\{\mu \in \text{cabv}\left(\mathbb{R}^n\right) \middle| |\mu|(T) \leq c, \text{ where } |\mu| \text{ is the variation of } \mu\right\}.$$

Theorem 6 *Let* $X = \mathbb{R}^n$. *Suppose that* $\int\limits_{[a,b]} \|R_\alpha\|_\circ (1 + r_\alpha) d\lambda < 1$. *Let* $\mu^\circ \in \text{cabv}\left(\mathbb{R}^n\right)$, *arbitrarily and* $P : \text{cabv}\left(\mathbb{R}^n\right) \to \text{cabv}\left(\mathbb{R}^n\right)$, $P(\mu) = H(\mu) + \mu^\circ$. *We consider* $\overline{B} \subset B_c\left(\mathbb{R}^n\right)$, *close in the weak–∗ topology, such that* $P(\overline{B}) \subset \overline{B}$. *Then*

(i) $\exists! \mu^* \in \overline{B}$ *such that* $P(\mu^*) = \mu^*$;
(ii) μ° *can be chosen such that* $\text{supp}(\mu^*) = A$ *(where, for a vector measure* μ, $\text{supp}(\mu) \stackrel{def}{=} \text{supp}(|\mu|)$*).*

Proof

(i) The set $\overline{B}$, being weak–∗ close in $B_c\left(\mathbb{R}^n\right)$ is compact, hence complete, in this topology. But the weak − ∗ topology is the same with the one given by $\|\cdot\|_{MK}$ on $B_c\left(\mathbb{R}^n\right)$ (see [2]). So, $\overline{B}$ is complete in the topology given by $\|\cdot\|_{MK}$. Let $\mu_1, \mu_2 \in \overline{B}$. We have:

$$\|P(\mu_1) - P(\mu_2)\| = \|H(\mu_1) - H(\mu_2)\| \leq \|H\|_\circ \|\mu_1 - \mu_2\|.$$

But, from Theorem 5, $\|H\|_\circ \leq \int\limits_{[a,b]} \|R_\alpha\|_\circ (1 + r_\alpha) d\lambda < 1$. So, the restriction of P on $\overline{B}$ is a contraction of ratio less or equal to $\int\limits_{[a,b]} \|R_\alpha\|_\circ (1 + r_\alpha) d\lambda$. Using the contraction principle, $\exists! \mu^* \in \overline{B}$ with $P(\mu^*) = \mu^*$.

(ii) Let $p : [a, b] \to [0, 1]$ be such that $\int\limits_{[a,b]} p(\alpha) d\lambda(\alpha) = 1$ (as before) and μ be the fractal measure associated to the U.I.F.S. $(\omega_\alpha)_{\alpha \in [a,b]}$ and to the function p. We know that $\text{supp}(\mu) = A$ (the attractor associated to the U.I.F.S.). For $x \in \mathbb{R}^n$, $x \neq 0$, fixed, we define $\tilde{\mu} = \mu \cdot x$. We will prove that $\tilde{\mu} \in \text{cabv}\left(\mathbb{R}^n\right)$ and $|\tilde{\mu}|(T) = \|x\|$. Let $(A_j)_{1 \leq j \leq m}$ be a partition of T with Borel sets. We have:

$$\sum_{j=1}^{m} \|\tilde{\mu}(A_j)\| = \sum_{j=1}^{m} \|\mu(A_j) \cdot x\| =$$

$$= \|x\| \sum_{j=1}^{m} \mu(A_j) = \|x\| \mu \left(\bigcup_{j=1}^{m} A_j\right) = \|x\| \mu(T) = \|x\|.$$

We deduce that $|\tilde{\mu}|(T) = \|x\|$, hence $\tilde{\mu} \in \text{cabv}(\mathbb{R}^n)$.

We define $\mu^\circ = (I - H)(\tilde{\mu})$ (I-the identity operator). Then, from $P(\mu^*) = \mu^*$, we have: $H(\mu^*) + \mu^\circ = \mu^*$ or, $\mu^\circ = (I - H)(\mu^*)$, that means: $(I - H)(\tilde{\mu}) = (I - H)(\mu^*)$. But $\|H\|_\circ < 1$. So, $I - H$ has the contiuous inverse $(I - H)^{-1}$. Thus, we obtain: $\mu^* = \tilde{\mu}$ and $\text{supp}(\mu^*) = \text{supp}(\tilde{\mu}) = A$.

Remark 1 The condition "$P(\overline{B}) \subset \overline{B}$" imposed in the enounce of the theorem is fullfilled, for example, if $c \geq \frac{1+\|H\|_\circ}{1-\|H\|_\circ} \cdot \|\tilde{\mu}\|$. To see this, we denote $\|\mu\| := |\mu|(T)$ (the variational norm of the measure μ). It is known (see [2]) that $\|\mu\|_{MK} \leq \|\mu\|$. We can write (for $\mu \in B_c(\mathbb{R}^n)$):

$$\|P(\mu)\|_{MK} \leq \|H(\mu)\|_{MK} + \left\|\mu^\circ\right\|_{MK} \leq \|H(\mu)\|_{MK} + \left\|\mu^\circ\right\| \leq$$

$$\leq \|H\|_\circ\|\mu\| + \|(I - H)(\tilde{\mu})\| = \|H\|_\circ\|\mu\| + (\|I\|_\circ + \|H\|_\circ)\,\|\tilde{\mu}\| =$$

$$= \|\tilde{\mu}\| + \|H\|_\circ\,(\|\mu\| + \|\tilde{\mu}\|) \leq \|\tilde{\mu}\|\,(1 + \|H\|_\circ) + c\|H\|_\circ$$
$$\leq c\,(1 - \|H\|_\circ) + c\|H\|_\circ = c.$$

So, in this case, we can take $\overline{B} = B_c(\mathbb{R}^n)$.

4 Sequences of U.I.F.S. Convergence Properties

Let X, Y be Banach spaces, $\omega : Y \to X$ be a contraction of ratio $r \in (0, 1)$ and $(T_n)_{n\geq 1} \subset \mathscr{L}(X, Y)$ such that $\alpha \overset{not}{=} \sup_n \|T_n\| < \frac{1}{r}$. For each $n \geq 1$, we define $U_n : Y \to Y$, $U_n = T_n \circ \omega$.

Lemma 1 (See [6]) *U_n is a contraction of ratio less or equal to αr.*

Lemma 2 (See [6]) *Suppose that there exists $T \in \mathscr{L}(X, Y)$ such that $T_n \xrightarrow{\|\cdot\|_\circ} T$. We denote:*

$$\mathcal{K}(Y) = \{K \subset Y | K \textit{ is compact and nonempty}\}.$$

Then, $U_n(K) \xrightarrow{\delta} U(K)$ (where $T = T \circ \omega$), $\forall K \in \mathcal{K}$.

Let, now, $\overline{Y}$ be a Banach space, $Y \in \mathcal{K}(\overline{Y})$, $\omega_\alpha : Y \to X$ be a contraction, for each $\alpha \in [a, b]$ and $r \overset{not}{=} \sup_\alpha r_\alpha < 1$. We consider the operators $T_n, T \in \mathcal{L}(X, \overline{Y})$ such that $\beta \overset{not}{=} \sup_n \|T_n\| < \frac{1}{r}$ and $T_n \xrightarrow{\|\cdot\|_\circ} T$. We also suppose that $T_n(\omega_\alpha(Y)) \subset Y$, $\forall n \geq 1$, $\forall \alpha \in [a, b]$. Let $U_\alpha^n = T_n \circ \omega_\alpha : Y \to Y$ and $U_\alpha = T \circ \omega_\alpha : Y \to Y$. Hence $\left(U_\alpha^n\right)_\alpha$ and $(U_\alpha)_\alpha$ are U.I.F.S. and we denote by K_n, respectively K their attractors.

Theorem 7 $\lim_{n\to\infty} \delta(K_n, K) = 0.$

Proof

$$\delta(K_n, K) = \delta\left(\overline{\bigcup_{\alpha\in[a,b]} U_\alpha^n(K_n)}, \overline{\bigcup_{\alpha\in[a,b]} U_\alpha(K)}\right) \le$$

$$\le \underbrace{\delta\left(\overline{\bigcup_{\alpha\in[a,b]} U_\alpha(K)}, \overline{\bigcup_{\alpha\in[a,b]} U_\alpha^n(K)}\right)}_{I} + \underbrace{\delta\left(\overline{\bigcup_{\alpha\in[a,b]} U_\alpha^n(K)}, \overline{\bigcup_{\alpha\in[a,b]} U_\alpha^n(K_n)}\right)}_{II}.$$

Let $\varepsilon > 0$, arbitrarily, fixed. We have:

$I \le \sup_{\alpha\in[a,b]} \delta\left(U_\alpha(K), U_\alpha^n(K)\right) \le \varepsilon$, for n big enough (using Lemma 2);

$II \le \sup_{\alpha\in[a,b]} \delta\left(U_\alpha^n(K), U_\alpha^n(K_n)\right) \le r\beta\delta(K_n, K)$ (using Lemma 1).

We obtain: $(1 - r\beta)\delta(K_n, K) \le \varepsilon$, for n big enough. So, $K_n \xrightarrow{\delta} K$.

For $N \in \mathbb{N}^*$, $\alpha \in [a, b]$, let $R_\alpha \in \mathscr{L}\left(\mathbb{R}^N\right)$, $H^n(\mu) = \int_{[a,b]} \left(R_\alpha \circ \mu \circ \left(U_\alpha^n\right)^{-1}\right) d\lambda$;

$H(\mu) = \int_{[a,b]} \left(R_\alpha \circ \mu \circ (U_\alpha)^{-1}\right) d\lambda$, $\forall \mu \in \text{cabv}\left(\mathbb{R}^N\right)$, $\forall n \ge 1$. Suppose that $\int_{[a,b]} \|R_\alpha\|_\circ(1 + \beta r)d\lambda < 1$. Let us define $P^n, P : \text{cabv}\left(\mathbb{R}^N\right) \to \text{cabv}\left(\mathbb{R}^N\right)$,

$P^n(\mu) = H^n(\mu) + \mu^\circ$, $P(\mu) = H(\mu) + \mu^\circ$, (where $\mu^\circ \in \text{cabv}\left(\mathbb{R}^N\right)$, arbitrarily, fixed). Let μ_n^*, μ^* be the fractal measures associated to P^n and P, according to Theorem 6.

Theorem 8 *If* $T_n \xrightarrow{\|\cdot\|_\circ} T$, *then* $\mu_n^* \xrightarrow{\|\cdot\|_{MK}} \mu^*$.

Proof Let $\varepsilon > 0$, arbitrarily, fixed. We will denote by q_n, respectively, q_{n_α} the ratios of the contractions P^n, respectively U_α^n. We have:

$$\left\|\mu^* - \mu\right\|_{MK} = \left\|P^n\left(\mu_n^*\right) - P\left(\mu^*\right)\right\|_{MK} \le \left\|P^n\left(\mu_n^*\right) - P^n\left(\mu^*\right)\right\|_{MK} +$$

$$+ \left\|P^n\left(\mu^*\right) - P\left(\mu^*\right)\right\|_{MK} \le q_n\|\mu_n^* - \mu^*\|_{MK} + \left\|P^n\left(\mu^*\right) - P\left(\mu^*\right)\right\|_{MK} \quad (3)$$

Denote $q = \sup_n q_n$. Using the conditions given above the enounce, we have:

$$q_n \le \int_{[a,b]} \|R_\alpha\|_\circ \left(1 + \left\|U_\alpha^n\right\|_\circ\right) d\lambda \le \int_{[a,b]} \|R_\alpha\|_\circ(1 + \beta r)d\lambda < 1.$$

So, $q < 1$. Hence, from (3), we deduce:

$$\left\|\mu_n^* - \mu^*\right\|_{MK} \leq q \left\|\mu_n^* - \mu^*\right\|_{MK} + \left\|P^n\left(\mu^*\right) - P\left(\mu^*\right)\right\|_{MK} \Longrightarrow$$

$$\Longrightarrow (1-q)\left\|\mu_n^* - \mu^*\right\|_{MK} \leq \left\|P^n\left(\mu^*\right) - P\left(\mu^*\right)\right\|_{MK} = \left\|H^n\left(\mu^*\right) - H\left(\mu^*\right)\right\|_{MK}.$$

But, it can be proved that, if $H(\mu) = R \circ \mu \circ \omega^{-1}(\mu \in \text{cabv}\left(\mathbb{R}^N\right), R \in \mathscr{L}\left(\mathbb{R}^N\right)$, $\omega : Y \to Y$ contraction), then: $\int f dH(\mu) = \int g d\mu$, where $g = R^* \circ f \circ \omega$, R^* being the adjoint of R. We can write:

$$\text{for any } f \in BL_1\left(\mathbb{R}^N\right), \left|\int f dH^n(\mu^*) - \int f dH(\mu^*)\right| =$$

$$= \left|\int_{[a,b]} \left(\int R_\alpha^* \circ f \circ U_\alpha^n d\mu^* - \int R_\alpha^* \circ f \circ U_\alpha d\mu^*\right) d\lambda\right| =$$

$$= \left|\int_{[a,b]} \left(\int R_\alpha^* \circ \left(f \circ U_\alpha^n - f \circ U_\alpha\right) d\mu^*\right) d\lambda\right| \leq$$

$$= \int_{[a,b]} \left(\int \left\|R_\alpha^*\right\|_\circ \left\|f \circ U_\alpha^n - f \circ U_\alpha\right\| d\left|\mu^*\right|\right) d\lambda \leq$$

$$\leq \int_{[a,b]} \left\|R_\alpha^*\right\|_\circ \left(\int \sup_\alpha \sup_{x \in Y} \left\|U_\alpha^n(x) - U_\alpha(x)\right\| d\left|\mu^*\right|\right) d\lambda \leq$$

$$\leq \underbrace{\left(\int_{[a,b]} \left\|R_\alpha\right\|_\circ d\lambda\right)}_{<1} \cdot \sup_\alpha \sup_{x \in Y} \left\|U_\alpha^n(x) - U_\alpha(x)\right\| \left|\mu^*\right|(Y) \leq \varepsilon, \text{ for } n \text{ big enough.}$$

Hence, we deduce that $\sup\limits_{f \in BL_1(\mathbb{R}^N)} \left|\int f dH^n\left(\mu^*\right) - \int f dH\left(\mu^*\right)\right| \leq \varepsilon.$

Consequently, $\left\|H^n\left(\mu^*\right) - H\left(\mu^*\right)\right\|_{MK} \leq \varepsilon$ and thus $(1-q)\left\|\mu_n^* - \mu^*\right\|_{MK} \leq$ $\leq \varepsilon$, for n big enough. It results: $\mu_n^* \overset{\|\cdot\|_{MK}}{\longrightarrow} \mu^*$.

5 Conclusions

In this paper, we gave three generalizations of the classical theory of iterated function system:

(a) Here we considered an uncountable set of contractions, unlike the classical frame, where an iterated function system is built with a finite set of contractions;
(b) The fractal measure, here, is a vector measure, not a positive normalized measure, as in the classical theory;
(c) We considered not only an iterated function system, but a sequence of (uncountable) iterated function systems, studying some convergence properties.

In the following we will be focused on two questions

(1°) The problem of convergence of the sequences of attractors and fractal vector measures if we have other type of convergence of the U.I.F.S.
(2°) How could apply these results to the theory of differential equations?

References

1. Chitescu, I., Ioana, L., Miculescu, R., Nita, L.: Sesquiliniar uniform vector integral. Proc. Indian Acad. Sci. (Math. Sci.) **125**(2), 229–236 (2015)
2. Chitescu, I., Ioana, L., Miculescu, R., Nita, L.: Monge-Kantorovich norms on spaces of vector measures. Results Math. **70**, 349–371 (2016)
3. Chitescu, I.: Spatii de Functii (in Romanian). Editura Stiintifica si Enciclopedica, Bucharest (1983)
4. Nita, L.: The attractor and the Hutchinson measure for an uncountable iterated function system. In: Proceedings of Mathematics and Educational Symposium of Department of Mathematics and Computer Science, Bucharest, 2nd edn., pp. 109–111 (2016)
5. Siretchi, G.: Spatii Concrete in Analiza Functionala (in Romanian). Centrul de Multiplicare al Universitatii Bucuresti, Bucharest (1982)
6. Constantinescu, S., Mierlus-Mazilu, I., Nita, L.: Families of iterated function systems. Convergences properties of the associated attractors and fractal measures. U.P.B. Sci. Bull. Ser. A **84**(2), 107–115 (2022)
7. D'Aniello, E., Steele, T.H.: Attractors for classes of iterated function systems. Eur. J. Math. **5**(1), 116–137 (2019)
8. Mihail, A.: Recurrent iterated function systems. Rev. Roumaine Math. Pures Appl. **53**(1), 43–53 (2008)

Method of Hydrodynamic Images and Quantum Calculus in Fock-Bargmann Representation of Quantum States

Oktay K. Pashaev

Abstract We propose a new approach to quantum states in Fock space in terms of classical hydrodynamics. By conformal mapping of complex analytic function, representing the wave function of quantum states in Fock-Bargmann representation, we define the complex potential, describing these quantum states by incompressible and irrotational classical hydrodynamic flow. In our approach, zeros of the wave function appear as a set of point vortices (sources) in plane with the same strength, allowing interpretation of them as images in a bounded domain. For the cat states we find fluid representation as descriptive of a point source in the oblique strip domain, with infinite number of periodically distributed images. For the annular domain, the infinite set of images is described by Jackson q-exponential functions. We show that these functions represent the wave functions of quantum coherent states of the q-deformed quantum oscillator in q-Fock-Bargmann representation and describe the infinite set of point vortices, distributed in geometric progression.

1 Introduction

The hydrodynamical analogy is a fruitful approach in many fields of physical science and mathematics, including electromagnetic theory, quantum mechanics, nuclear physics, dynamical systems and complex analysis. In quantum mechanics, the hydrodynamic representation of wave function proposed by Madelung [1], gives intuitive picture of probability flow propagating in quantum fluid. The argument of wave function is identified with the velocity potential of the fluid and every eigenfunction of the Schrödinger equation can be interpreted as a special type of stationary flow. By expressing the velocity of fluid as an analytic function and applying the residue theorem, WKB got semiclassical quantization rule, which in fact is known in complex analysis as the argument principle, counting number of zeros

O. K. Pashaev (✉)
Izmir Institute of Technology, Urla, Izmir, Turkey
e-mail: oktaypashaev@iyte.edu.tr

V. Gayoso Martínez et al. (eds.), *Mathematical Methods for Engineering Applications*, Springer Proceedings in Mathematics & Statistics 439,
https://doi.org/10.1007/978-3-031-49218-1_5

and poles. The hydrodynamic interpretation gives meaning to an analytic function as descriptive of incompressible and irrotational fluid in plane and classification of singularities of the function by point vortices, sinks and sources, etc. Conformal mappings and method of images are important technical tools to solve boundary value problems in bounded domains. The method of images was developed by W. Thomson as powerful method of solving electrostatic problems and it was applied to many problems in Maxwell's Treatise on Electricity and Magnetism [2]. For simple planar domains like the circle, the wedge, the strip and the annular domain [3], the distribution of the hydrodynamic images is determined by the circle theorem [4], the wedge theorem [5], the strip theorem [6] and the two circle theorem [7], respectively. For the strip domain, the infinite set of periodic images is described by elliptic functions. For the wedge and the annular domain it can be represented naturally by the q-periodic functions of the quantum or the q-calculus [8]. For real q, the stretching from the origin results from inversion in two circles [9], while for q as a root of unity, $q^{2n} = 1$, the rotation corresponds to reflection in two lines. This is why the geometric inversion in complex plain becomes the origin of the method of images. For the wedge domain the kaleidoscope of finite number of images is located at vertices of regular polygon [10, 11], and possess the quantum group symmetry [12].

In this paper we propose a new approach to quantum states in Fock space in terms of classical hydrodynamics. By conformal mapping of complex analytic function, representing the wave function of quantum states in Fock-Bargmann representation, we define the complex potential, describing these quantum states by incompressible and irrotational classical hydrodynamic flow. In our approach, zeros of the wave function appear as a set of point vortices in plane with the same strength. This allows us in some cases to interpret these vortices as images of the one real vortex in a bounded domain, with vanishing normal velocity across the boundary curve. For the cat states we find fluid representation as descriptive of point source in the oblique strip domain, with infinite number of periodically distributed images. For the annular domain with real q, the infinite set of images is described by Jackson q-exponential functions. We show that these functions represent the wave functions of quantum coherent states of the q-deformed quantum oscillator in q-Fock-Bargmann representation and describe the infinite set of point vortices, distributed along a half-line in geometric progression.

2 Holomorphic Wave Functions and Hydrodynamic Flow

The wave function in quantum mechanics has appeared in Schrödinger first communication on wave mechanics [13]. By the Log transformation $S = \hbar \ln \Psi$ in the Hamilton-Jacobi equation with action variable S he introduced unknown function Ψ , the logarithmic derivative of which then becomes complex velocity vector of Madelung fluid [1]. This suggests us to introduce the following hydrodynamic representation for holomorphic wave functions.

Definition 1 The analytic functions

$$f(z) = \frac{i\Gamma}{2\pi} \ln \Psi(z), \qquad \bar{v}(z) = \frac{df}{dz} = \frac{i\Gamma}{2\pi} \frac{\Psi'(z)}{\Psi(z)}, \tag{1}$$

where $z = x + iy$, represent the vortex type complex potential $f(z)$ of two dimensional incompressible and irrotational fluid flow and corresponding complex velocity $\bar{v}(z)$.

The function f determines a conformal mapping from z-plane to $f(z) = \varphi + i\psi$ plane, and expression for velocity $\bar{v}(z)$ is holomorphic version of the Cole-Hopf transformation from the heat equation to nonlinear Burgers equation [14]. The holomorphic wave function in terms of complex potential f, for the vortex representation (1) is

$$\Psi(z) = |\Psi(z)|e^{i \arg \Psi} = \exp\left(-i\frac{2\pi}{\Gamma} f(z)\right) = \exp\left(\frac{2\pi}{\Gamma}\psi\right) \exp\left(-i\frac{2\pi}{\Gamma}\varphi\right). \tag{2}$$

In this representation, the velocity potential curves $\varphi = constant$ are orthogonal to the velocity field and preserve the constant argument of $\Psi(z)$. From another side, the curves $\psi = constant$, along which the modulus of the wave function Ψ is constant, are lines of the stream. Such interpretation of the hydrodynamic flow in terms of the modulus and argument of some complex wave function $\Psi(z)$ was indicated in Poincare lectures on vortex theory in 1893 [3], much early before discovery of quantum mechanics and the wave function.

Proposition 1 *Every zero $z = z_0$ of the wave function (1), $\Psi(z_0) = 0$, describes the point vortex with complex potential f and corresponding complex velocity $\bar{v}$,*

$$f(z) = \frac{i\Gamma}{2\pi} \ln(z - z_0), \qquad \bar{v}(z) = \frac{i\Gamma}{2\pi} \frac{1}{z - z_0}, \tag{3}$$

where Γ characterises the strength (intensity) of the vortex.

The wave function $w = w(z) = \Psi(z)$ gives conformal mapping from complex plane $z \in C$ to complex plane $w = u + iv \in C$. Then, (1) represents the point vortex at the origin in w plane, $f(z(w)) = \frac{i\Gamma}{2\pi} \ln w$. The stream function of this flow

$$\psi(u, v) = \frac{\Gamma}{2\pi} \ln |w| = \frac{\Gamma}{2\pi} \ln |\Psi(z)|, \tag{4}$$

describes the stream lines $\psi(u, v) = c_0$ as circles of radius

$$|\Psi(z)| = e^{\frac{2\pi}{\Gamma} c_0}. \tag{5}$$

For the flow in a domain, bounded by curve C, the boundary condition on normal velocity to the curve is zero, which implies that the stream function $\psi = 0$, and as follows, the wave function $\Psi(z)$ on boundary curve is unimodular, $|\Psi(z)| = 1$.

Definition 2 The analytic functions

$$f(z) = \frac{N}{2\pi} \ln \Psi(z), \qquad \bar{v}(z) = \frac{df}{dz} = \frac{N}{2\pi} \frac{\Psi'(z)}{\Psi(z)}, \tag{6}$$

represent the source/sink type complex potential $f(z)$ of the two dimensional flow and corresponding complex velocity $\bar{v}(z)$.

The wave function for the source/sink field representation is

$$\Psi(z) = \exp \frac{2\pi}{N} f = \exp\left(\frac{2\pi}{N}\varphi(x, y)\right) \exp\left(i\frac{2\pi}{N}\psi(x, y)\right). \tag{7}$$

The function gives conformal mapping to complex plane $w = \Psi(z)$, so that in this plane (6) represents the point source/sink at the origin $f(z(w)) = \frac{N}{2\pi} \ln w$. The stream lines of this flow in w plane correspond to constant argument of the wave function $\frac{N}{2\pi} \arg \Psi(z) = \psi(u, v) = c_0$.

Proposition 2 *Every zero $z = z_0$ of the wave function (6), $\Psi(z_0) = 0$, corresponds to the point source/sink with complex potential f and complex velocity $\bar{v}$,*

$$f(z) = \frac{N}{2\pi} \ln(z - z_0), \qquad \bar{v}(z) = \frac{N}{2\pi} \frac{1}{z - z_0}, \tag{8}$$

where N characterizes the strength (abundance) of the source ($N > 0$) or the sink ($N < 0$).

2.1 Boundary Conditions

For a domain bounded by curve C the normal velocity of the flow vanishes at the boundary. This implies the boundary condition as the reality condition for complex potential $\Im f(z)|_C = \psi|_C = 0$. For the vortex flow (1) on C, this gives $\psi(x, y) = \frac{\Gamma}{4\pi} \ln |\Psi(z)| = 0$ or the unimodularity condition on the wave function $|\Psi(z)|^2 = 1$, which describes the unit circle in w plane. For the source/sink flow (6) on C, $\psi(x, y) = \frac{N}{4\pi i} \ln \frac{\Psi(z)}{\overline{\Psi(z)}} = 0$, this gives the reality condition on the wave function $\Psi(z) = \overline{\Psi(z)}$. As a result we have following proposition.

Proposition 3 *The boundary condition for the two dimensional flow, as vanishing of normal velocity v_n on the boundary curve C, corresponds to boundary condition on the wave function:*

(1) for the vortex flow

$$|\Psi(z)|^2 = 1, \tag{9}$$

(2) for the source/sink flow

$$\Psi(z) = \overline{\Psi(z)}. \tag{10}$$

2.2 Normalization Freedom

The complex potential $f(z)$ of the flow with complex velocity $\bar{v}(z) = df/dz$ is determined up to an arbitrary complex constant: $f(z) \to f(z) + c_0$. This implies that the wave function corresponding to the given flow $\bar{v}(z)$ is defined up to an arbitrary (frequently an infinite) multiplication constant B: $\Psi(z) \to B\,\Psi(z)$.

3 The Wedge Theorem and Wave Function

The wedge theorem [12] describes the flow in the wedge domain.

Theorem 1 *For a given in plane flow $f(z)$, introduction of the boundary wedge domain with angle $\alpha = \frac{\pi}{n}$, produces the flow*

$$F_q(z) = \sum_{k=0}^{n-1} f(q^{2k} z) + \sum_{k=0}^{n-1} \bar{f}(q^{2k} z). \tag{11}$$

where $q = e^{i\frac{\pi}{n}}$ is primitive root of unity, so that $q^{2n} = 1$. The flow is q^2-periodic (rotation invariant): $F_q(q^2 z) = F_q(z)$.

This theorem implies following representation for the wave function in the wedge domain.

Theorem 2 *The wave function Ψ_q in the wedge domain with angle $\alpha = \frac{\pi}{n}$ and $q = e^{i\frac{\pi}{n}}$, in vortex representation (1) is*

$$\Psi_q(z) = \prod_{k=0}^{n-1} \frac{\Psi(q^{2k} z)}{\bar{\Psi}(q^{2k} z)}, \tag{12}$$

and in source/sink representation (6) it is

$$\Psi_q(z) = \prod_{k=0}^{n-1} \Psi(q^{2k} z)\, \overline{\Psi}(q^{2k} z), \tag{13}$$

where function $\Psi(z)$ corresponds to the whole complex plane. These functions are q^2-periodic: $\Psi_q(q^2 z) = \Psi_q(z)$.

Proof The wave function (12) on boundaries of the wedge domain $z = x$ and $z = qx$, satisfy following equations, respectively,

$$\overline{\Psi_q(x)} = \frac{1}{\Psi_q(x)}, \qquad \overline{\Psi_q(qx)} = \frac{1}{\Psi_q(qx)},$$

which means that the function on the wedge boundaries is the unimodular

$$|\Psi_q(x)|^2 = 1, \qquad |\Psi_q(qx)|^2 = 1.$$

The wave function (13) on boundaries of the wedge domain $z = x$ and $z = qx$, is equal, respectively,

$$\Psi_q(x) = \prod_{k=0}^{n-1} |\Psi(q^{2k} x)|^2, \qquad \Psi_q(qx) = \prod_{k=0}^{n-1} |\Psi(q^{2k+1} x)|^2. \tag{14}$$

It is the real and non-negative valued function on the boundaries.

4 The Strip Theorem and Wave Function

For the hydrodynamic flow in a strip domain we have the strip theorem [6].

Theorem 3 *The complex potential for incompressible and irrotational flow in the strip domain $S : \{z = x + iy;\ -h/2 \leq y \leq h/2\}$ is*

$$F(z) = \sum_{n=-\infty}^{\infty} f(z + (2n)ih) + \sum_{n=-\infty}^{\infty} \bar{f}(z + (2n-1)ih), \tag{15}$$

where $f(z)$ is potential of the flow in whole plane. It satisfies the boundary conditions: $\Im F(z)|_{z=x\pm i\frac{h}{2}} = 0$, periodicity and the combined periodicity conditions, respectively: $F(z + 2ih) = F(z)$, $\bar{F}(z + ih) = F(z)$.

This theorem implies the form of wave functions in the strip domain according to following theorem.

Theorem 4 *The wave function corresponding to the strip domain S in vortex representation (1) is*

$$\Psi_h(z) = \prod_{n=-\infty}^{\infty} \frac{\Psi(z + (2n)ih)}{\overline{\Psi}(z + (2n-1)ih)}, \tag{16}$$

and in source/sink representation (6) it is

$$\Psi_h(z) = \prod_{n=-\infty}^{\infty} \Psi(z + (2n)ih)\,\overline{\Psi}(z + (2n-1)ih), \tag{17}$$

where function $\Psi(z)$ corresponds to the whole complex plane. The function (16) is inverse combined periodic and the one in (17) is combined periodic, respectively,

$$\bar{\Psi}_h(z + ih) = \frac{1}{\Psi_h(z)}, \qquad \bar{\Psi}_h(z + ih) = \Psi_h(z). \tag{18}$$

Both functions are periodic: $\Psi_h(z + 2ih) = \Psi_h(z)$.

Example 1 For the point vortex at the origin in the strip domain S,

$$\Psi_h(z) = \prod_{n=-\infty}^{\infty} \frac{(z + (2n)ih)}{(z + (2n-1)ih)} = \tanh \frac{\pi}{2h} z. \tag{19}$$

Example 2 For the point source at the origin in S,

$$\Psi_h(z) = \prod_{n=-\infty}^{\infty} (z + (2n)ih)(z + (2n-1)ih) = \prod_{n=-\infty}^{\infty} (z + nih) = z \prod_{n=1}^{\infty} (z^2 + n^2h^2)$$

it gives (here we neglect irrelevant multiplicative constant)

$$\Psi_h(z) = \sinh \frac{\pi}{h} z. \tag{20}$$

As we show in next section, this wave function describes the coherent cat state (35) in Fock-Barmann representation (37), where complex parameter of the coherent state is fixed as $\alpha = \frac{\pi}{h}$.

4.1 *Oblique Strip Theorem*

For upcoming applications we need a more general version of the strip theorem, with the oblique strip domain.

Theorem 5 *The complex potential*

$$F_\beta(z) = \sum_{-\infty}^{\infty} f(z + (2n)ihe^{i\beta}) + \sum_{-\infty}^{\infty} \bar{f}(z + (2n-1)ihe^{i\beta}), \tag{21}$$

for the flow in oblique strip domain:

$$S_\beta : \{z = x + iy; x \tan\beta - \frac{h}{2\cos\beta} \leq y \leq x \tan\beta + \frac{h}{2\cos\beta}\},$$

between two oblique lines ($-\infty < x < \infty$)*:* $z = x + i(x \tan\beta \pm \frac{h}{2\cos\beta})$, *is real on these lines and satisfies periodicity and combined periodicity conditions*

$$F_\beta(z + 2ihe^{i\beta}) = F_\beta(z), \qquad \bar{F}(z + ihe^{i\beta}) = F(z). \tag{22}$$

5 Fock-Bargmann Representation

Now we are going to construct the hydrodynamic flow for several types of quantum coherent states. The pair of bosonic operators a and a^+ satisfy the commutation relation $[a, a^+] = 1$, and determines the Fock states

$$|n\rangle = \frac{(a^+)^n}{\sqrt{n!}}|0\rangle, \qquad a|0\rangle = 0. \tag{23}$$

Definition 3 The Glauber coherent states [19] (not normalized) are vectors in Fock space

$$|\alpha\rangle = \sum_{n=0}^{\infty} \frac{\alpha^n}{\sqrt{n!}}|n\rangle, \tag{24}$$

where $\alpha \in C$ is an arbitrary complex number.

Proposition 4 *The inner product of two coherent states is* $\langle\beta|\alpha\rangle = e^{\bar{\beta}\alpha}$, *and normalization factor is* $\langle\alpha|\alpha\rangle = e^{|\alpha|^2}$.

A representation of quantum states in Fock space by complex analytic functions is called the Fock-Bargmann representation [15, 16]. Let $|\Psi\rangle = \sum_{n=0}^{\infty} c_n|n\rangle$ is an arbitrary vector in Fock space with normalization condition $\langle\Psi|\Psi\rangle = \sum_{n=0}^{\infty} |c_n|^2 = 1$.

Definition 4 The wave function

$$\langle z|\Psi\rangle = \Psi(\bar{z}), \tag{25}$$

where $|z\rangle$ is the Glauber coherent state (24), determines the entire complex analytic function

$$\Psi(z) = \sum_{n=0}^{\infty} c_n \frac{z^n}{\sqrt{n!}}. \tag{26}$$

Proposition 5 *The wave function of the Fock state $|n\rangle$ in Fock-Bargmann representation is monomial*

$$\Psi(z) = \frac{z^n}{\sqrt{n!}}. \tag{27}$$

It represents the point vortex and point source/sink flow at origin, with strength $n\Gamma$ and nN, correspondingly,

$$f_v(z) = \frac{in\Gamma}{2\pi} \ln z, \qquad f_s(z) = \frac{nN}{2\pi} \ln z. \tag{28}$$

Proposition 6 *The wave function of the Glauber coherent state $|\alpha\rangle$ in Fock-Bargmann representation is entire function*

$$\Psi(z) = e^{\alpha z}. \tag{29}$$

It describes homogeneous flow in plane with linear complex potential and constant velocity, determined by complex number α,

$$f(z)_{vortex} = \frac{i\Gamma}{2\pi}\alpha\, z, \qquad \bar{v}(z) = \frac{i\Gamma}{2\pi}\alpha, \tag{30}$$

$$f(z)_{source/sink} = \frac{N}{2\pi}\alpha\, z, \qquad \bar{v}(z) = \frac{N}{2\pi}\alpha. \tag{31}$$

Displaced Coherent State The displaced coherent states are defined as

$$|n; \alpha\rangle = e^{\alpha a^+ - \bar{\alpha} a}|n\rangle, \tag{32}$$

and in Fock-Bargman representation their representatives are complex wave functions with translated origin

$$\Psi_n(z; \alpha) = \frac{(z - \bar{\alpha})^n}{\sqrt{n!}} e^{\alpha z}. \tag{33}$$

The corresponding homogeneous vortex flow

$$f(z) = \frac{i\Gamma n}{2\pi} \ln(z - \bar{\alpha}) + \frac{i\Gamma}{2\pi}\alpha z \tag{34}$$

includes also point vortex with strength Γn at position $z = \bar{\alpha}$.

5.1 Cat States Flow

The even and odd superpositions of two Glauber coherent states

$$|0\rangle_\alpha = \frac{\cosh \alpha\, a^+}{\sqrt{\cosh |\alpha|^2}}|0\rangle, \qquad |1\rangle_\alpha = \frac{\sinh \alpha\, a^+}{\sqrt{\sinh |\alpha|^2}}|0\rangle, \tag{35}$$

are called the cat states. An arbitrary state $|\Psi\rangle$ in the cat state basis is

$$_z\langle 0|\Psi\rangle = \frac{\Psi_0(\bar z)}{\sqrt{\cosh |z|^2}} = \frac{\Psi(\bar z) + \Psi(-\bar z)}{2\sqrt{\cosh |z|^2}}, \; _z\langle 1|\Psi\rangle = \frac{\Psi_1(\bar z)}{\sqrt{\sinh |z|^2}} = \frac{\Psi(\bar z) - \Psi(-\bar z)}{2\sqrt{\sinh |z|^2}}.$$

These wave functions determine even and odd entire analytic functions

$$\Psi_0(z) = \frac{1}{2}(\Psi(z) + \Psi(-z)), \qquad \Psi_1(z) = \frac{1}{2}(\Psi(z) - \Psi(-z)). \tag{36}$$

For Glauber coherent state $|\alpha\rangle$, this gives

$$\Psi_0(z) = \frac{e^{\alpha z} + e^{-\alpha z}}{2} = \cosh \alpha z, \qquad \Psi_1(z) = \frac{e^{\alpha z} - e^{-\alpha z}}{2} = \sinh \alpha z. \tag{37}$$

The corresponding complex potential and complex velocity of the vortex flow are

$$f_0(z) = \frac{i\Gamma}{2\pi} \ln \cosh \alpha z, \qquad \bar v_0(z) = \frac{i\Gamma}{2\pi}\alpha \tanh \alpha z, \tag{38}$$

$$f_1(z) = \frac{i\Gamma}{2\pi} \ln \sinh \alpha z, \qquad \bar v_1(z) = \frac{i\Gamma}{2\pi}\alpha \coth \alpha z. \tag{39}$$

By using infinite product representation of hyperbolic functions

$$\sinh \alpha z = \alpha z \prod_{n=1}^{\infty}\left(1 + \frac{\alpha^2 z^2}{(\pi n)^2}\right), \quad \cosh \alpha z = \prod_{n=1}^{\infty}\left(1 + \frac{\alpha^2 z^2}{(\pi n - \frac{\pi}{2})^2}\right), \tag{40}$$

and omitting irrelevant constants we have following proposition.

Proposition 7 *The complex potentials of vortex flow in cat states are*

$$f_0(z) = \frac{i\Gamma}{2\pi} \sum_{n=-\infty}^{\infty} \ln\left(z + i\frac{\pi}{\alpha}(n + \frac{1}{2})\right), \tag{41}$$

$$f_1(z) = \frac{i\Gamma}{2\pi} \sum_{n=-\infty}^{\infty} \ln\left(z - i\frac{\pi}{\alpha} n\right), \tag{42}$$

with corresponding velocities

$$\bar{v}_0(z) = \frac{i\Gamma}{2\pi} \sum_{n=-\infty}^{\infty} \frac{1}{z + i\frac{\pi}{\alpha}(n+\frac{1}{2})}, \quad \bar{v}_1(z) = \frac{i\Gamma}{2\pi} \sum_{n=-\infty}^{\infty} \frac{1}{z - i\frac{\pi}{\alpha}n}. \tag{43}$$

These formulas show that both flows are periodic $f_0\left(z + i\frac{\pi}{\alpha}\right) = f_0(z)$, $f_1\left(z + i\frac{\pi}{\alpha}\right) = f_1(z)$, and describe one dimensional lattice with equal strength vortices, distributed with period $\frac{\pi}{|\alpha|}$ along the line from origin, inclined by angle $\beta = \frac{\pi}{2} - \arg\alpha$.

Proposition 8 *The source/sink flows corresponding to cat states are*

$$f_0(z) = \frac{N}{2\pi} \sum_{n=-\infty}^{\infty} \ln\left(z + i\frac{\pi}{\alpha}(n+\frac{1}{2})\right), \tag{44}$$

$$f_1(z) = \frac{N}{2\pi} \sum_{n=-\infty}^{\infty} \ln\left(z + i\frac{\pi}{\alpha}n\right). \tag{45}$$

The flows are also periodic and like in the vortex representation, describe one dimensional lattice of sources/sinks. But in contrast to the vortex case, these flows allow interpretation in terms of the oblique strip theorem (21), as just one source/sink in bounded domain in form of the oblique strip, which is determined by complex parameter α of the coherent state.

Theorem 6 *The source/sink flow corresponding to the cat state $|1\rangle_\alpha$ represents one point source/sink at origin in the oblique strip domain ($-\infty < x < \infty$),*

$$S_\alpha : \{z = x + iy;\ -x\tan\arg\alpha - \frac{\pi}{2|\alpha|\cos\arg\alpha} \leq y \leq -x\tan\arg\alpha + \frac{\pi}{2|\alpha|\cos\arg\alpha}\}, \tag{46}$$

the symmetric strip on interval $[-\frac{\pi}{2|\alpha|}, \frac{\pi}{2|\alpha|}]$, with width $h = \frac{\pi}{|\alpha|}$, and inclined by angle $\beta = -\arg\alpha$.

Theorem 7 *The source/sink flow corresponding to the cat state $|0\rangle_\alpha$ represents one point source/sink in the shifted oblique strip domain (46), with interval $[0, \frac{\pi}{|\alpha|}]$, which implies translation $z \to z + i\frac{\pi}{2|\alpha|}$.*

The cat states represent orthogonal superposition of coherent states and have many applications from quantum optics to quantum information theory. The above theorems provide interpretation of these states as one point source/sink for the hydrodynamic flow in the strip domain.

6 q-Deformed Coherent States

The pair of q-bosonic operators a_q and a_q^+ satisfy the commutation relation $a_q a_q^+ - q\, a_q^+ a_q = 1$, and determines the q-Fock states [17],

$$|n\rangle_q = \frac{(a_q^+)^n}{\sqrt{[n]_q!}}|0\rangle_q, \qquad a_q|0\rangle_q = 0, \tag{47}$$

where non-symmetric q-number is $[n]_q = 1 + q + q^2 + ...q^{n-1} = \frac{q^n-1}{q-1}$.

Definition 5 The q-deformed coherent states (not normalized) are vectors in q-Fock space

$$|\alpha\rangle_q = \sum_{n=0}^{\infty} \frac{\alpha^n}{\sqrt{[n]_q!}}|n\rangle_q, \tag{48}$$

where $\alpha \in C$ is an arbitrary complex number.

Proposition 9 *The wave function of the q-deformed coherent state $|\alpha\rangle_q$ in q-Fock-Bargmann representation is*

$$_q\langle z|\alpha\rangle_q = \Psi_q(\bar{z}) \quad \rightarrow \quad \Psi_q(z) = e_q^{\alpha z}, \tag{49}$$

where e_q^z is the Jackson exponential function [8].

For $|q| > 1$ this function is entire and can be represented by infinite product

$$e_q^{\alpha z} = \prod_{k=0}^{\infty} \left(1 + \frac{\alpha z}{q^k}\left(1 - \frac{1}{q}\right)\right). \tag{50}$$

For $|q| < 1$ the function is analytic in the disk $|z| < \frac{1}{\sqrt{1-q}}$, and meromorphic, with infinite number of poles

$$e_q^{\alpha z} = \prod_{k=0}^{\infty} \frac{1}{(1 - q^k(1-q)\alpha z)}. \tag{51}$$

Theorem 8 *The vortex flow for q-deformed coherent states, for $|q| > 1$*

$$f_q(z) = \frac{i\Gamma}{2\pi} \sum_{k=0}^{\infty} \ln\left(z + \frac{q^k}{\alpha(q-1)}\right) \tag{52}$$

represents an infinite set of point vortices with positions at geometric progression $z_k = -\frac{q^k}{\alpha(q-1)}$, *and for* $|q| < 1$

$$f_q(z) = -\frac{i\Gamma}{2\pi}\sum_{k=0}^{\infty}\ln\left(z - \frac{1}{\alpha(1-q)q^k}\right) \tag{53}$$

by the anti-vortices at positions $z_k = \frac{1}{\alpha(1-q)q^k}$.

An infinite set of point vortex images, located along the line in geometric progression appears in method of images for hydrodynamic flow applied to two concentric circles, the problem discussed seems the first time in lectures of Poincare [3]. These images were calculated by the two circle theorem [7] and represented in terms of q-calculus by Jackson's exponential function in [18]. These results were motivation for us to search the relation between hydrodynamic flow in bounded domain and quantum states, representable by complex analytic functions with quantum group symmetry.

7 Conclusions

The Madelung fluid representation provides a basis for several interpretations of quantum mechanics and becomes fundamental tool in description of quantum fluids and nonlinear evolution equations. It has been applied as computational tool for solving non-stationary quantum problems, when instead of following quantum trajectories of individual particles, both trajectories and hydrodynamic fields are computed by fluid dynamics computational techniques. In present paper a new approach is proposed, to relate quantum theory with fluid dynamics. The key point of this approach is Fock-Bargmann representation of the wave function by complex analytic function, which provides a bridge between these two subjects. This work was supporting by BAP project 2022IYTE-1-0002.

References

1. Madelung, E.: Quantentheorie in hydrodynamischer form. Zeitschr. f. Phys. **40**, 322–326 (1926). https://doi.org/10.1007/BF01400372
2. Maxwell, J.C.: A Treatise on Electricity and Magnetism, vol. 1. Clarendon Press, Oxford (1873). https://doi.org/10.1017/CBO9780511709333
3. Poincare, H.: Theorie des Tourbillions. Georges Carre, Paris (1893)
4. Milne-Thomson, L.M.: Theoretical Hydrodynamics. Macmillan, London (1968)
5. Pashaev, O.K.: Variations on a theme of q-oscillator. Phys. Scr. **90**, 074010 (2015). https://doi.org/10.1088/0031-8949/90/7/074010

6. Pashaev, O.K.: Quantum calculus of classical vortex images, integrable models and quantum states. J. Phys.: Conf. Ser. **766**, 012015 (2016). https://doi.org/10.1088/1742-6596/766/1/012015
7. Pashaev, O.K.: Two-circles theorem, q-periodic functions and entangled qubit states. J. Phys.: Conf. Ser. **482**, 012033 (2014). https://doi.org/10.1088/1742-6596/482/1/012033
8. Kac, V., Cheung, P.: Quantum Calculus. Springer, New York (2002). https://doi.org/10.1007/978-1-4613-0071-7
9. Ford, L.R.: Automorphic Functions. McGraw-Hill Book Company, New York (1929)
10. Pashaev, O.K., Kocak, A.: Kaleidoscope of classical vortex images and quantum coherent states. In: Kac, V.G., et al. (eds.) Symmetries, Differential Equations and Applications, vol. 266, pp. 179–199. Springer (2018). https://doi.org/10.1007/978-3-030-01376-9_10
11. Kocak, A., Pashaev, O.K.: Special functions with mod n symmetry and kaleidoscope of quantum coherent states. J. Phys.: Conf. Ser. **1194**, 012059 (2019). https://doi.org/10.1088/1742-6596/1194/1/012059
12. Pashaev, O.K.: Quantum group symmetry for kaleidoscope of hydrodynamic images and quantum states. J. Phys.: Conf. Ser. **1194**, 012087 (2019). https://doi.org/10.1088/1742-6596/1194/1/012087
13. Schrödinger, E.: Quantisierung als Weigenwertproblem (I Mitt.). Ann. Phys. **79**, 361 (1926). https://doi.org/10.1002/andp.19263861802
14. Pashaev, O.K., Gurkan, Z.: Abelian Chern-Simons vortices and holomorphic Burgers hierarchy. Theor. Math. Phys. **152**, 1017–1029 (2007). https://doi.org/10.1007/s11232-007-0086-0
15. Fock, V.: Verallgemeinerung und Lösung der Diracschen statistischen Gleichung. Z. Phys. **49**, 339–357 (1928). https://doi.org/10.1007/BF01337923
16. Bargmann, V.: On a Hilbert space of analytic functions and an associated integral transform Part I. Commun. Pure Appl. Math. **XIV**, 187–214 (1961). https://doi.org/10.1002/cpa.3160140303
17. Arik, M., Coon, D.D.: Hilbert spaces of analytic functions and generalized coherent states. J. Math. Phys. **17**, 524–527 (1976). https://doi.org/10.1063/1.522937
18. Pashaev, O.K., Yilmaz, O.: Vortex images and q-elementary functions. J. Phys. A: Math. Theor. **41**, 135207 (2008). https://doi.org/10.1088/1751-8113/41/13/135207
19. Perelomov, A.: Generalized Coherent States and Their Applications. Springer, Berlin (1986). https://doi.org/10.1007/978-3-642-61629-7

On Leonardo Numbers and Fibonacci Fundamental System

Elen Viviani Pereira Spreafico and Paula Maria Machado Cruz Catarino

Abstract Our goal is to explore the generalized Leonardo numbers, through the properties of the Fibonacci fundamental systems related to the elements of this sequence. We explicitly describe the closed connection between the sequences of the Fibonacci fundamental system and the generalized Leonardo numbers. Moreover, the matrix approach is considered for studying the combinatorial identities and the generalized Cassini identity for the generalized Leonardo numbers. In addition, the analytical aspect about each sequence of the generalized Leonardo numbers is elaborated. Finally, the combinatorial and the analytical formula of the generalized Cassini identity are investigated.

1 Introduction

In recent times, numerous sets of number sequences have been researched, including but not limited to Fibonacci, Pell, Lucas, Jacobsthal, Padovan, and Perrin numbers, along with their generalizations. One of these interesting sequences is that it is consisted of the Leonardo numbers, $\{Le_n\}_{n\geq 0}$. The Leonardo numbers was introduced in the mathematics literature by Catarino and Borges in [5] and it is defined by the recurrence relation

$$Le_n = Le_{n-1} + Le_{n-2} + 1, \ (n \geq 2), \tag{1}$$

with initial conditions $Le_0 = Le_1 = 1$. This sequence of numbers is studied in several aspects, as analytic and matrix perspectives. Also we find connections to

E. V. P. Spreafico (✉)
Institute of Mathematics, Federal University of Mato Grosso do Sul, Campo Grande, Brazil
e-mail: elen.spreafico@ufms.br

P. M. M. C. Catarino
Department of Mathematics, University of Trás-os-Montes e Alto Douro, Vila Real, Portugal
e-mail: pcatarin@utad.pt

V. Gayoso Martínez et al. (eds.), *Mathematical Methods for Engineering Applications*, Springer Proceedings in Mathematics & Statistics 439,
https://doi.org/10.1007/978-3-031-49218-1_6

various areas of mathematics. They appear in the study of recursive sequences, combinatorial problems, and in the analysis of certain algorithms and number systems. Additionally, they have applications in computer science, particularly in areas such as data structures and algorithm design (see more in [1, 2, 6, 10, 11], and references therein).

Similar to the Fibonacci numbers, the Leonardo numbers exhibit a remarkable pattern in nature. They can be found in the branching of trees, the arrangement of leaves and petals, the spirals of shells, and many other natural phenomena. This pattern is known as phyllotaxis and is closely related to the Fibonacci and Leonardo sequences.

In summary, the Leonardo numbers form a fascinating sequence of integers that share similarities with the Fibonacci numbers. They arise from a simple recurrence relation and have connections to various branches of mathematics. Moreover, their presence in natural structures adds to their significance and intrigue.

Since the recurrence relation (1) can be write under the form

$$Le_{n+1} = 2Le_n - Le_{n-2},\ (n \geq 2),$$

in the present study we discuss properties of a generalized model of Leonardo numbers, defined by a general linear recursive relation,

$$L_{n+1} = 2^d L_n - kL_{n-2},\ (n \geq 2), \tag{2}$$

with $d, k \geq 1$ and initial conditions $L_0 = \alpha_0$, $L_1 = \alpha_1$ and $L_2 = \alpha_2$, called generalized Leonardo numbers. Our goal is explore the generalized Leonardo numbers defined by the recurrence relation (2), through the properties of the Fibonacci fundamental systems related to the elements of this sequence.

The fundamental Fibonacci system, which encompasses sequences defined by linear recursive relations of the Fibonacci type, serves as a valuable tool for studying such sequences. Its general form was extensively explored by Rachidi et al. in [3, 9]. Furthermore, it has been employed in other studies such as [12] for investigating generalized Fibonacci numbers, in [13] for deriving results on generalized Pell numbers, in [7] for establishing properties of a generalized Pell model and in [8] where the generalized Cassini identities were established.

Therefore, several new properties and identities of this general model are established, and many results of the literature are recovered. More precisely, we obtain the combinatorial expressions and the analytic formulas of sequences defined by (2).

The content of this paper is as follows. Section 3 concerns the generalized Leonardo numbers and its associated Fibonacci fundamental system. Section 4 is devoted to study the Casoratian matrix and to provide identities for the generalized Leonardo numbers. The combinatorial aspects of generalized Leonardo numbers are given in Sect. 5. In Sect. 6 the analytical formulas for the generalized Leonardo numbers are established. Finally, Sect. 7 is devoted to the discuss of the Cassini identity for Leonardo numbers, and also the combinatorial and the analytical formula of the generalized Cassini identity are investigated.

2 The Generalized Leonardo Numbers and Fibonacci Fundamental System

As mentioned before, the Fibonacci fundamental system is an interesting tool for analyzing sequences defined by linear recursive relations of the Fibonacci type. This approach has been applied for models of generalized sequences, as generalized Fibonacci numbers and the generalized Pell numbers (see more details in [7, 12] and [13]), and several results were established. In this section, we establish the Fibonacci fundamental system associated to the generalized Leonardo numbers (2) and explicitly describe the closed connection between the elements, which are sequences, of the Fibonacci fundamental system.

2.1 *Fibonacci Fundamental System Related to (2)*

Given a non negative integer d and a positive integer k, we consider the sequences of generalized Leonardo numbers $\{L_n^{(1)}\}_{n\geq 0}$, $\{L_n^{(2)}\}_{n\geq 0}$ and $\{L_n^{(3)}\}_{n\geq 0}$, defined by,

$$\begin{aligned} L_{n+1}^{(j)} &= 2^d L_n^{(j)} - kL_{n-2}^{(j)}, \text{ for } n \geq 3, \\ L_n^{(j)} &= \delta_{n+1}^{(j)}, \text{ for } n = 0,\ 1,\ 2, \end{aligned} \tag{3}$$

where $\delta_i^{(j)} = 1$ if $i = j$ and $\delta_i^{(j)} = 0$, otherwise.

As an example, for $d = 1$ and $k = 3$, Expression (3) is given by

$$\begin{aligned} L_{n+1}^{(j)} &= 2L_n^{(j)} - 3L_{n-2}^{(j)}, \text{ for } n \geq 3, \\ L_n^{(j)} &= \delta_{n+1}^{(j)}, \text{ for } n = 0,\ 1,\ 2. \end{aligned}$$

for every j, $1 \leq j \leq 3$.

Table 1 describes the values of these sequences for $0 \leq n \leq 11$.

Let $\mathcal{E}_L$ be the set of sequences $\{L_n\}_{n\geq 0}$ satisfying the recurrence relation (2) and the set $\mathcal{S}_L = \left\{\{L_n^{(j)}\}_{n\geq 0};\ 1 \leq j \leq 3\right\}$. We can prove that the set $\mathcal{E}_L$ of sequences satisfying the recursive relation (2) is a real vector space. Moreover, the next result shows us that the set $\mathcal{S}_L = \{\{L_n^{(j)}\}_{n\geq 0};\ 1 \leq j \leq 3\}$ is a generator system of $\mathcal{E}_L$.

Proposition 1 *Let $\{L_n\}_{n\geq 0}$ be the sequence defined by the recursive relation (2) and with initial conditions $(\alpha_0, \alpha_1, \alpha_2)$. Then, for every $n \geq 0$, we have,*

Table 1 The generalized Leonardo numbers $\{L_n^{(j)}\}_{n\geq 0}$ with $d=1$ and $k=3$

n	0	1	2	3	4	5	6	7	8	9	10	11
$L_n^{(1)}$	1	0	0	−3	−6	−12	−15	−12	12	69	174	312
$L_n^{(2)}$	0	1	0	0	−3	−6	−12	−15	−12	12	69	174
$L_n^{(3)}$	0	0	1	2	4	5	4	−4	−23	−58	−104	−139

$$L_n = \sum_{j=0}^{2} \alpha_j L_n^{(j+1)}.$$

Proof Let $\{v_n\}_{n\geq 0}$ be the sequence defined by, $v_n = \alpha_0 L_n^{(1)} + \alpha_1 L_n^{(2)} + \alpha_2 L_n^{(3)}$, for every $n \geq 0$, with initial conditions $(\alpha_0, \alpha_1, \alpha_2)$. For every $n \geq 3$, we have,

$$\begin{aligned} v_n &= \sum_{j=0}^{2} \alpha_j L_n^{(j+1)} = \sum_{j=0}^{2} \alpha_j [2^d L_{n-1}^{(j+1)} - k L_{n-3}^{(j+1)}] \\ &= 2^d v_{n-1} - k v_{n-3}. \end{aligned}$$

Therefore, the sequence $\{v_n\}_{n\geq 0}$ satisfies the recursive relation (2). In addition, we get,

$$\begin{aligned} v_0 &= \alpha_0 v_0^{(1)} + \alpha_1 v_0^{(2)} + \alpha_2 v_0^{(3)} = \alpha_0, \\ v_1 &= \alpha_0 v_1^{(1)} + \alpha_1 v_1^{(2)} + \alpha_2 v_1^{(3)} = \alpha_1, \\ v_2 &= \alpha_0 v_2^{(1)} + \alpha_2 v_2^{(2)} + \alpha_2 v_2^{(3)} = \alpha_2, \end{aligned}$$

because $v_d^{(j+1)} = 1$ if $j = d$, and $v_d^{(j+1)} = 0$ if $j \neq d$. Hence, the two sequences $\{v_n\}_{n\geq 0}$ and $\{L_n\}_{n\geq 0}$ satisfy the recursive relation (2) and own the same initial conditions $(\alpha_0, \alpha_1, \alpha_2)$. Therefore, for every $n \geq 0$, we have $L_n = v_n = \sum_{j=0}^{2} \alpha_j L_n^{(j+1)}$. □

By Proposition 1, the set $\mathcal{S}_L = \{\{L_n^{(j)}\}_{n\geq 0};\ 1 \leq j \leq 3\}$ is the *Fibonacci fundamental system* of the generalized Leonardo numbers defined by the recurrence relation (2).

2.2 *Properties of the Fibonacci Fundamental System Related to Generalized Leonardo Numbers*

In this section we explicitly describe the closed connection between the sequences $\{L_n^{(1)}\}_{n\geq 0}$, $\{L_n^{(2)}\}_{n\geq 0}$ and the called fundamental sequence of generalized Leonardo numbers $\{L_n^{(3)}\}_{n\geq 0}$. More precisely, we have the following result.

Proposition 2 *Let* $\mathcal{S}_L = \left\{\{L_n^{(j)}\}_{n\geq 0};\ 1 \leq j \leq 3\right\}$ *be the Fibonacci fundamental system of the generalized Leonardo numbers (2). Then, for every* j*, we have,*

$$L_n^{(j)} = -kL_{n-j}^{(3)} \tag{4}$$

for $n \geq j$.

Proof First we prove the proposition for $j = 1$, namely, $L_n^{(1)} = -kL_{n-1}^{(3)}$. Let w_n be the sequence defined by $w_n = -kL_{n-1}^{(3)}$. Note that for $n = 1, 2, 3$, we have $w_n = L_n^{(1)}$. Moreover, $2^d w_{n+2} - kw_n = -k(2^d L_{n+1}^{(3)} - kL_{n-1}^{(3)}) = -kL_{n+2}^{(3)} = w_{n+3}$, i.e., the sequence w_n satisfies the recurrence relation (2). Then, $L_n^{(1)} = w_n = -kL_{n-1}^{(3)}$, for $n \geq 1$.

Similarly, for $j = 2$, consider the sequence u_n defined by $u_n = -kL_{n-2}^{(3)}$. For $n = 2, 3, 4$, we have $u_n = L_n^{(2)}$. Moreover, $2^d u_{n+2} - ku_n = -k(2^d L_n^{(3)} - kL_{n-2}^{(3)}) = -kL_{n+1}^{(3)} = u_{n+3}$, i.e., the sequence u_n satisfies the recurrence relation (2). Then, $L_n^{(2)} = u_n = -kL_{n-2}^{(3)}$, for $n \geq 2$. □

Combining the Propositions 1 and 2 we obtain the following general result, which the proof is omitted.

Proposition 3 *Let* $\{L_n\}_{n\geq 0}$ *be the sequence defined by the recursive relation (2) and with initial conditions* $(\alpha_0, \alpha_1, \alpha_2)$*. Then, for every* $n \geq 0$*, we have,*

$$L_n = -k\alpha_0 L_{n-1}^{(3)} - k\alpha_1 L_{n-2}^{(3)} + \alpha_2 L_n^{(3)}. \tag{5}$$

As an interesting example by replacing $d = k = 1$ and initial conditions $\alpha_0 = 1, \alpha_1 = 1, \alpha_2 = 3$ in Eq. (5), we obtain the Leonardo numbers $\{Le_n\}_{n\geq 0}$, in terms of the fundamental solution, $\{L_n^{(3)}\}_{n\geq 0}$.

Corollary 1 *Let* $\{Le_n\}_{n\geq 0}$ *be the sequence of Leonardo numbers. Then, for every* $n \geq 0$*, we have,*

$$Le_n = 3L_n^{(3)} - L_{n-1}^{(3)} - L_{n-2}^{(3)}.$$

3 Casoratian Matrix and Identities for the Generalized Leonardo Numbers

In this section we will establish some identities of the generalized Leonardo numbers. For this purpose we will start by exploring the matrix formulation of Expression (2) and its relation with Fibonacci fundamental system. The relation with matrices and the Fibonacci fundamental system provided interesting results (see more in [7, 8, 12, 13]). The matrix formulation of the recursive relation (2) is given by,

$$V_n = AV_{n-1}, \text{ for every } n \geq 3,$$

where

$$A = \begin{pmatrix} 2^d & 0 & -k \\ 1 & 0 & 0 \\ 0 & 1 & 0 \end{pmatrix} \text{ and } V_n = \begin{pmatrix} L_n \\ L_{n-1} \\ L_{n-2} \end{pmatrix}, \tag{6}$$

It is possible gives an explicit form of the entries of the matrices powers of A in terms of the elements of the Fibonacci fundamental system (3). This will allow to obtain properties of the generalized Leonardo numbers.

To reach our goal, we consider the well-known Casoratian matrix, $\widehat{C}(n)$, related to the family of sequences $\{L_n^{(1)}\}_{n\geq 0}$, $\{L_n^{(2)}\}_{n\geq 0}$, $\{L_n^{(3)}\}_{n\geq 0}$, defined by

$$\widehat{C}(n) = \begin{pmatrix} L_n^{(1)} & L_n^{(2)} & L_n^{(3)} \\ L_{n+1}^{(1)} & L_{n+1}^{(2)} & L_{n+1}^{(3)} \\ L_{n+2}^{(1)} & L_{n+2}^{(2)} & L_{n+2}^{(3)} \end{pmatrix}. \tag{7}$$

A direct verification shows that the Casoratian matrix can be written under the form,

$$\widehat{C}(n) = J \times M_n \times J,$$

where $J = (b_{i,j})_{1\leq i,j\leq 3}$ is the anti-diagonal unit matrix, namely, $b_{i,j} = 1$, for $i + j = 4$, and $b_{i,j} = 0$, otherwise and $M_n = (L_{n+3-i-1}^{(j)})_{1\leq i,j\leq 3}$. We show that the matrix M_n, can be written under the form $M_n = A^n$, where A is the classical companion matrix given in (6), (for more details see, [3, 9] and references therein). Hence, we get the following property stated in the next result.

Theorem 1 *Let* $S_L = \left\{\{L_n^{(j)}\}_{n\geq 0};\ 1 \leq j \leq 3\right\}$ *be the Fibonacci fundamental system of the generalized Leonardo numbers defined by the recurrence relation (2). Then, for every* $n \geq 0$, *the entries of the powers* $A^n = (a_{ij}^{(n)})_{1\leq i,j\leq 3}$ *are given under the form,*

$$a_{ij}^{(n)} = L_{n+3-i}^{(3-j+1)}. \tag{8}$$

Identity $\widehat{C}(n) = J \times M_n \times J$ implies the matrix identity $\widehat{C}(n+m) = \widehat{C}(n).\widehat{C}(m)$, for every n and m. Hence, the entries of the matrix $\widehat{C}(n+m) = (c_{ij}^{(n+m)})_{1\le i,\, j\le r}$, are expressed in terms of those of the matrices $\widehat{C}(m) = (c_{ij}^{(m)})_{1\le i,\, j\le r}$ and $\widehat{C}(n) = (c_{ij}^{(n)})_{1\le i,\, j\le 3}$ as follows,

$$c_{ij}^{(n+m)} = \sum_{k=1}^{r} c_{ik}^{(n)} c_{kj}^{(m)} = \sum_{k=1}^{r} c_{ik}^{(m)} c_{kj}^{(n)}, \text{ for every } n,\ m \ge 0, \tag{9}$$

where $1 \le i,\ j \le 3$.

Combining the Expression (9) and the Theorem 1 we already get the next result,

Proposition 4 *Let $\mathcal{S}_L = \left\{\{L_n^{(j)}\}_{n\ge 0};\ 1 \le j \le 3\right\}$ be the Fibonacci fundamental system of the generalized Leonardo numbers defined by the recurrence relation (2). Then, for each $m, n \ge 0$, the following identity is verified,*

$$L_{n+m+3-i}^{(3-j+1)} = \sum_{k=1}^{3} L_{m+3-i}^{(4-k)} L_{n+3-k}^{(4-j)}. \tag{10}$$

In special, for $i = 3$ and $j = 1$, we obtain

$$L_{n+m}^{(3)} = \sum_{k=1}^{3} L_m^{(4-k)} L_{n+3-k}^{(3)} = L_m^{(3)} L_{n+2}^{(3)} + L_m^{(2)} L_{n+1}^{(3)} + L_m^{(1)} L_n^{(3)}.$$

By Expression (4) of Proposition 2, namely, $L_n^{(j)} = -kL_{n-j}^{(3)}$, we obtain the identity,

$$L_{n+m}^{(3)} = L_m^{(3)} L_{n+2}^{(3)} - k(L_{m-2}^{(3)} L_{n+1}^{(3)} + L_{m-1}^{(3)} L_n^{(3)}).$$

Now, under the previous discussion and the Corollary 1, the next identity involving the Leonardo numbers and the fundamental solution $\{L_n^{(3)}\}_{n\ge 0}$ is provide.

Corollary 2 *Let $\{Le_n\}_{n\ge 0}$ be the sequence of Leonardo numbers. Then, for every $n \ge 0$, we have,*

$$Le_{n+m} = L_m^{(3)} Le_{n+2} - L_{m-1}^{(3)} Le_{n+1} - L_{m-2}^{(3)} Le_n.$$

4 Combinatorial Formulas for the Generalized Leonardo Numbers

The combinatorial expressions of Fibonacci type sequences $\{u_n\}_{n\geq 0}$ defined by recurrence relation $u_{n+1} = \sum_{i=0}^{r-1} a_i u_{n-i-1}$, for $n \geq r$, have been largely studied and some interesting results for special cases as generalized Fibonacci and generalized Pell numbers were provide (see more in [7, 12, 13], and references therein). This section is devoted to discuss the combinatorial expressions of the generalized Leonardo numbers. Consider the following combinatorial expression,

$$\rho(n,3) = \sum_{m_0+2m_1+3m_2=n-3} \frac{(m_0+m_1+m_2)!}{m_0!m_1!m_2!} a_0^{m_0} a_1^{m_1} a_2^{m_2}, \text{ for every } n \geq 3, \tag{11}$$

where a_0, a_1, a_2 are real or complex numbers, $\rho(j,3) = 0$, for $0 \leq j \leq 2$, and $\rho(3,3) = 1$. Since

$$a_0\rho(n,3) + a_1\rho(n-1,3) + a_2\rho(n-2,3) =$$
$$= \sum_{m_0+2m_1+3m_2=n-2} \left(\frac{(m_0+m_1+m_2-1)!}{(m_0-1)!m_1!m_2!} + \frac{(m_0+m_1+m_2-1)!}{m_0!(m_1-1)!m_2!} + \frac{(m_0+m_1+m_2-1)!}{m_0!m_1!(m_2-1)!} \right)$$
$$\times a_0^{m_0} a_1^{m_1} a_2^{m_2}$$
$$= \sum_{m_0+2m_1+3m_2=n-2} \frac{(m_0+m_1+m_2)!}{m_0!m_1!m_2!} \times a_0^{m_0} a_1^{m_1} a_2^{m_2} = \rho(n+1,3),$$

then sequence $\rho(n,3)$ satisfies the following linear difference equation $\rho(n+1,3) = a_0\rho(n,3) + a_1\rho(n-1,3) + a_2\rho(n-2,3)$, for every $n \geq 3$. Specially, for $a_0 = 2^d$, $a_1 = 0$ and $a_2 = -k$, for every $n \geq 3$, we get

$$\rho(n+1,3) = (-1)^{m_2} \sum_{m_0+2m_1+3m_2=n-2} \frac{(m_0+m_1+m_3)!}{m_0!m_1!m_2!} 2^{dm_0} k^{m_2}, \tag{12}$$

with $\rho(0,3) = \rho(1,3) = \rho(2,3) = 0$ and $\rho(3,3) = 1$. Therefore, the sequence $\{\rho(n+1,r)\}_{n\geq 0}$ satisfies the Expression (2) and its initial conditions are given by $\rho(1,3) = \rho(2,3) = 0$ and $\rho(3,3) = 1$. In addition, we formulate the combinatorial expression of the fundamental solution $(L_n(3))_{n\geq 0}$.

Proposition 5 *The combinatorial expression of* $(L_n(3))_{n\geq 0}$ *is given by*

$$L_n^{(3)} = \rho(n+1,3) = (-1)^{m_2} \sum_{m_0+2m_1+3m_2=n-2} \frac{(m_0+m_1+m_3)!}{m_0!m_1!m_2!} 2^{dm_0} k^{m_2}, \tag{13}$$

for every $n \geq 3$, *where* $L_j(3) = \rho(j+1,3) = 0$, *for* $j = 0, 1$, *and* $L_2(3) = \rho(3,3) = 1$.

Combining the Propositions 3 and 5 we obtain the following general result.

Proposition 6 *Let* $\{L_n\}_{n\geq 0}$ *be the sequence defined by the recursive relation* (2) *and with initial conditions* $(\alpha_0, \alpha_1, \alpha_2)$. *Then, for every* $n \geq 0$, *we have,*

$$L_n = -k\alpha_0\rho(n,3) - k\alpha_1\rho(n-1,3) + \alpha_2\rho(n+1,3). \tag{14}$$

Similarly, by replacing $d = k = 1$ and initial conditions $\alpha_0 = 1, \alpha_1 = 1, \alpha_2 = 3$ in Eq. (14), we obtain a combinatorial expression for Leonardo numbers $\{Le_n\}_{n\geq 0}$, stated in the following corollary.

Corollary 3 *Let* $\{Le_n\}_{n\geq 0}$ *be the sequence of Leonardo numbers. Then, for every* $n \geq 0$, *we have,*

$$Le_n = 3\rho(n+1,3) - \rho(n,3) - \rho(n-1,3).$$

5 Analytical Aspect of the Generalized Leonardo Numbers

It is well known that the analytic formula for linear recursive sequences is related to the roots of the associated characteristic polynomial equation. In this section, we study the roots of the characteristic polynomials of the generalized recursive equation (2) given by $P(z) = z^3 - 2^d z^2 - k$, using the Sylvester matrix approach (see more in [7]).

In summary, the Sylvester matrix is a matrix associated with two univariate polynomials $P(z)$ and $Q(z)$, whose entries are given by coefficients of these two polynomials with the following property : when the determinant of the Sylvester matrix $S_{P,Q}$, called the resultant, is zero, then the two polynomials have a common root. Considering the polynomial $P(z)$ and its derivative $P'(z)$, if the determinant of the Sylvester matrix $S_{P,P'}$ is different from 0, then the polynomials $P(z)$ and $P'(z)$ do not have common roots. This means that if $det(S_{P,P'}) \neq 0$, then the roots of $P(z)$ are simple. In this case, the associated characteristic polynomial is $P(z) = z^3 - 2^d z^2 + k$, with derivative $P'(z) = 3z^2 - 2^{d+1}z$, and its Sylvester matrix associated is given by,

$$S_{P,P'} = \begin{pmatrix} 1 & -2^d & 0 & k & 0 \\ 0 & 1 & -2^d & 0 & k \\ 3 & -2^{d+1} & 0 & 0 & 0 \\ 0 & 3 & -2^{d+1} & 0 & 0 \\ 0 & 0 & 3 & -2^{d+1} & 0 \end{pmatrix}.$$

Hence, $det(S_{P,P'}) = -4k2^{3d} + 18k^2$. Suppose that $det(S_{P,P'}) = 0$. Then $-4k2^{3d} + 18k^2 = k(-2^{3d+2} + 18k) = 0$. Since k is a positive integer, then $k \neq 0$ and $k = \frac{2^{3d+1}}{3^2}$. Thus, $det(S_{P,P'}) \neq 0$, for all non negative integer d and positive integer k. Under the preceding discussion, we get the follow analytical result.

Lemma 1 *Consider a non negative integer d and a positive integer k. Then, the roots of the polynomial*

$$P(z) = z^3 - 2^d z^2 + k,$$

are simple.

In addition, the sequence $\{\rho(n,3)\}_{n\geq 0}$ defined by (11) is expressed in an analytic form in terms of the roots of characteristic polynomial. The result is given in the following lemma (see [4] and references therein).

Lemma 2 *Let $\{\rho(n,3)\}_{n\geq 0}$ be the sequence defined by (11). Suppose that the roots $\lambda_1,\lambda_2,\lambda_3$ of its characteristic polynomial $P(z) = z^3 - 2^d z^2 + k$, satisfying $\lambda_i \neq \lambda_j$ for $i \neq j$. Then, we have*

$$\rho(n,3) = \sum_{i=1}^{3} \frac{\lambda_i^{n-1}}{P'(\lambda_i)} = \sum_{i=1}^{3} \frac{\lambda_i^{n-1}}{\prod\limits_{k\neq i}(\lambda_i - \lambda_k)} \text{ for every } n \geq 3,$$

otherwise $\rho(3,3) = 1$, $\rho(i,3) = 0$ for $i \leq 2$, where $P'(z) = \frac{dP}{dz}(z)$.

By Lemma 2 and Proposition 5, we obtain the analytic form for the fundamental solution numbers $L_n^{(3)}$.

Proposition 7 *The analytical expression of $(L_n(3))_{n\geq 0}$ is given by*

$$L_n^{(3)} = \sum_{i=1}^{3} \frac{\lambda_i^{n}}{\prod\limits_{k\neq i}(\lambda_i - \lambda_k)}, \tag{15}$$

for every $n \geq 3$, where $\lambda_1,\lambda_2,\lambda_3$ are roots of the characteristic polynomial $P(z) = z^3 - 2^d z^2 + k$, satisfying $\lambda_i \neq \lambda_j$ for $i \neq j$.

By extending the Proposition 7 for any sequence $\{L_n\}_{n\geq 0}$ defined by the recursive relation (2), we get the following general result,

Proposition 8 *Let $\{L_n\}_{n\geq 0}$ be the sequence defined by the recursive relation (2) and with initial conditions $(\alpha_0, \alpha_1, \alpha_2)$. Then, for every $n \geq 3$, we have,*

$$L_n = -k\alpha_0 \sum_{i=1}^{3} \frac{\lambda_i^{n-1}}{\prod\limits_{k\neq i}(\lambda_i - \lambda_k)} - k\alpha_1 \sum_{i=1}^{3} \frac{\lambda_i^{n-2}}{\prod\limits_{k\neq i}(\lambda_i - \lambda_k)} + \alpha_2 \sum_{i=1}^{3} \frac{\lambda_i^{n}}{\prod\limits_{k\neq i}(\lambda_i - \lambda_k)}, \tag{16}$$

where λ_1,λ_2, λ_3 *are roots of the characteristic polynomial* $P(z) = z^3 - 2^d z^2 + k$, *satisfying* $\lambda_i \neq \lambda_j$ *for* $i \neq j$.

Proof The proof is given by direct application of the Lemma 2 to Proposition 6. □

As a special case, doing $d = k = 1$ and initial conditions $\alpha_0 = 1, \alpha_1 = 1, \alpha_2 = 3$ in Expression (16) , we obtain an analytical formula for Leonardo numbers $\{Le_n\}_{n\geq 0}$, given in the following corollary.

Corollary 4 *Let* $\{Le_n\}_{n\geq 0}$ *be the sequence of Leonardo numbers. Then, for every* $n \geq 3$, *it is verified,*

$$Le_n = 3\sum_{i=1}^{3} \frac{\lambda_i^n}{\prod\limits_{k\neq i}(\lambda_i - \lambda_k)} - \sum_{i=1}^{3} \frac{\lambda_i^{n-1}}{\prod\limits_{k\neq i}(\lambda_i - \lambda_k)} - \sum_{i=1}^{3} \frac{\lambda_i^{n-2}}{\prod\limits_{k\neq i}(\lambda_i - \lambda_k)},$$

where $\lambda_1 = 1$, $\lambda_2 = \frac{1+\sqrt{5}}{2}$, $\lambda_3 = \frac{1-\sqrt{5}}{2}$ *are roots of the characteristic polynomial* $P(z) = z^3 - 2z^2 + 1$.

6 The Cassini Identity for the Generalized Leonardo Numbers

Recently the generalization of the Cassini identity based on the generalized Fibonacci fundamental system was established in [8]. Moreover, the authors explore the generalized Cassini identities for the weighted generalized Fibonacci sequences, through the associated generalized Fibonacci fundamental system, providing the algebraic, combinatoric and analytic formulations of the Cassini identities. This section is devoted to provide the Cassini identity for the generalized Leonardo numbers via the Fibonacci fundamental system, which constitutes an application among the results of the article [8].

Recall the matrix formulation of the recurrence relation (2), namely,

$$V_n = AV_{n-1}, \text{ for every } n \geq 3,$$

where A is the classical companion matrix given in (6), namely,

$$A = \begin{pmatrix} 2^d & 0 & -k \\ 1 & 0 & 0 \\ 0 & 1 & 0 \end{pmatrix} \text{ and } V_n = \begin{pmatrix} L_n \\ L_{n-1} \\ L_{n-2} \end{pmatrix}.$$

Then $V_{n+2} = AV_{n-1} = A^n V_2$. By considering the Fibonacci fundamental system $\{\{L_n^{(s)}\}_{n\geq 0},\ s = 1, 2, 3\}$, each entries of the vector $V_{n+2} = (L_{n+2}, L_{n+1}, L_n)^t$, can be written under the form $L_{n+j} = \sum_{s=1}^{3} \alpha_s L_{n+j}^{(s)}$, where $0 \le j \le 2$. On the other hand, the Casoratian matrix, $\widehat{C}(n)$, related to the family of sequences $\{L_n^{(1)}\}_{n\ge 0}$, $\{L_n^{(2)}\}_{n\ge 0}$, $\{L_n^{(3)}\}_{n\ge 0}$, satisfies the relation

$$\widehat{C}(n) = J \times M_n \times J,$$

where $J = (b_{i,j})_{1\le i,j\le 3}$ is the anti-diagonal unit matrix, namely, $b_{i,j} = 1$, for $i + j = 4$, and $b_{i,j} = 0$, otherwise and $M_n = (L_{n+3-i-1}^{(j)})_{1\le i,j\le 3}$. Since, it is verify $M_n = A^n$, where A is the classical companion matrix given in (6), then we show that

$$det[\widehat{C}(n)] = det[A^n] = det[A]^n = (-k)^n.$$

Also,

$$\widehat{C}(n) = \begin{pmatrix} L_n^{(1)} & L_n^{(2)} & L_n^{(3)} \\ L_{n+1}^{(1)} & L_{n+1}^{(2)} & L_{n+1}^{(3)} \\ L_{n+2}^{(1)} & L_{n+2}^{(2)} & L_{n+2}^{(3)} \end{pmatrix} = \begin{pmatrix} -kL_{n-1}^{(3)} & -kL_{n-2}^{(3)} & L_n^{(3)} \\ -kL_n^{(3)} & -kL_{n-1}^{(3)} & L_{n+1}^{(3)} \\ -kL_{n+1}^{(3)} & -kL_n^{(3)} & L_{n+2}^{(3)} \end{pmatrix}, \tag{17}$$

and,

$$det[\widehat{C}(n)] = k^2 \left(det \begin{bmatrix} L_{n-1}^{(3)} & L_{n-2}^{(3)} & L_n^{(3)} \\ L_n^{(3)} & L_{n-1}^{(3)} & L_{n+1}^{(3)} \\ L_{n+1}^{(3)} & L_n^{(3)} & L_{n+2}^{(3)} \end{bmatrix} \right)$$

$$= k^2((L_{n-1}^{(3)})^2 L_{n+2}^{(3)} + (L_{n+1}^{(3)})^2 L_{n-2}^{(3)} + (L_n^{(3)})^3) - 2L_n^{(3)} L_{n-1}^{(3)} L_{n+1}^{(3)} - L_{n-2}^{(3)} L_n^{(3)} L_{n+2}^{(3)}$$

Under the previous discussion, we obtain the following result,

Theorem 2 *The generalized Cassini identity, related to the Fibonacci fundamental system (3), is given under the following form,*

$$(L_{n-1}^{(3)})^2 L_{n+2}^{(3)} + (L_{n+1}^{(3)})^2 L_{n-2}^{(3)} + (L_n^{(3)})^3)$$
$$- 2L_n^{(3)} L_{n-1}^{(3)} L_{n+1}^{(3)} - L_{n-2}^{(3)} L_n^{(3)} L_{n+2}^{(3)} = (-1)^n k^{n-2}.$$

In addition, combining Expression 13 and Theorem 2 we obtain the combinatorial generalized Cassini identity related to the Fibonacci fundamental system (3),

Proposition 9 *The combinatorial generalized Cassini identity, related to the Fibonacci fundamental system (3), is given under the following form,*

$$(\rho(n,3))^2\rho(n-1,3)+(\rho(n+2,3))^2\rho(n-1,3)+(\rho(n+1,3))^3)$$

$$-2\rho(n+1,3)\rho(n,3)\rho(n+2,3)-\rho(n-1,3)\rho(n+1,3)\rho(n+3,3)=(-1)^n k^{n-2}.$$

As an interesting example, we can derive the Cassini identity for Leonardo numbers, $\{Le_n\}_{n\geq 0}$. In fact, since

$$\begin{pmatrix} Le_{n+1} & Le_n & Le_{n-1} \\ Le_n & Le_{n-1} & Le_{n-2} \\ Le_{n-1} & Le_{n-2} & Le_{n-3} \end{pmatrix} = \begin{pmatrix} 2 & 0 & -1 \\ 1 & 0 & 0 \\ 0 & 1 & 0 \end{pmatrix}^{n-3} \begin{pmatrix} Le_4 & Le_3 & Le_2 \\ Le_3 & Le_2 & Le_1 \\ Le_2 & Le_1 & Le_0 \end{pmatrix},$$

then

$$\begin{vmatrix} Le_{n+1} & Le_n & Le_{n-1} \\ Le_n & Le_{n-1} & Le_{n-2} \\ Le_{n-1} & Le_{n-2} & Le_{n-3} \end{vmatrix} = (-1)^{n-3} \begin{vmatrix} Le_4 & Le_3 & Le_2 \\ Le_3 & Le_2 & Le_1 \\ Le_2 & Le_1 & Le_0 \end{vmatrix}$$

Thus, a long direct computation, allows us to obtain the following result,

Corollary 5 *Let $\{Le_n\}_{n\geq 0}$ be the sequence of Leonardo numbers. Then, for every $n \geq 3$, it is verified,*

$$(Le_n)^2 - Le_{n+1}Le_{n-1} =$$

$$2(Le_{n-2})^3+2(Le_{n-2})^2(Le_{n-1}+1)+Le_{n-2}(2(Le_{n-1})^2+Le_{n-1})+Le_{n-1}+4(-1)^n.$$

7 Conclusions

In this paper we study the generalized Leonardo numbers based in properties of the Fibonacci fundamental system. Moreover, some identities and combinatorial identities for the generalized Leonardo numbers are provided. On the other hand, we presented a study of the characteristic polynomial associated to the generalized Leonardo numbers and provided analytic formulas. Also, as an application, the Cassini identity for the Fibonacci fundamental system associated to generalized Leonardo numbers are established.

It seems for us that several results of our study is new in literature.

Acknowledgments The first author is CNPq scholarship holder Process 2007770/2022-5 and express their sincere thanks to the Brazilian National Council for Scientific and Technological Development- CNPq- Brazil and Federal University of Mato Grosso do Sul – UFMS/MEC – Brazil for their valuable support. The second author is member of the Research Centre CMAT-UTAD (Polo of Research Centre CMAT - Centre of Mathematics of University of Minho) and thanks the Portuguese Funds through FCT – Fundação para a Ciência e a Tecnologia, within the Projects UIDB/00013/2020 and UIDP/00013/2020.

References

1. Alp, Y., Koçer, E.G.: Some properties of Leonardo numbers. Konuralp J. Math. **9**(1), 183–189 (2021)
2. Alves, F.R.V, Vieira, R.P.M.: The Newton fractal's Leonardo sequence study with the Google colab. Int. Electron. J. Math. Educ. **15**, 1–9 (2020)
3. Ben Taher, R., Rachidi, M.: On the matrix powers and exponential by r-generalized Fibonacci sequences methods: the companion matrix case. Linear Algebra Appl. **370**, 341–353 (2003)
4. Ben Taher, R., Rachidi, M.: Solving some generalized Vandermonde systems and inverse of their associate matrices via new approaches for the Binet formula. Appl. Math. Comput. **290**, 267–280 (2016)
5. Catarino, P., Borges, A.: On Leonardo numbers. Acta Math. Univ. Comenianae **89**(1), 75–86 (2019)
6. Catarino, P., Borges, A.: A note on incomplete Leonardo numbers. Integers **20**, A43 (2020)
7. Craveiro, I.M., Pereira Spreafico, E.V., Rachidi, M.: On a model of generalized Pell numbers. Int. J. Adv. Eng. Res. Sci. **8**(9), 527–550 (2021)
8. Craveiro, I.M., Pereira Spreafico, E.V., Rachidi, M.: Generalized Cassini identities via the generalized Fibonacci fundamental system. Applications. Indian J. Pure Appl. Math. (2023). https://doi.org/10.1007/s13226-023-00430-1
9. El Wahbi, B., Mouline, M., Rachidi, M.: Solving nonhomogeneous recurrences relations by matrix method. Fibonacci Quart. **40**(2), 109–117 (2002)
10. Kara, N., Yilmaz, F.: On hybrid numbers with Gaussian Leonardo coefficients. Mathematics **11**, 1551 (2023). https://doi.org/10.3390/math11061551
11. Kuhapatanakul, K., Chobsorn, J.: On the generalized Leonardo numbers. Integers **22**, A48 (2022)
12. Spreafico, E.V.P., Rachidi, M.: Fibonacci fundamental system and generalized Cassini identity. Fibonacci Quart. **57**(2), 155–157 (2019)
13. Spreafico, E.V.P., Rachidi, M.: On generalized Pell numbers of order $r \geq 2$. Trends Comput. Appl. Math. **22**(1), 125–138 (2021)

A Quadratic Estimation Approach from Fading Measurements Subject to Deception Attacks

Raquel Caballero-Águila and Josefa Linares-Pérez

Abstract In this paper, using covariance information, the least-squares quadratic filtering and fixed-point smoothing problems are addressed under the assumption that the measurements are perturbed by both a multiplicative noise and a time-correlated additive noise. Additionally, they are affected by the fading phenomena and exposed to random deception attacks. In the least-squares quadratic estimation approach, the signal and observation vectors are augmented by combining the original vectors with their second-order Kronecker powers. Then, using the Kronecker algebra rules and under an innovation approach, the linear estimators of the original signal based on the augmented observations are obtained. These linear estimators provide the required quadratic estimators. A simulation example shows the feasibility of the proposed quadratic estimation algorithms; also, the superiority of the quadratic estimators over the conventional linear ones is illustrated and the influence of the deception attack success probabilities on the estimation accuracy is analyzed.

1 Introduction

The signal estimation problem in networked systems typically involves measurements that are subject to various sources of uncertainty. These uncertainties arise not only from additive noises but also from other stochastic disturbances inherent to the network itself (network-induced uncertainties), which can result from physical limitations, environmental complexities, and other factors. Multiplicative noises—arising, for example, from random failures and repairs of system components—are

R. Caballero-Águila (✉)
Departamento de Estadística e I.O., Universidad de Jaén, Jaén, Spain
e-mail: raguila@ujaen.es

J. Linares-Pérez
Departamento de Estadística e I.O., Universidad de Granada, Granada, Spain

V. Gayoso Martínez et al. (eds.), *Mathematical Methods for Engineering Applications*, Springer Proceedings in Mathematics & Statistics 439,
https://doi.org/10.1007/978-3-031-49218-1_7

a common source of random uncertainties. Another prevalent type of uncertainty is the degradation of measurements, which can occur due to limitations of physical equipment, aging of sensors, or inaccuracies in measurement instruments. In such scenarios, the network communication may not always be perfect, leading to random fading or degradation of system measurements. This phenomenon covers the case of missing measurements as a particular case. These stochastic perturbations need to be carefully considered when designing observation models and estimation algorithms to ensure accurate signal estimation. Consequently, the investigation of estimation problems in stochastic systems involving multiplicative noises and/or fading measurements has emerged as a prominent research field in recent years. Several studies, including references [1–7] and related work, have addressed this topic and contributed to its development.

The presence of non-white additive noises affecting measurements is another relevant issue when dealing with the estimation problem. Traditional estimation algorithms typically yield accurate estimations when the additive noise is either white or correlated within a finite-time interval. However, real-world scenarios often involve sequentially time-correlated measurement noise, which arises as the output of a linear system driven by white noise. In the context of systems with time-correlated noise, the design of recursive estimation algorithms becomes a relevant topic of discussion. Previous works, such as [8–10], have addressed this issue, using the time-differencing approach to handle noise correlation. This approach utilizes the differences between consecutive measurements. More recently, alternative methods have been proposed in [11, 12] to address the signal estimation problem in linear systems with time-correlated additive noises and random packet dropouts. These methods provide non-augmentation approaches that do not require the availability of consecutive measurements.

While sensor networks offer numerous advantages, it is important to acknowledge their potential vulnerabilities when addressing the estimation problem. Ensuring the accuracy of designed estimators requires considering these vulnerabilities. One of the most common vulnerabilities that sensor networks face is the risk of cyber attacks. In particular, the estimation problem in networked systems susceptible to deception attacks—which involve intentionally altering the measurement signal to compromise data integrity—has inspired many significant research efforts. Noteworthy studies, such as [13–17] and the related literature, have explored various approaches to address this issue, contributing to the understanding and development of techniques that safeguard the quality of estimators in the presence of deception attacks.

The estimation problem of random signals from noisy measurements using the least-squares (LS) method has attracted significant research attention over the last decades, due to its diverse and compelling applications across various fields, including engineering and economics, among others. The well-known Kalman filter provides the LS optimal signal estimator for linear systems with Gaussian and mutually independent initial signal and noise processes. This estimator is linear in nature and, therefore, under ideal conditions of Gaussianity and independence, the LS linear estimator is equal to the optimal one. However, in scenarios involving non-Gaussian disturbances, the LS linear estimator generally becomes suboptimal and computationally burdensome. This drawback becomes more pronounced when dealing with complex systems featuring diverse uncertainties, as the LS linear estimator usually experiences considerable degradation in performance. As a result, alternative suboptimal estimators—such as polynomial and quadratic estimators—have gained significant attention. These estimators strike a favorable trade-off between estimation accuracy and computational complexity, rendering them valuable in practical applications. The relevance and significance of these suboptimal estimators are evident in the existing body of literature, exemplified by [18–24] and the comprehensive references cited therein.

Despite extensive research on the LS quadratic estimation problem in various classes of stochastic systems with known signal evolution models, practical situations often make it difficult to access such models. In these cases, alternative types of information, such as covariance information, become necessary. Within this framework, LS quadratic estimation algorithms have been proposed in different scenarios, based on the knowledge of statistical moments—including up to the fourth-order moments—of the stochastic processes involved in the observation equation. Notably, LS quadratic estimation algorithms have been developed in scenarios such as measurements with one-step correlated random delays [25] and measurements with random packet dropouts [26]. However, to the best of the authors' knowledge, the quadratic estimation problem has not yet been investigated within the class of systems where measurements are perturbed by time-correlated noise, affected by fading phenomena and exposed to stochastic deception attacks. This represents a significant gap in the existing literature, highlighting the need for further research and development of estimation methods in these complex scenario.

Bearing these considerations in mind, this paper addresses the LS quadratic filtering and fixed-point smoothing problems of discrete-time stochastic signals under the following conditions:

Covariance information. It is assumed that the evolution model generating the signal process is not necessarily known and, instead, only information on the statistical moments—up to the fourth order—of the stochastic processes involved in the observation equation is required.

Multiplicative noise and fading phenomena. The measurements are affected by the fading phenomena and perturbed by a multiplicative noise.

Time-correlated measurement noise. The additive noise in the measurements is described by a discrete-time linear model driven by white noise.

Stochastic deception attacks. Deception attacks are assumed to be launched by potential adversaries at the measured outputs, injecting false information that randomly perturbs the real measurements.

The paper is structured as follows. The observation model is described in Sect. 2, together with the required assumptions about the processes involved. Section 3 deals with the formulation of the quadratic estimation problem under an orthogonal projection approach. After defining the augmented vectors, we study their dynamics and second-order statistical properties (Sect. 3.1) and we describe the innovation approach (Sect. 3.2), which will be used to derive the recursive quadratic estimation algorithms in Sect. 4. Section 5 presents a simulation study to assess the effectiveness of the proposed filtering and fixed-point smoothing estimators. The simulation results highlight their superiority over conventional linear estimators and illustrate the effect of the probability of successful attacks on the estimation accuracy. Furthermore, Sect. 6 provides some concluding remarks that summarize the findings and implications of the study.

Notations Standard mathematical notation will be used throughout the paper. If not explicitly stated, the dimension of all vectors and matrices is assumed to be compatible with algebraic operations. $\mathbb{R}^n$ represents the n-dimensional Euclidean space. M^T and M^{-1} depict the transpose and inverse of the matrix M. $(M_1 \mid \dots \mid M_k)$ denotes a partitioned matrix whose blocks are the submatrices $M_1, \dots, M_k$, and 0 represents zero scalar or matrix with proper dimension. $\otimes$ denotes the Kronecker product and K_{n^2} an $n^2 \times n^2$ matrix such that $K_{n^2}(z \otimes v) = v \otimes z + z \otimes v$, $\forall z, v \in \mathbb{R}^n$. $M^{[2]} = M \otimes M$ represents the second-order Kronecker power of the vector or the matrix M.

2 Observation Model

Let x_k be an n_x-dimensional discrete-time random signal to be estimated and let us assume that, at each time $k \geq 1$, the measurement output, $z_k \in \mathbb{R}^{n_z}$, is described by:

$$z_k = \theta_k \big(G_{a,k} + \beta_k G_{b,k}\big) x_k + v_k, \quad k \geq 1, \tag{1}$$

where θ_k is a [0, 1]-valued random variable, β_k is a scalar white multiplicative noise, $G_{a,k}$ and $G_{b,k}$ are known time-varying matrices, and v_k is a time-correlated additive noise, which is assumed to be generated by the following linear evolution model:

$$v_k = D_{k-1} v_{k-1} + u_{k-1}, \quad k \geq 1, \tag{2}$$

where D_k are non-singular known deterministic matrices and u_k is a white noise.

Deception attacks are considered to be launched by potential adversaries, who inject false information into the measured outputs. This false information will randomly perturb the real measurements, depending on whether the attack is successful or not. Specifically, the injected deception signal is modeled as a sum of two components; the first component, $-z_k$, neutralizes the actual measurement, while the second one, w_k, is a noise representing the deceptive information added by the adversary. So, the injected deception signal is modeled by:

$$\breve{z}_k = -z_k + w_k, \quad k \geq 1. \tag{3}$$

The attacks launched by adversaries may be successful or not and this uncertainty about success or failure is incorporated to the model by means of Bernoulli random variables. Therefore, the mathematical model describing the potentially attacked measurements, y_k, that will be processed in the signal estimation, is formulated as:

$$y_k = z_k + \lambda_k \breve{z}_k, \ k \geq 1. \tag{4}$$

Clearly, $\lambda_k = 0$ means that the attack at time k has failed and the measurement to be processed is the actual one ($y_k = z_k$), while $\lambda_k = 1$ means that the attack has been successful and the processed measurement is the deceptive one ($y_k = w_k$). So, using (3), the available observations (4), that will be used to estimate the signal, are rewritten as:

$$y_k = (1 - \lambda_k) z_k + \lambda_k w_k, \ \ k \geq 1. \tag{5}$$

2.1 Model Assumptions

The goal of this paper is finding the least-squares (LS) quadratic estimator of the signal x_k given the observations $y_1, \ldots, y_L$. These estimators can be obtained as the orthogonal projection of the vector x_k onto the space of n_x-dimensional linear transformation of the observations $y_1, \ldots, y_L$ and their second-order Kronecker powers $y_1^{[2]}, \ldots, y_L^{[2]}$. Therefore, in order to address the LS quadratic estimation problem, the second-order moments of the Kronecker powers $y_1^{[2]}, \ldots, y_L^{[2]}$ must exist. To guarantee their existence, it is required that the second-, third- and fourth-order moments of the random processes involved in the observation model (5) exist and are known. More specifically, throughout this paper, the following assumptions are required:

Assumption (A1) The signal $\{x_k\}_{k \geq 1}$ is a zero-mean process whose covariance function, $\Sigma^x_{k,s} = Cov[x_k, x_s]$, as well as the covariance function of its second-order powers, $\Sigma^{x^{[2]}}_{k,s} = Cov\left[x_k^{[2]}, x_s^{[2]}\right]$, can be factorized as follows:

$$\Sigma^x_{k,s} = A_k B_s^T, \quad \Sigma^{x^{[2]}}_{k,s} = \breve{A}_k \breve{B}_s^T, \ s \leq k,$$

where the $n_x \times M_1$ matrices A_k, B_s and the $n_x^2 \times M_2$ matrices $\breve{A}_k$, $\breve{B}_s$ are known for all $k, s \geq 1$. Also, the cross-covariance function of the signal and its second-order powers, $\Sigma^{xx^{[2]}}_{k,s} = Cov\left[x_k, x_s^{[2]}\right]$, can be expressed as:

$$\Sigma^{xx^{[2]}}_{k,s} = \begin{cases} \dot{A}_k \dot{B}_s^T, & s \leq k, \\ \ddot{B}_k \ddot{A}_s^T, & k \leq s, \end{cases}$$

where, for all $k, s \geq 1$, $\dot{A}_k$, $\dot{B}_s$, $\ddot{B}_k$ and $\ddot{A}_s$ are $n_x \times P_1, n_x^2 \times P_1, n_x \times P_2$ and $n_x^2 \times P_2$ known matrices, respectively.

Remark 1 The derivation of the proposed quadratic estimation algorithms will not require the evolution model of the signal; instead, we will use a covariance-based estimation approach. In this approach, although the signal evolution model is not necessary, a zero-mean signal is required and the covariance and cross-covariance functions of the signal and its second-order powers must be expressed in a separable form. These separability property of the signal autocovariance function required in Assumption (A1) covers different types of stationary and non-stationary signals and, as a result, the estimation problem approach based on such assumption provides a unifying methodology to obtain general algorithms that are applicable to a wide range of practical situations, regardless of whether or not the state-space model is fully known.

Assumption (A2) The process $\{\theta_k\}_{k\geq 1}$ is a sequence of independent random variables with arbitrary distribution over the interval [0, 1], whose moments up to the fourth-order ones are known. The sequence $\{\beta_k\}_{k\geq 1}$ is a scalar white process with zero mean and known moments (up to the fourth-order ones).

Assumption (A3) v_0 is a zero-mean random vector whose moments up to the fourth-order one are known. The process $\{u_k\}_{k\geq 0}$ is a zero-mean white sequence with known moments up to the fourth-order one. We will denote

$$
\begin{aligned}
\Sigma_0^v &= Cov[v_0], & \Sigma_0^{vv^{[2]}} &= Cov[v_0, v_0^{[2]}], & \Sigma_0^{v^{[2]}} &= Cov[v_0^{[2]}], \\
\Sigma_k^u &= Cov[u_k], & \Sigma_k^{uu^{[2]}} &= Cov[u_k, u_k^{[2]}], & \Sigma_k^{u^{[2]}} &= Cov[u_k^{[2]}].
\end{aligned}
$$

Remark 2 In order to deal with the time-correlated noise, we will not use the measurement differencing method, but the direct estimation of the noise. Since the derivation of the estimation algorithms will be carried out under a covariance-based approach, it is necessary to express the noise covariance function in a separable form. As a consequence of Assumption (A3), the measurement noise $\{v_k\}_{k\geq 1}$ is a zero-mean second-order process, and the non-singularity of the matrices D_k allows us to factorise the covariance matrices $\Sigma_{k,s}^v$ in a similar way to those of the signal in (A1); specifically, $\Sigma_{k,s}^v = \boldsymbol{D}_k \boldsymbol{F}_s^T$, in which $\boldsymbol{D}_k = D_{k,0}$, $\boldsymbol{F}_s^T = D_{s,0}^{-1}\Sigma_s^v$, where $D_{k,0} = D_{k-1}\cdots D_0$ and the matrices Σ_s^v are recursively computed by $\Sigma_s^v = D_{s-1}\Sigma_{s-1}^v D_{s-1}^T + \Sigma_{s-1}^u$, $s \geq 1$, with initial condition Σ_0^v.

Assumption (A4) The noise process $\{w_k\}_{k\geq 1}$ is a zero-mean white sequence with known moments up to the fourth-order one. We will denote

$$
\Sigma_k^w = Cov[w_k], \qquad \Sigma_k^{ww^{[2]}} = Cov[w_k, w_k^{[2]}], \qquad \Sigma_k^{w^{[2]}} = Cov[w_k^{[2]}].
$$

Assumption (A5) $\{\lambda_k\}_{k\geq 1}$ is a sequence of independent Bernoulli random variables with known probabilities $P(\lambda_k = 1) = \overline{\lambda}_k$, $k \geq 1$.

Assumption (A6) The signal process $\{x_k\}_{k\geq 1}$, the vector v_0 and the processes $\{\theta_k\}_{k\geq 1}$, $\{\beta_k\}_{k\geq 1}$, $\{u_k\}_{k\geq 0}$, $\{w_k\}_{k\geq 1}$ and $\{\lambda_k\}_{k\geq 1}$ are mutually independent.

3 Quadratic Estimators and Orthogonal Projections

In this section, the LS quadratic estimation problem of the signal is addressed. Our aim is to derive recursive algorithms for the LS quadratic filter, $\widehat{x}_{k/k}$, and the quadratic estimators $\widehat{x}_{k/L}$ at the fixed-point k, for any $L > k$, that is, the LS quadratic fixed-point smoother. As indicated above, such quadratic estimators, $\widehat{x}_{k/L}$, $k \leq L$, can be obtained as the orthogonal projection of x_k on the space of n_x-dimensional linear transformations of $y_1, \ldots, y_L$ and $y_1^{[2]}, \ldots, y_L^{[2]}$. From assumptions (A1)-(A6), it is clear that $E\big[y_k^{[2]T} y_k^{[2]}\big] < \infty$ and, then, the existence of the LS quadratic estimators is guaranteed.

If we define the augmented observation vectors by stacking the original vectors and their second-order Kronecker powers, $\mathcal{Y}_k = \left(y_k^T \mid y_k^{[2]T}\right)^T$, it is clear that the space of n_x-dimensional linear transformations of $y_1, \ldots, y_L$ and $y_1^{[2]}, \ldots, y_L^{[2]}$, is equal to the space of n_x-dimensional linear transformations of $\mathcal{Y}_1, \ldots, \mathcal{Y}_L$. As a consequence, the LS quadratic estimator $\widehat{x}_{k/L}$ is the orthogonal projection of x_k on the space of n_x-dimensional linear transformations of $\mathcal{Y}_1, \ldots, \mathcal{Y}_L$; that is, the LS quadratic estimator $\widehat{x}_{k/L}$ is the LS linear estimator of x_k based on the augmented observation vectors $\mathcal{Y}_1, \ldots, \mathcal{Y}_L$. Since the augmented observation vectors have non-zero mean, for simplicity in the derivation of estimation algorithms, we consider the centered vectors $\mathbf{y}_k = \mathcal{Y}_k - E[\mathcal{Y}_k]$; clearly, the LS linear estimator of x_k based on $\mathcal{Y}_k, \ldots, \mathcal{Y}_L$ is equal to the one based on $\mathbf{y}_1, \ldots, \mathbf{y}_L$.

3.1 Second-Order Statistical Properties of the Augmented Processes

To obtain LS linear estimator of x_k based on $\mathbf{y}_1, \ldots, \mathbf{y}_L$, the second-order statistical properties of the random processes involved should be studied. For this purpose, let us consider the following augmented vectors:

$$\mathbf{x}_k = \mathcal{X}_k - E[\mathcal{X}_k], \ \mathbf{z}_k = \mathcal{Z}_k - E[\mathcal{Z}_k], \ \mathbf{v}_k = \mathcal{V}_k - E[\mathcal{V}_k], \ \mathbf{w}_k = \mathcal{W}_k - E[\mathcal{W}_k],$$

where

$$\mathcal{X}_k = \begin{pmatrix} x_k \\ x_k^{[2]} \end{pmatrix}, \ \mathcal{Z}_k = \begin{pmatrix} z_k \\ z_k^{[2]} \end{pmatrix}, \ \mathcal{V}_k = \begin{pmatrix} v_k \\ v_k^{[2]} \end{pmatrix}, \ \mathcal{W}_k = \begin{pmatrix} w_k \\ w_k^{[2]} \end{pmatrix}.$$

We start by obtaining the dynamics of the vectors $\mathbf{z}_k$, $\mathbf{v}_k$ and $\mathbf{y}_k$.

Dynamics of the augmented measurement output vectors $\mathbf{z}_k$. To simplify further mathematical discussions, we denote $G_k = \theta_k(G_{a,k} + \beta_k G_{b,k})$. The Kronecker product properties then lead to

$$G_k^{[2]} = \theta_k^2 \left(G_{a,k}^{[2]} + \beta_k^2 G_{b,k}^{[2]} + \beta_k K_{n_z^2}(G_{a,k} \otimes G_{b,k}) \right).$$

Now, by using (1), and taking again into account the Kronecker product properties, we have:

$$z_k^{[2]} = G_k^{[2]} x_k^{[2]} + K_{n_z^2}(G_k x_k \otimes v_k) + v_k^{[2]}, \quad k \geq 1.$$

Then, by defining

$$\mathcal{G}_k = \begin{pmatrix} G_k & 0 \\ 0 & G_k^{[2]} \end{pmatrix}, \quad \mathcal{V}_k^* = \begin{pmatrix} 0 \\ K_{n_z^2}(G_k x_k \otimes v_k) \end{pmatrix},$$

it is clear that the augmented measurement vectors $\mathcal{Z}_k$ satisfy:

$$\mathcal{Z}_k = \mathcal{G}_k \mathcal{X}_k + \mathcal{V}_k + \mathcal{V}_k^*, \quad k \geq 1.$$

Next, we denote

$$\dot{\mathbf{v}}_k = \mathbf{v}_k + \begin{pmatrix} 0 \\ K_{n_z^2}(\overline{\theta}_k G_{a,k} x_k \otimes v_k) \end{pmatrix}, \quad \ddot{\mathbf{v}}_k = \begin{pmatrix} 0 \\ \widetilde{G}_k^{[2]} vec(A_k B_k^T) + K_{n_z^2}(\widetilde{G}_k x_k \otimes v_k) \end{pmatrix},$$

with $\widetilde{G}_k = G_k - E[G_k]$ and $\widetilde{G}_k^{[2]} = G_k^{[2]} - E[G_k^{[2]}]$ verifying

$$\widetilde{G}_k = (\theta_k - \overline{\theta}_k) G_{a,k} + \theta_k \beta_k G_{b,k},$$

$$\widetilde{G}_k^{[2]} = (\theta_k^2 - m_{\theta_k}^{(2)}) G_{a,k}^{[2]} + (\theta_k^2 \beta_k^2 - m_{\theta_k}^{(2)} m_{\beta_k}^{(2)}) G_{b,k}^{[2]} + \theta_k^2 \beta_k K_{n_z^2}(G_{a,k} \otimes G_{b,k}).$$

We then conclude that the centered augmented measurement process $\{\mathbf{z}_k\}_{k \geq 1}$ obey the following equation:

$$\mathbf{z}_k = \mathcal{G}_k \mathbf{x}_k + \dot{\mathbf{v}}_k + \ddot{\mathbf{v}}_k, \quad k \geq 1. \tag{6}$$

Dynamics of the augmented noise vectors $\mathbf{v}_k$. From (2) and using again the Kronecker product properties, we obtain:

$$v_k^{[2]} = D_{k-1}^{[2]} v_{k-1}^{[2]} + K_{n_z^2}(D_{k-1} v_{k-1} \otimes u_{k-1}) + u_{k-1}^{[2]}, \quad k \geq 1.$$

Now, defining

$$\mathcal{D}_k = \begin{pmatrix} D_k & 0 \\ 0 & D_k^{[2]} \end{pmatrix}, \qquad \mathcal{U}_k = \begin{pmatrix} u_k \\ K_{n_z^2}(D_k v_k \otimes u_k) + u_k^{[2]} \end{pmatrix},$$

the following equation for the augmented noise $\mathcal{V}_k$ holds:

$$\mathcal{V}_k = \mathcal{D}_{k-1}\mathcal{V}_{k-1} + \mathcal{U}_{k-1}, \quad k \geq 1.$$

Then, defining the centered vectors $\mathbf{u}_k = \mathcal{U}_k - E[U_k]$, it is clear that the dynamics of the centered augmented noise process $\{\mathbf{v}_k\}_{k\geq 0}$ are given by the following equation:

$$\mathbf{v}_k = \mathcal{D}_{k-1}\mathbf{v}_{k-1} + \mathbf{u}_{k-1}, \quad k \geq 1. \tag{7}$$

Dynamics of the augmented observation vectors $\mathbf{y}_k$. Using (5) and taking into account that $\lambda_k(1-\lambda_k) = 0$, it is clear that the augmented observation vectors, $\mathcal{Y}_k$, verify the following equation:

$$\mathcal{Y}_k = (1-\lambda_k)\mathcal{Z}_k + \lambda_k \mathcal{W}_k, \quad k \geq 1.$$

Therefore, denoting

$$\mathbf{h}_k = \begin{pmatrix} 0 \\ m_{\theta_k}^{(2)}(G_{a,k}^{[2]} + V_k^{\beta} G_{b,k}^{[2]})vec(A_k B_k^T) + vec(\mathbf{D}_k \mathbf{F}_k^T) - vec(\Sigma_k^w) \end{pmatrix},$$

we have that the centered augmented observation process $\{\mathbf{y}_k\}_{k\geq 1}$ satisfies:

$$\mathbf{y}_k = (1-\lambda_k)\mathbf{z}_k + \lambda_k \mathbf{w}_k - (\lambda_k - \overline{\lambda}_k)\mathbf{h}_k, \quad k \geq 1. \tag{8}$$

The second-order statistical properties of the augmented signal and processes involved in Eqs. (6)–(8) are set out in the following propositions, the proofs of which are omitted, as they follow quite easily from the model assumptions.

Proposition 1 *Under assumptions (A1)–(A6), the augmented signal process* $\{\mathbf{x}_k\}_{k\geq 1}$ *has zero mean and its covariance function admits the following factorization:*

$$\Sigma_{k,s}^{\mathbf{x}} = \mathbb{A}_k \mathbb{B}_s^T, \quad s \leq k,$$

where

$$\mathbb{A}_k = \begin{pmatrix} A_k & \dot{A}_k & 0 & 0 \\ 0 & 0 & \ddot{A}_k & \breve{A}_k \end{pmatrix}, \quad \mathbb{B}_k = \begin{pmatrix} B_k & 0 & \ddot{B}_k & 0 \\ 0 & \dot{B}_k & 0 & \breve{B}_k \end{pmatrix}.$$

Also, the expectations $E[x_k \mathbf{x}_s^T]$ *are factorized as follows:*

$$E[x_k \mathbf{x}_s^T] = \begin{cases} \breve{\mathbb{A}}_k \mathbb{B}_s^T, & s \leq k, \\ \breve{\mathbb{B}}_k \mathbb{A}_s^T, & k \leq s, \end{cases}$$

where

$$\breve{\mathbb{A}}_k = \left(A_k \mid \dot{A}_k \mid 0 \mid 0\right), \quad \breve{\mathbb{B}}_k = \left(B_k \mid 0 \mid \ddot{B}_k \mid 0\right).$$

Proposition 2 *Under assumptions (A1)–(A6), the additive noise processes involved in (6) satisfy the following properties:*

(a) The noise process $\{\dot{\mathbf{v}}_k\}_{k\geq 0}$ is a sequence of zero-mean random vectors and its covariance function $\Sigma_{k,s}^{\dot{\mathbf{v}}}$ can be factorized as follows:

$$\Sigma_{k,s}^{\dot{\mathbf{v}}} = \Delta_k \Upsilon_s^T, \; s \leq k,$$

with $\Delta_k = (\mathbb{C}_k \mid \mathbb{D}_k)$, $\Upsilon_k = (\mathbb{E}_k \mid \mathbb{F}_k)$, $k \geq 1$, in which

$$\mathbb{C}_k = \begin{pmatrix} 0 \\ K_{n_z^2}\left(\theta_k G_{a,k} A_k \otimes \mathbf{D}_k\right) \end{pmatrix}, \qquad \mathbb{D}_k = \mathcal{D}_{k,0},$$

$$\mathbb{E}_k = \begin{pmatrix} 0 \\ K_{n_z^2}\left(\theta_k G_{a,k} B_k \otimes \mathbf{F}_k\right) \end{pmatrix}, \qquad \mathbb{F}_k = \Sigma_k^{\mathbf{v}T} \mathcal{D}_{k,0}^{-1T},$$

where $\mathcal{D}_{k,0} = \mathcal{D}_{k-1} \cdots \mathcal{D}_0$ and $\Sigma_s^{\mathbf{v}}$ is recursively computed as follows:

$$\Sigma_s^{\mathbf{v}} = \mathcal{D}_{s-1} \Sigma_{s-1}^{\mathbf{v}} \mathcal{D}_{s-1}^T + \Sigma_{s-1}^{\mathbf{u}}, \quad s \geq 1; \qquad \Sigma_0^{\mathbf{v}} = \begin{pmatrix} \Sigma_0^{v} & \Sigma_0^{vv^{[2]}} \\ (\Sigma_0^{vv^{[2]}})^T & \Sigma_0^{v^{[2]}} \end{pmatrix}.$$

(b) The noise process $\{\ddot{\mathbf{v}}_k\}_{k\geq 1}$ is a sequence of zero-mean mutually uncorrelated random vectors with covariance

$$\Sigma_k^{\ddot{\mathbf{v}}} = \begin{pmatrix} 0 & 0 \\ 0 & (\Sigma_k^{\ddot{\mathbf{v}}})^{(22)} \end{pmatrix},$$

in which

$$\Sigma_k^{\ddot{\mathbf{v}}(22)} = E\left[\widetilde{G}_k^{[2]} vec(A_k B_k^T) vec^T(A_k B_k^T) \widetilde{G}_k^{[2]T}\right] + K_{n_z^2}\left(E\left[\widetilde{G}_k A_k B_k^T \widetilde{G}_k\right] \otimes \mathbf{D}_k \mathbf{F}_k^T\right) K_{n_z^2}.$$

Proposition 3 *Under assumptions (A1)–(A6), the processes $\{\mathbf{u}_k\}_{k\geq 0}$ and $\{\mathbf{w}_k\}_{k\geq 1}$, involved in (7) and (8), respectively, are sequences of zero-mean mutually uncorre-*

lated random vectors with covariance matrices given by

$$\Sigma_k^{\mathbf{w}} = \begin{pmatrix} \Sigma_k^{w} & \Sigma_k^{ww^{[2]}} \\ (\Sigma_k^{ww^{[2]}})^T & \Sigma_k^{w^{[2]}} \end{pmatrix}, \ k \geq 1; \quad \Sigma_k^{\mathbf{u}} = \begin{pmatrix} \Sigma_k^{u} & \Sigma_k^{uu^{[2]}} \\ (\Sigma_k^{uu^{[2]}})^T & (\Sigma_k^{\mathbf{u}})^{(22)} \end{pmatrix}, \ k \geq 0,$$

in which $(\Sigma_k^{\mathbf{u}})^{(22)} = K_{n_z^2}\left(D_k vec(\mathbf{D}_k \mathbf{F}_k^T) D_k^T \otimes \Sigma_k^u\right) K_{n_z^2} + \Sigma_k^{u^{[2]}}$.

In addition, the stochastic processes $\{\mathbf{x}_k\}_{k\geq 1}$, $\{\mathbf{u}_k\}_{k\geq 0}$, $\{\mathbf{w}_k\}_{k\geq 1}$, $\{\dot{\mathbf{v}}_k\}_{k\geq 0}$ *and* $\{\ddot{\mathbf{v}}_k\}_{k\geq 1}$ *are uncorrelated to each other.*

Proposition 4 *Under assumptions (A1)–(A6),* $\Sigma_k^{\mathbf{z}}$ *and* $\Sigma_k^{\mathbf{y}}$*, the covariance matrices of the processes* $\{\mathbf{z}_k\}_{k\geq 1}$ *and* $\{\mathbf{y}_k\}_{k\geq 1}$ *involved in Eqs. (6) and (8), respectively, are given by:*

$$\Sigma_k^{\mathbf{z}} = E[\mathcal{G}_k \mathbb{A}_k \mathbb{B}_k^T \mathcal{G}_k^T] + \Delta_k \Upsilon_k^T + \Sigma_k^{\ddot{\mathbf{v}}}, \quad k \geq 1,$$

$$\Sigma_k^{\mathbf{y}} = (1 - \overline{\lambda}_k)\Sigma_k^{\mathbf{z}} + \overline{\lambda}_k \Sigma_k^{\mathbf{w}} + \overline{\lambda}_k(1 - \overline{\lambda}_k)\mathbf{h}_k \mathbf{h}_k^T, \quad k \geq 1.$$

3.2 Innovation Approach

As already indicated, the quadratic estimator, $\widehat{x}_{k/L}$, of the signal x_k based on the observations $y_1, \ldots, y_L$ is equal to the LS linear estimator of x_k based on the augmented observations $\mathbf{y}_1, \ldots, \mathbf{y}_L$. To address this problem, we use an innovation approach, which simplifies the derivation of the LS linear estimation algorithms, since the innovation process is a white process. According to the innovation approach, the observation process, $\{\mathbf{y}_k\}_{k\geq 1}$, is replaced by the innovation process, $\{\mu_k\}_{k\geq 1}$, and, hence, the LS linear estimator of x_k based on the observations $\mathbf{y}_1, \ldots, \mathbf{y}_L$ can be expressed as a linear combination of the innovations $\mu_1, \ldots, \mu_L$; namely,

$$\widehat{x}_{k/L} = \sum_{h=1}^{L} \breve{\mathcal{S}}_{k,h}^{x} \Pi_h^{-1} \mu_h, \quad k, L \geq 1, \tag{9}$$

where $\breve{\mathcal{S}}_{k,h}^{x} = E[x_k \mu_h^T]$ and $\Pi_h = E[\mu_h \mu_h^T]$.

The innovation at time k is defined by $\mu_k = \mathbf{y}_k - \widehat{\mathbf{y}}_{k/k-1}$, where $\widehat{\mathbf{y}}_{k/k-1}$ is the LS linear one-stage predictor of the observation $\mathbf{y}_k$. Since the estimators are expressed in terms of the innovations, we must begin by determining a proper expression for the predictors, $\widehat{\mathbf{y}}_{k/k-1}$, that allows us to calculate the innovations μ_k. Taking into account Eqs. (6) and (8), together with the incorrelation properties established in Proposition 3 and the Orthogonal Projection Lemma (OPL), the predictors $\widehat{\mathbf{y}}_{k/k-1}$ are expressed in terms of the signal and noise predictors as follows:

$$\widehat{\mathbf{y}}_{k/k-1} = (1-\overline{\lambda}_k)\big(\mathcal{H}_k\widehat{\mathbf{x}}_{k/k-1} + \widehat{\dot{\mathbf{v}}}_{k/k-1}\big), \quad k \geq 1. \tag{10}$$

This expression for the one-stage observation predictor is the starting point for obtaining the quadratic estimation algorithms presented in the following section.

4 Recursive Quadratic Estimation Algorithms

Recursive algorithms for the LS quadratic filter, $\widehat{x}_{k/k}$, and fixed-point smoother, $\widehat{x}_{k/k+N}$, $N \geq 1$, are presented in Theorems 1 and 2, respectively.

Theorem 1 *The LS quadratic filtering estimators, $\widehat{x}_{k/k}$, and the error covariance matrices, $\widehat{\Sigma}_{k/k} = E[(x_k - \widehat{x}_{k/k})(x_k - \widehat{x}_{k/k})^T]$, are calculated by:*

$$\widehat{x}_{k/k} = (\breve{\mathbb{A}}_k \mid 0)\mathbf{e}_k, \quad k \geq 1, \tag{11}$$

$$\widehat{\Sigma}_{k/k} = A_k B_k^T - (\breve{\mathbb{A}}_k \mid 0)\mathbf{T}_k(\breve{\mathbb{A}}_k \mid 0)^T, \quad k \geq 1. \tag{12}$$

The vectors $\mathbf{e}_k$ and the matrices $\mathbf{T}_k$ are recursively obtained by:

$$\mathbf{e}_k = \mathbf{e}_{k-1} + \boldsymbol{\Psi}_k \Pi_k^{-1}\mu_k, \quad k \geq 1; \quad \mathbf{e}_0 = 0, \tag{13}$$

$$\mathbf{T}_k = \mathbf{T}_{k-1} + \boldsymbol{\Psi}_k \Pi_k^{-1}\boldsymbol{\Psi}_k^T, \quad k \geq 1; \quad \mathbf{T}_0 = 0, \tag{14}$$

where the matrices $\boldsymbol{\Psi}_k$ are given by the following expression:

$$\boldsymbol{\Psi}_k = (1-\overline{\lambda}_k)\Big(\big(\overline{\mathcal{H}}_k\mathbb{B}_k \mid \Upsilon_k\big) - \big(\overline{\mathcal{H}}_k\mathbb{A}_k \mid \Delta_k\big)\mathbf{T}_{k-1}\Big)^T, \quad k \geq 1. \tag{15}$$

The innovations μ_k and their covariance matrices, Π_k, are calculated by:

$$\mu_k = \mathbf{y}_k - (1-\overline{\lambda}_k)\big(\overline{\mathcal{H}}_k\mathbb{A}_k \mid \Delta_k\big)\mathbf{e}_{k-1}, \quad k \geq 1, \tag{16}$$

$$\Pi_k = \Sigma_k^{\mathbf{y}} - (1-\overline{\lambda}_k)^2\big(\overline{\mathcal{H}}_k\mathbb{A}_k \mid \Delta_k\big)\mathbf{T}_{k-1}\big(\overline{\mathcal{H}}_k\mathbb{A}_k \mid \Delta_k\big)^T, \quad k \geq 1, \tag{17}$$

where the covariance matrices $\Sigma_k^{\mathbf{y}}$ are given in Proposition 4.

Proof According to expression (9) for the filter $\widehat{x}_{k/k}$, the matrices $\breve{\mathcal{S}}_{k,h}^{x} = E[x_k\mu_h^T]$, $h \leq k$, must be calculated. Similarly to (9), denoting $\mathcal{S}_{k,h}^{\mathbf{x}} = E[\mathbf{x}_k\mu_h^T]$ and $\mathcal{S}_{k,h}^{\dot{\mathbf{v}}} = E[\dot{\mathbf{v}}_k\mu_h^T]$, the one-stage predictor of both the augmented signal, $\widehat{\mathbf{x}}_{k/k-1}$, and the augmented noise, $\widehat{\dot{\mathbf{v}}}_{k/k-1}$, are written as:

$$\widehat{\mathbf{x}}_{k/k-1} = \sum_{h=1}^{k-1} \mathcal{S}^{\mathbf{x}}_{k,h} \Pi_h^{-1} \mu_h, \quad \widehat{\dot{\mathbf{v}}}_{k/k-1} = \sum_{h=1}^{k-1} \mathcal{S}^{\dot{\mathbf{v}}}_{k,h} \Pi_h^{-1} \mu_h; \quad k \geq 1.$$

Then, using (10) together with Propositions 1 and 3, we obtain:

$$\breve{\mathcal{S}}^{x}_{k,h} = (1-\overline{\lambda}_h)\Big[\breve{\mathbb{A}}_k \mathbb{B}_h^T \overline{\mathcal{H}}_h^T - (1-\delta_{h,1})\sum_{j=1}^{h-1} \breve{\mathcal{S}}^{x}_{k,j} \Pi_j^{-1} \big(\overline{\mathcal{H}}_h \mathcal{S}^{\mathbf{x}}_{h,j} + \mathcal{S}^{\dot{\mathbf{v}}}_{h,j}\big)^T\Big], \ h \geq 1.$$

Analogously, the matrices $\mathcal{S}^{\mathbf{x}}_{k,h} = E[\mathbf{x}_k \mu_h^T]$ and $\mathcal{S}^{\dot{\mathbf{v}}}_{k,h} = E[\dot{\mathbf{v}}_k \mu_h^T]$ are written as:

$$\mathcal{S}^{\mathbf{x}}_{k,h} = (1-\overline{\lambda}_h)\Big[\mathbb{A}_k \mathbb{B}_h^T \overline{\mathcal{H}}_h^T - (1-\delta_{h,1})\sum_{j=1}^{h-1} \mathcal{S}^{\mathbf{x}}_{k,j} \Pi_j^{-1} \big(\overline{\mathcal{H}}_h \mathcal{S}^{\mathbf{x}}_{h,j} + \mathcal{S}^{\dot{\mathbf{v}}}_{h,j}\big)^T\Big], \ h \geq 1,$$

$$\mathcal{S}^{\dot{\mathbf{v}}}_{k,h} = (1-\overline{\lambda}_h)\Big[\Delta_k \Upsilon_h^T - (1-\delta_{h,1})\sum_{j=1}^{h-1} \mathcal{S}^{\dot{\mathbf{v}}}_{k,j} \Pi_j^{-1} \big(\overline{\mathcal{H}}_h \mathcal{S}^{\mathbf{x}}_{h,j} + \mathcal{S}^{\dot{\mathbf{v}}}_{h,j}\big)^T\Big], \ h \geq 1.$$

From these expressions, by denoting

$$\begin{aligned} \Psi^{\mathbf{x}}_h &= (1-\overline{\lambda}_h)\Big[\mathbb{B}_h^T \overline{\mathcal{H}}_h^T - (1-\delta_{h,1})\sum_{j=1}^{h-1} \Psi^{\mathbf{x}}_j \Pi_j^{-1} \big(\overline{\mathcal{H}}_h \mathcal{S}^{\mathbf{x}}_{h,j} + \mathcal{S}^{\dot{\mathbf{v}}}_{h,j}\big)^T\Big], \ h \geq 1, \\ \Psi^{\dot{\mathbf{v}}}_h &= (1-\overline{\lambda}_h)\Big[\Upsilon_h^T - (1-\delta_{h,1})\sum_{j=1}^{h-1} \Psi^{\dot{\mathbf{v}}}_j \Pi_j^{-1} \big(\overline{\mathcal{H}}_h \mathcal{S}^{\mathbf{x}}_{h,j} + \mathcal{S}^{\dot{\mathbf{v}}}_{h,j}\big)^T\Big], \ h \geq 1, \end{aligned} \tag{18}$$

the following factorizations are established:

$$\breve{\mathcal{S}}^{x}_{k,h} = \breve{\mathbb{A}}_k \Psi^{\mathbf{x}}_h, \quad \mathcal{S}^{\mathbf{x}}_{k,h} = \mathbb{A}_k \Psi^{\mathbf{x}}_h, \quad \mathcal{S}^{\dot{\mathbf{v}}}_{k,h} = \Delta_k \Psi^{\dot{\mathbf{v}}}_h, \ h \leq k. \tag{19}$$

So, defining

$$e^{\mathbf{x}}_k = \sum_{h=1}^{k} \Psi^{\mathbf{x}}_h \Pi_h^{-1} \mu_h, \ e^{\dot{\mathbf{v}}}_k = \sum_{h=1}^{k} \Psi^{\dot{\mathbf{v}}}_h \Pi_h^{-1} \mu_h, \ k \geq 1, \tag{20}$$

we conclude that

$$\widehat{x}_{k/s} = \breve{\mathbb{A}}_k e^{\mathbf{x}}_s, \ \widehat{\mathbf{x}}_{k/s} = \breve{\mathbb{A}}_k e^{\mathbf{x}}_s, \ \widehat{\dot{\mathbf{v}}}_{k/s} = \Delta_k e^{\dot{\mathbf{v}}}_s, \ 1 \leq s \leq k. \tag{21}$$

Now, denoting $\mathbf{e}_k = \big(e^{\mathbf{x}T}_k \mid e^{\dot{\mathbf{v}}T}_k\big)^T$ and $\mathbf{\Psi}_k = \big(\Psi^{\mathbf{x}T}_k \mid \Psi^{\dot{\mathbf{v}}T}_k\big)^T$, expression (11) is straightforward, just using that—from (21)—the quadratic filter of the signal is $\widehat{x}_{k/k} = \breve{\mathbb{A}}_k e^{\mathbf{x}}_k$. The filtering error covariances are expressed as $\widehat{\Sigma}_{k/k} = E[x_k x_k^T] -$

$E[\widehat{x}_{k/k}\widehat{x}^T_{k/k}]$; hence, defining $\mathbf{T}_k = E[\mathbf{e}_k\mathbf{e}_k^T]$, expression (12) is derived by using Assumption (A1).

The recursive relation (13) for $\mathbf{e}_k$ is clear from (20), and the formula (14) for $\mathbf{T}_k$ is easily obtained using that $\mathbf{e}_{k-1}$ is orthogonal to μ_k.

Next, by substituting (19) into (18), and taking into account that $\mathbf{T}_k = E[\mathbf{e}_k\mathbf{e}_k^T] = \sum_{h=1}^{k}\boldsymbol{\Psi}_h\Pi_h^{-1}\boldsymbol{\Psi}_h^T,\ k \geq 1$, expression (15) is derived.

Finally, in view of (10) and (21), it is clear that

$$\widehat{\mathbf{y}}_{k/k-1} = (1-\overline{\lambda}_k)\left(\mathcal{H}_k\mathbb{A}_k \mid \Delta_k\right)\mathbf{e}_{k-1},\quad k \geq 1, \tag{22}$$

and expression (16) for the innovation $\mu_k = \mathbf{y}_k - \widehat{\mathbf{y}}_{k/k-1}$ is straightforward. To obtain the innovation covariance matrix, $\Pi_k = E[\mu_k\mu_k^T]$, we write $\Pi_k = \Sigma_k^{\mathbf{y}} - E[\widehat{\mathbf{y}}_{k/k-1}\widehat{\mathbf{y}}^T_{k/k-1}]$, and using (22), expression (17) is easily deduced.

Theorem 2 *At any fixed sampling time $k \geq 1$, the LS quadratic fixed-point smoothers, $\widehat{x}_{k/k+N}$, and the error covariances, $\widehat{\Sigma}_{k/k+N}$, are obtained by starting from the filter, $\widehat{x}_{k/k}$, and the filtering error covariance matrix, $\widehat{\Sigma}_{k/k}$, respectively, by the following recursive relations:*

$$\widehat{x}_{k/k+N} = \widehat{x}_{k/k+N-1} + \breve{\mathcal{S}}^x_{k,k+N}\Pi^{-1}_{k+N}\mu_{k+N},\quad N \geq 1, \tag{23}$$

$$\widehat{\Sigma}_{k/k+N} = \widehat{\Sigma}_{k/k+N-1} - \breve{\mathcal{S}}^x_{k,k+N}\Pi^{-1}_{k+N}\breve{\mathcal{S}}^{xT}_{k,k+N},\quad N \geq 1, \tag{24}$$

where

$$\breve{\mathcal{S}}^x_{k,k+N} = (1-\overline{\lambda}_{k+N})\Big((\breve{\mathbb{B}}_k \mid 0) - \mathbf{M}_{k,k+N-1}\Big)\left(\mathcal{H}_{k+N}\mathbb{A}_{k+N} \mid \Delta_{k+N}\right)^T,\quad N \geq 1, \tag{25}$$

and

$$\begin{aligned}\mathbf{M}_{k,k+N} &= \mathbf{M}_{k,k+N-1} + \breve{\mathcal{S}}^x_{k,k+N}\Pi^{-1}_{k+N}\boldsymbol{\Psi}^T_{k+N},\quad N \geq 1,\\ \mathbf{M}_k &= (\breve{\mathbb{A}}_k \mid 0)\mathbf{T}_k,\quad k \geq 1.\end{aligned} \tag{26}$$

Proof Recursive formula (23) it is immediately derived using (9). From it, the recursion for the error covariance matrices (24) is easily obtained using the OPL.

In order to calculate $\breve{\mathcal{S}}^x_{k,k+N} = E[x_k\mu^T_{k+N}] = E[x_k\mathbf{y}^T_{k+N}] - E[x_k\widehat{\mathbf{y}}^T_{k+N/k+N-1}]$, let us note that, from Propositions 1–3,

$$E\left[x_k\mathbf{y}^T_{k+N}\right] = (1-\overline{\lambda}_{k+N})\breve{\mathbb{B}}_k\mathbb{A}^T_{k+N}\overline{\mathcal{H}}^T_{k+N}.$$

On the other hand, using (22) and denoting $\mathbf{M}_{k,k+N} = E[x_k\mathbf{e}^T_{k+N}],\ N \geq 0$, it is clear that

$$E\big[x_k\widehat{\mathbf{y}}^T_{k+N/k+N-1}\big] = (1-\overline{\lambda}_{k+N})\mathbf{M}_{k,k+N-1}\left(\mathcal{H}_{k+N}\mathbb{A}_{k+N} \mid \Delta_{k+N}\right)^T.$$

Now, combining both expressions, (25) is straightforward.

Finally, expression (26) is deduced from the recursive relation (13) for $\mathbf{e}_{k+N}$; its initial condition is derived by observing that, from the OPL, $\mathbf{M}_k = E[\widehat{x}_{k/k}\mathbf{e}_k^T]$ and using (11) for $\widehat{x}_{k/k}$.

5 Simulation Study

A numerical simulation example is presented in order to: *(a)* Analyze the feasibility and effectiveness of the proposed quadratic estimators. *(b)* Illustrate the superior performance of the quadratic estimators over the linear ones. *(c)* Examine the effect of the probability of successful deception attacks, $\overline{\lambda}$, on the performance of the estimators.

Scalar AR(1) signal model. We consider a zero-mean scalar signal $\{x_k\}_{k\geq 1}$, such that

$$\Sigma^x_{k,s} = 0.95^{k-s}\,\Sigma^x_s, \quad \Sigma^{x^{[2]}}_{k,s} = 0.95^{2(k-s)}\,\Sigma^{x^{[2]}}_s, \ s \leq k; \quad \Sigma^{xx^{[2]}}_{k,s} = 0, \ \forall k, s,$$

where the functions Σ^x_s and $\Sigma^{x^{[2]}}_s$ are recursively obtained by

$$\Sigma^x_s = 0.9025\,\Sigma^x_{s-1} + 0.1, \ s \geq 1,$$

$$\Sigma^{x^{[2]}}_s = 0.8145\,\Sigma^{x^{[2]}}_{s-1} + 0.361\,\Sigma^x_{s-1} + 0.02, \ s \geq 1.$$

According to Assumption (A1), these covariance functions can be expressed in a separable form defining, for example:

$$A_k = 0.95^k, \ B_k = 0.95^{-k}\,\Sigma^x_k; \quad \breve{A}_k = 0.95^{2k}, \ \breve{B}_k = 0.95^{-2k}\,\Sigma^{x^{[2]}}_k;$$
$$\dot{A}_k = \dot{B}_k = 0; \quad \ddot{A}_k = \ddot{B}_k = 0.$$

For the simulations, the signal is assumed to be generated by the following AR(1) model:

$$x_k = 0.95x_{k-1} + \varepsilon_{k-1}, \quad k \geq 1,$$

where x_0 is a zero-mean Gaussian variable with $\Sigma^x_0 = 0.1$, and $\{\varepsilon_k\}_{k\geq 0}$ is a zero-mean white Gaussian noise with $\Sigma^\varepsilon_k = 0.1, \ \forall k \geq 0$.

Actual measurements. Assume that the real measurements of the signal, z_k, are described by

$$z_k = \theta_k(0.9 + 0.5\beta_k)x_k + v_k, \quad k \geq 1,$$

in which

- $\{\theta_k\}_{k\geq 1}$ is a sequence of independent random variables uniformly distributed over $[0.5, 0.9]$ and the white multiplicative noise $\{\beta_k\}_{k\geq 1}$ is a standard Gaussian process.
- The noise process $\{v_k\}_{k\geq 0}$ is generated by

$$v_k = 0.8v_{k-1} + u_{k-1}, \quad k \geq 1,$$

where $\{u_k\}_{k\geq 0}$ is a zero-mean white Gaussian noise with $\Sigma_k^u = 0.01, \ \forall k \geq 0$, and v_0 is a zero-mean Gaussian variable with $\Sigma_0^v = 1$; hence,

$$\Sigma_0^{vv^{[2]}} = \Sigma_k^{uu^{[2]}} = 0, \quad \Sigma_0^{v^{[2]}} = 2, \quad \Sigma_k^{u^{[2]}} = 0.0002.$$

Let us suppose that the measurements are subject to deception attacks and the signal injected by the adversary is given by

$$\breve{z}_k = -z_k + w_k, \quad k \geq 1.$$

The attack noise $\{w_k\}_{k\geq 1}$ is assumed to be a white non-Gaussian sequence with distribution

$$P(w_k = -8) = 1/8, \quad P(w_k = 8/7) = 7/8, \quad \forall k \geq 1.$$

Hence, $E[w_k] = 0, \ \Sigma_k^w = 9.1429, \ \Sigma_k^{ww^{[2]}} = -62.6939, \ \Sigma_k^{w^{[2]}} = 429.9009.$

Observations available for estimation. Finally, we suppose that the observations available for estimation are given by

$$y_k = (1 - \lambda_k)z_k + \lambda_k w_k, \quad k \geq 1,$$

where the white sequence of Bernoulli random variables, $\{\lambda_k\}_{k\geq 1}$, modeling whether the deception attacks actually succeed or not, are identically distributed with probabilities $P(\lambda_k = 1) = \overline{\lambda}$.

First, in order to show the feasibility and effectiveness of the proposed quadratic estimators, as well as to illustrate their performance and their superiority over the conventional linear estimators, the error variances of both linear and quadratic estimators are calculated considering the value 0.5 for the probability $\overline{\lambda}$. The results are displayed in Fig. 1, which shows that:

- The quadratic estimation error variances are lower than the linear ones (superiority of quadratic estimators over linear estimators).

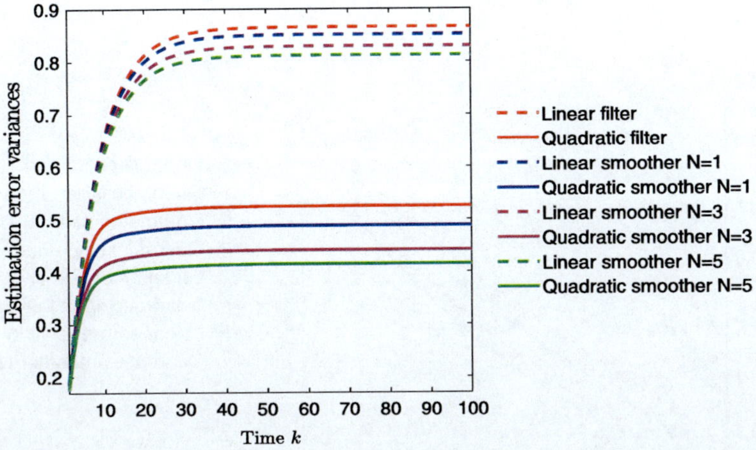

Fig. 1 Estimation error variance comparison of the linear and quadratic filtering and smoothing estimators when $\overline{\lambda} = 0.5$

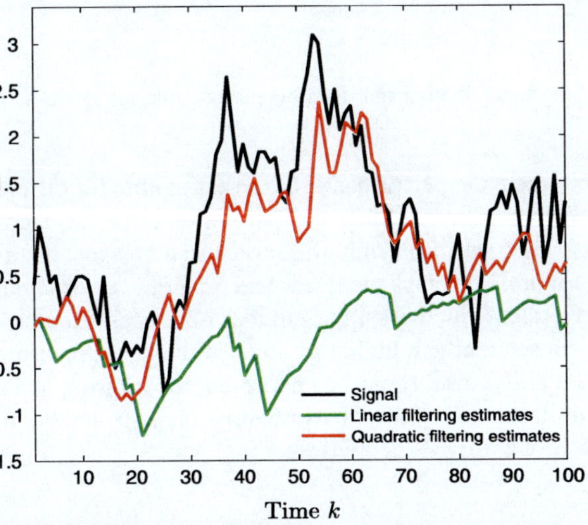

Fig. 2 Simulated signal, linear and quadratic filtering estimates when $\overline{\lambda} = 0.5$

- The estimation accuracy of the smoothers is superior to that of the filters and, also, the performance of the fixed-point smoothers improves as the number of available observations increases.
- Both the quadratic and linear estimation error variances stabilize, approximately from the 50th iteration onwards.

Figure 2 displays a simulated signal trajectory and their corresponding linear and quadratic filtering estimates for $\overline{\lambda} = 0.5$. Agreeing with the comments made about Fig. 1, it is observed that the quadratic filtering estimates track the signal evolution better than the linear ones.

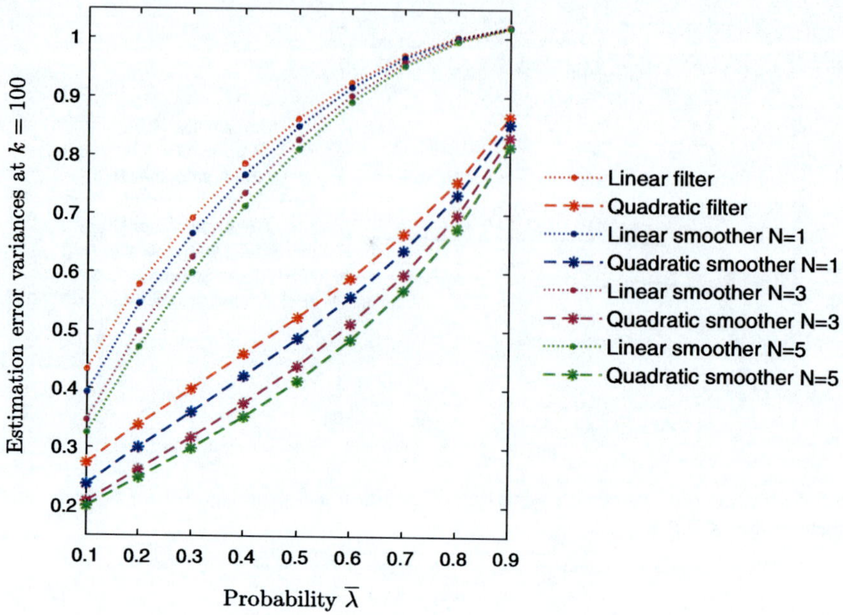

Fig. 3 Linear and quadratic filtering and smoothing error variances versus $\overline{\lambda}$

Finally, we compare the performance of the estimators for different values of the successful attack probability, $\overline{\lambda}$. Since the estimation error variances show a similar behaviour from $k = 50$ onwards, only the variances at the specific iteration $k = 100$ are considered. Figure 3 displays the linear and quadratic error variances at $k = 100$ for $\overline{\lambda} = 0.1$ to 0.9. As $\overline{\lambda}$ increases, the number of successful attack is expected to be greater and, consequently, a higher number of available measurements used for estimation will be only noise. Hence, as expected, the filtering and smoothing error variances of both linear and quadratic estimators, become smaller as the successful deception attack probability, $\overline{\lambda}$, decreases.

6 Conclusions

Recursive algorithms for the LS quadratic filtering and fixed-point smoothing problems have been derived in the presence of different perturbations that include multiplicative noise, time-correlated additive noise, fading phenomena and random deception attacks. The authors have proposed a new approach—based on the use of covariance information—to deal with these challenges.

Simulation results validate the feasibility of the proposed quadratic estimation algorithms and highlight the superiority of quadratic estimators over linear estimators. The analysis of the effect of deception attack success probabilities on the estimation accuracy further emphasizes the importance of considering such attacks in the estimation process.

Overall, the findings of this research show the potential of using covariance information to deal with the quadratic estimation problem from fading measurements affected by multiplicative noise, time-correlated additive noise and random deception attacks. The quadratic estimation approach opens up new horizons for improving estimation accuracy in real-world scenarios where these challenges are prevalent.

Acknowledgments This research was supported by the "Ministerio de Ciencia e Innovación, Agencia Estatal de Investigación" of Spain and the European Regional Development Fund [grant number PID2021-124486NB-I00].

References

1. Li, W., Jia, Y., Du, J.: Distributed filtering for discrete-time linear systems with fading measurements and time-correlated noise. Digit. Signal Process. **60**, 211–219 (2017). https://doi.org/10.1016/j.dsp.2016.10.003
2. Caballero-Águila, R., Hermoso-Carazo, A., Linares-Pérez, J.: Fusion estimation from multi-sensor observations with multiplicative noises and correlated random delays in transmission. Mathematics **2017**, 45 (2017). https://doi.org/10.3390/math5030045
3. Lin, H., Sun, S.: Optimal sequential fusion estimation with stochastic parameter perturbations, fading measurements, and correlated noises. IEEE Trans. Signal Process. **66**, 3571–3583 (2018). https://doi.org/10.1109/TSP.2018.2831642
4. Li, Y., Liu, S., Zhong, M., Ding, S.X.: State estimation for stochastic discrete-time systems with multiplicative noises and unknown inputs over fading channels. Appl. Math. Comput. **320**, 116–130 (2018). https://doi.org/10.1016/j.amc.2017.09.008
5. Liu, W., Shi, P.: Optimal linear filtering for networked control systems with time-correlated fading channels. Automatica **101**, 345–353 (2019). https://doi.org/10.1016/j.automatica.2018.11.042
6. Wang, X., Sun, S.: Optimal recursive estimation for networked stochastic uncertain systems with fading measurements and time-correlated channel noises. J. Comput. Appl. Math. **346**, 549–565 (2019). https://doi.org/10.1016/j.cam.2018.07.026.
7. Ma, J., Sun, S.: Optimal linear recursive estimators for stochastic uncertain systems with time-correlated additive noises and packet dropout compensations. Signal Process. **176**, 107704 (2020). https://doi.org/10.1016/j.sigpro.2020.107704
8. Li, W., Jia, Y., Du, J.: Distributed filtering for discrete-time linear systems with fading measurements and time-correlated noise. Digit. Signal Process. **60**, 211–219 (2017). https://doi.org/10.1016/j.dsp.2016.10.003
9. Geng, H., Wang, Z., Cheng, Y., Alsaadi, F., Dobaie, A.M.: State estimation under non-Gaussian Lévy and time-correlated additive sensor noises: a modified Tobit Kalman filtering approach. Signal Process. **154**, 120–128 (2019). https://doi.org/10.1016/j.sigpro.2018.08.005
10. Caballero-Águila, R., Hermoso-Carazo, A., Linares-Pérez, J.: Networked fusion estimation with multiple uncertainties and time-correlated channel noise. Inf. Fusion **54**, 161–171 (2020). https://doi.org/10.1016/j.inffus.2019.07.008
11. Ma, J., Sun, S.: Optimal linear recursive estimators for stochastic uncertain systems with time-correlated additive noises and packet dropout compensations. Signal Process. **176**, 107704 (2020). https://doi.org/10.1016/j.sigpro.2020.107704

12. Caballero-Águila, R., Hu, J., Linares-Pérez, J.: Two compensation strategies for optimal estimation in sensor networks with random matrices, time-correlated noises, deception attacks and packet losses. Sensors **22**, 8505 (2022). https://doi.org/10.3390/s22218505
13. Wang, Z., Wang, D., Shen, B., Alsaadi, F.E.: Centralized security-guaranteed filtering in multirate-sensor fusion under deception attacks. J. Frankl. Inst. **355**, 406–420 (2018). https://doi.org/10.1016/j.jfranklin.2017.11.010
14. Yang, W., Zhang, Y., Chen, G., Yang, C., Shi, L.: Distributed filtering under false data injection attacks. Automatica **102**, 34–44 (2019). https://doi.org/org/10.1016/j.automatica.2018.12.027
15. Caballero-Águila, R., Hermoso-Carazo, A., Linares-Pérez, J.: A two-phase distributed filtering algorithm for networked uncertain systems with fading measurements under deception attacks. Sensors **20**, 6445 (2020). https://doi.org/10.3390/s20226445
16. Xiao, S., Han, Q., Ge, X., Zhang, Y.: Secure distributed finite-time filtering for positive systems over sensor networks under deception attacks. IEEE Trans. Cybern. **50**, 1200–1228 (2020). https://doi.org/10.1109/tcyb.2019.2900478
17. Ma, Y., Sun, S.: Distributed optimal and self-tuning filters based on compressed data for networked stochastic uncertain systems with deception attacks. Sensors **23**, 335 (2023). https://doi.org/10.3390/s23010335
18. Zhao, H., Zhang, C.: Non-Gaussian noise quadratic estimation for linear discrete-time time-varying systems. Neurocomputing **174**, 921–927 (2016). https://doi.org/10.1016/j.neucom.2015.10.015
19. Caballero-Águila, R., García-Garrido, I., Linares-Pérez, J.: Quadratic estimation problem in discrete-time stochastic systems with random parameter matrices. Appl. Math. Comput. **273**, 308–320 (2016). https://doi.org/10.1016/j.amc.2015.10.005
20. Cacace, F., Conte, F., Germani, A., Palombo, G.: Feedback quadratic filtering. Automatica **82**, 158–164 (2017). https://doi.org/10.1016/j.automatica.2017.04.046
21. Li, L., Tan, L., Song, X., Yan, X.: Quadratic filtering for discrete-time systems with measurement delay and packet dropping. Complexity **2020**, 1725121 (2020). https://doi.org/10.1155/2020/1725121
22. Liu, Q., Wang, Z., Han, Q., Jiang, C.: Quadratic estimation for discrete time-varying non-Gaussian systems with multiplicative noises and quantization effects. Automatica **113**, 108714 (2020). https://doi.org/10.1016/j.automatica.2019.108714
23. Wang, S., Wang, Z., Dong, H., Chen, Y., Alsaadi, F.E.: Recursive quadratic filtering for linear discrete non-Gaussian systems over time-correlated fading channels. IEEE Trans. Signal Process. **70**, 3343–3356 (2022). https://doi.org/10.1109/TSP.2022.3182511
24. Wang, S., Wang, Z., Dong, H., Chen, Y.: Recursive state estimation for stochastic nonlinear non-Gaussian systems using energy-harvesting sensors: a quadratic estimation approach. Automatica **147**, 110671 (2023). https://doi.org/10.1016/j.automatica.2022.110671
25. Hermoso-Carazo, A., Linares-Pérez, J.: Linear and quadratic least-squares estimation using measurements with correlated one-step random delay. Digit. Signal Process. **18**, 450–464 (2008). https://doi.org/10.1016/j.dsp.2007.04.007
26. Caballero-Águila, R., Hermoso-Carazo, A., Linares-Pérez, J.: Recursive least-squares quadratic smoothing from measurements with packet dropouts. Signal Process. **92**, 931–938 (2012). https://doi.org/10.1016/j.sigpro.2011.10.006

SOLO Taxonomy in the Evaluation of Engineering Students: A Case Study in Mathematics

Cristina M. R. Caridade and Verónica Pereira

Abstract This paper presents the results of a study whose main objective was to analyze the assessments by exam in higher education according to the SOLO Taxonomy (Structure of the Observed Learning Outcome) over the last five years, in particular the cognitive complexity required in each question and the "quality" of the exam. The data were collected in the Mathematical Analysis exams of the Degree in Electrical and Computer Engineering. In total, 130 items presented in 10 exams (2 per year) from 2019 to 2023 were surveyed. The SOLO index of these exams was always higher than 16, with the exception of one exam from 2023 with an index of 14.55. The analysis reveals that the exams are composed of items classified at the highest levels of cognitive complexity, according to the SOLO Taxonomy, and associated with the student's deep learning. Thus the authors were able to study their pedagogical practices in relation to the assessment and, consequently, to consciously prepare the assessment.

1 Introduction

The assessment of learning and competencies in higher education has been much discussed, not only due to political issues associated with the Bologna process, but also due to the development of different assessment methods. Assessment is a crucial element in the entire teaching-learning process [1, 2] and how students understand assessment determines how they learn [3].

C. M. R. Caridade (✉)
Polytechnic Institute of Coimbra, Coimbra Institute of Engineering, Coimbra, Portugal

CICGE, DGAOT, FCUP, Vila Nova de Gaia, Portugal
e-mail: caridade@isec.pt

V. Pereira
Polytechnic Institute of Coimbra, Coimbra Institute of Engineering, Coimbra, Portugal
e-mail: veronica.pereira@isec.pt

V. Gayoso Martínez et al. (eds.), *Mathematical Methods for Engineering Applications*, Springer Proceedings in Mathematics & Statistics 439,
https://doi.org/10.1007/978-3-031-49218-1_8

The SOLO Taxonomy (Structure of the Observed Learning Outcome) model was developed by Biggs and Collis (1992) [4], who proposed a system of categories to identify the levels of thought formalization, which considers the amount of knowledge involved in the response to each item, the complexity of reasoning and type of solution required [5]. Piaget's model of cognitive development and the stages of development of the SOLO model are similar [6]. There are already several authors who have published studies on the SOLO taxonomy in the academic world, such as [5, 7–11] although in higher education, more specifically in the area of mathematics, these studies have been rare [7, 12, 13].

SOLO Taxonomy is a classification or categorization methodology, ordered according to certain specific characteristics, which involves five stages: "Pre-Structural", "Unistructural", "Multistructural", "Relational" and "Abstract" (Table 1 left).

SOLO Taxonomy contributions to educational assessment are numerous, for example to assess the quality of learning, curricular objectives, categorization of questions, or even responses. At classroom level, teachers can have more control over their students' development or learning by designing and grading assignments or activities and questions [15]. In the evaluation process, taxonomy in addition to contributing to exam items elaboration, provides more complete results (involving the level of cognitive complexity of the students' thinking) to teachers [10]. Thus, the SOLO Taxonomy is an educational tool that can be applied in assessments. It can help in the preparation of items for exams, which allows the teacher to balance superficial and deep issues, and can also be used for analysis of test results (pedagogical interpretation). By exploring the SOLO taxonomy, it is possible to identify the level of cognitive complexity of learning that a student must present in relation to the evaluated contents (Table 1). The levels of complexity of the SOLO Taxonomy are ordered according to understanding, from the understanding of concrete to abstract elements, through an increasing process of organization of the number and dimension of relationships between concepts, consistency between these relationships and generalization of the principles used [7] as described in Table 1. For example, an item in the "Abstract" category should involve two or more topics of the higher level syllabus, using innovative strategies for its resolution while applying various concepts and information interconnected simultaneously. If an item belongs to the "Prestructural" category, it only involves a unique basic topic of a lower level of the syllabus, it does not involve any replication or innovation, or any type of attachment between concepts.

This paper displays the results of a study whose main objective was to analyze the assessments by exam in higher education according to the SOLO Taxonomy over the last five years, in particular the cognitive complexity required in each question and the "quality" of the exam.

Table 1 Description of SOLO taxonomy levels relating to response indicators adapted from [4, 14]

	Analysis parameters: topics and procedures			
	Topics		Procedures	
Category	#	Level	Innovation degree	Integration
Abstract	≥ 2	>syllabus **(Higher)**	Inedited: It involves the development of working hypotheses and innovative strategies.	Interconnected: It shows the application of several concepts and information in an integrated and simultaneous way.
Relational	≥ 2	= syllabus **(Adequate)**	Replica: It involves hypotheses of work and strategies described in the programs.	Interconnected and/or Compartmented
Multistructural	≥ 2	= syllabus **(Adequate)**	Replica: It involves hypotheses of work and strategies described in the programs.	Compartmented: It shows the application of several concepts and information in an isolated and successive way.
Unistructural	1	= syllabus **(Adequate)**	Replica: It involves hypotheses of work and strategies described in the programs.	(Not applicable).
Prestructural	1 or 0	< syllabus **(Inferior)**	It does not involve any type of replica (or unprecedented situation).	(Not applicable).

2 Developed Research

This study, in continuity with the study presented in [16], intends to analyze the exams of Mathematical Analysis in Electrical and Computer Engineering degree at the Coimbra Institute of Engineering during the last five years. Consequently, the aim is to verify which contents are evaluated in the exams and at what percentage, as well as the level of difficulty of the exams over time. In fact, with this study, it is possible to give teachers an idea about the importance and "quality" of their students' learning measured by the evaluation of the proposed exams. It is possible

Table 2 Syllabus of mathematical analysis by topics

Program topics	Included content
Topic 1: Functions	Inverse trigonometric functions; exponential functions; logarithm functions; hyperbolic functions; inverse functions: domains and ranges.
Topic 2: Differential calculus	Derivative; derivatives rules; Rolles's theorem; Lagrange's theorem; Cauchy theorem; successive derivatives; derivative of implicit function.
Topic 3: Primitivation	Definition and properties; primitive of a function; primitivation by decomposition; primitivation by parts; primitivation of rational functions; primitivation by substitution.
Topic 4: Integral calculus	Defined integral; fundamental theorem of calculus; applications: areas, volumes and curves lengths; indefinite integral; improper integrals.
Topic 5: Numerical analysis	Error theory; differential of a function; nonlinear equations: Bisection method; Newton-Raphson method; numerical integration: trapezoidal rule; Simpson's rule.

for the teacher to analyze and evaluate the level of reasoning and the learning capacity of his students. Monitoring the evaluation of students' competencies and their reflection in the sense of a possible adjustment is an important means of diagnosis for the higher education teacher.

Through the application of the SOLO taxonomy theory to the exams of the regular period (E1) and the recourse period (E2) proposed by the teacher, it is possible to analyze the key points of the exam through the research questions and to give teachers relevant suggestions for the teaching and learning of their students.

According to the Mathematical Analysis curriculum, there are five different topics (see Table 2) to be addressed. The first two topics, with exception of hyperbolic functions, are addressed as a review of secondary school subjects, as a basic support for solving the following contents. Therefore, these topics do not carry significant weight in the exams.

In agreement with the SOLO taxonomy theory, at the pre-structural level, students' knowledge of content is vague and they fail to understand and solve problems, giving logically confused answers with no basis to support [17]. At the "Unistructural" level, students deal with only a single aspect, they can only follow simple instructions, identify, name or count. Hence, these two levels are not suitable for a basic education in mathematics in engineering, and, for that reason, are not identified in the items of the Mathematical Analysis exams.

3 Analysis and Discussion

In the Degree in Electrotechnical and Computer Engineering, students of Mathematical Analysis are subject to two exams per year (E1—exam in the normal period, E2—exam in the recourse period) with score 200. Each exam is divided into two parts. Part 1 (Part 1) evaluates the first four topics (Table 2) and is solved without a calculator, while part 2 (Part 2) evaluates the topic five and is solved with a calculator. In Fig. 1, the numbers of items of the E1 exam are distributed, concerning Part 1 (on the left) and Part 2 (on the right) according to the SOLO levels during the years 2019 to 2023.

The "Unistructural" and "Prestructural" levels are not portrayed in the exams, as previously mentioned. The displayed items are centered on the "Relational" level, since, in Mathematical Analysis, the goal is to make students develop abilities (skills) to understand relationships between different contents, analytical reasoning and even the ability to apply mathematical concepts in solving practical problems. At the "Multistructural" level there are only five items in the E1 exam (1 in 2022 and 4 in 2023) and two items in the E2 exam (1 in 2022 and 1 in 2023) composed by the direct application of the methods and procedures. Items involving the "Abstract" level are rare (5 items in four years in the E1 exam and 4 items in three years in the E2 exam) as they assess the students' ability to transfer their knowledge to new areas, that is, the most abstract level of knowledge not attained by all.

In the same way, the distribution of the number of items, related to Part 2 according to the SOLO levels during the years 2019 to 2023, is represented in Fig. 2, E1 exam (on the left) and the E2 exam (on the right).

In this part of the syllabus, it is possible to develop items at the "Abstract" level (1 item in the E1 exam and 4 items in the E2 exams) more frequently once students sometimes have to deal with real problems. It should be noted that there are a

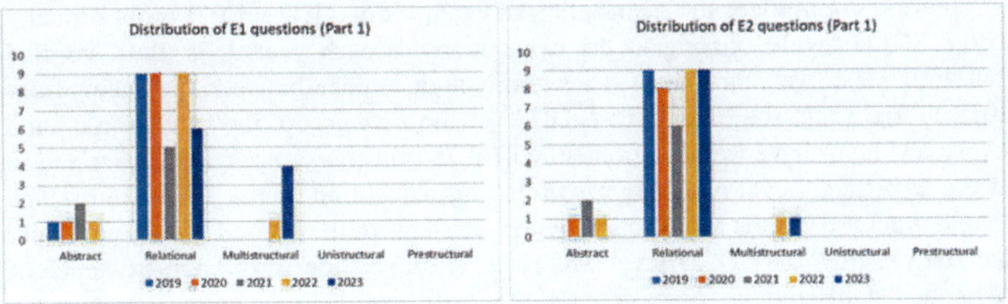

Fig. 1 Distribution of the number of items related to Part 1 according to SOLO levels: E1 (on the left) and E2 (on the right)

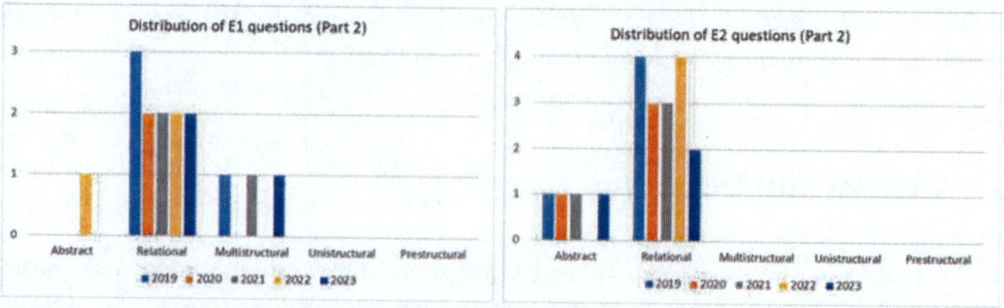

Fig. 2 Distribution of the number of items related to Part 2 according to SOLO levels: E1 (on the left) and E2 (on the right)

maximum of 2 "Abstract" level items per test, with the exception of E2 exam in 2021, which has 3 "Abstract" level items. Most exam questions are classified at the "Relational" level. Only one question classified at the "Multistructural" level appears in the E1 exams of the years 2019, 2021 and 2023.

The topics addressed in the exam questions regarding the SOLO level are highlighted in Fig. 3.

Only topics related to the "Multistructural", "Relational" and "Abstract" levels are shown while the remaining levels are not found in any question of the exams. On the left, there are the scores given to each of the 5 topics in the E1 exam and in the E2 exam, on the right. It appears that topic 3 and 4 (in the two exams E1 and E2) are the topics with the highest weight in the exam, because they are the ones with greater importance. Topics 1 and 2 do not always appear in the exams, for instance, in the 2019 to 2021 exams, this topic was not directly included in the exam items. All other topics are included in all exams with greater or lesser weight. Topic 4 and 5 are the topics with the highest weight in the "Abstract" (in blue) level. At the SOLO level "Multistructural" (in gray) in the E1 exam, only 1 item from topic 5 appears (2019 and 2021), 1 item from topic 1 (2022) and 4 items from topic 2 to 5 (2023). At the same time, in the E2 exam, it only appears 1 item from topic 1 (2022 and 2023). The "Relational" level (in orange) is the one that is staged in all questions of all topics and exam types with the exception of E1 exams in topic 1 (2021) and topic 2 (from 2019 to 2021).

Figure 4 summarizes the distribution of exam scores (E1 and E2) proportionally to the SOLO levels existing in the exam items of each year. It is, thus, verified that, in the last five years, there is a significant difference between the exams from 2019 to 2022 and the exams from 2023. In the first group, the "Relational" and "Abstract" items have weights greater than 93.75% in all exams and the appreciation of the "Abstract" level has been growing, with exception of the 2023 exams. The "Multistructural" items are practically non-existent. In the second group, with only the 2023 exams, the SOLO level was reduced, in relation to the non-existence of "Abstract" items and the increase in the "Multistructural" items score. However, the exams are balanced (the SOLO levels between them are similar) and aim to assess the knowledge of higher education students of engineering courses who must have

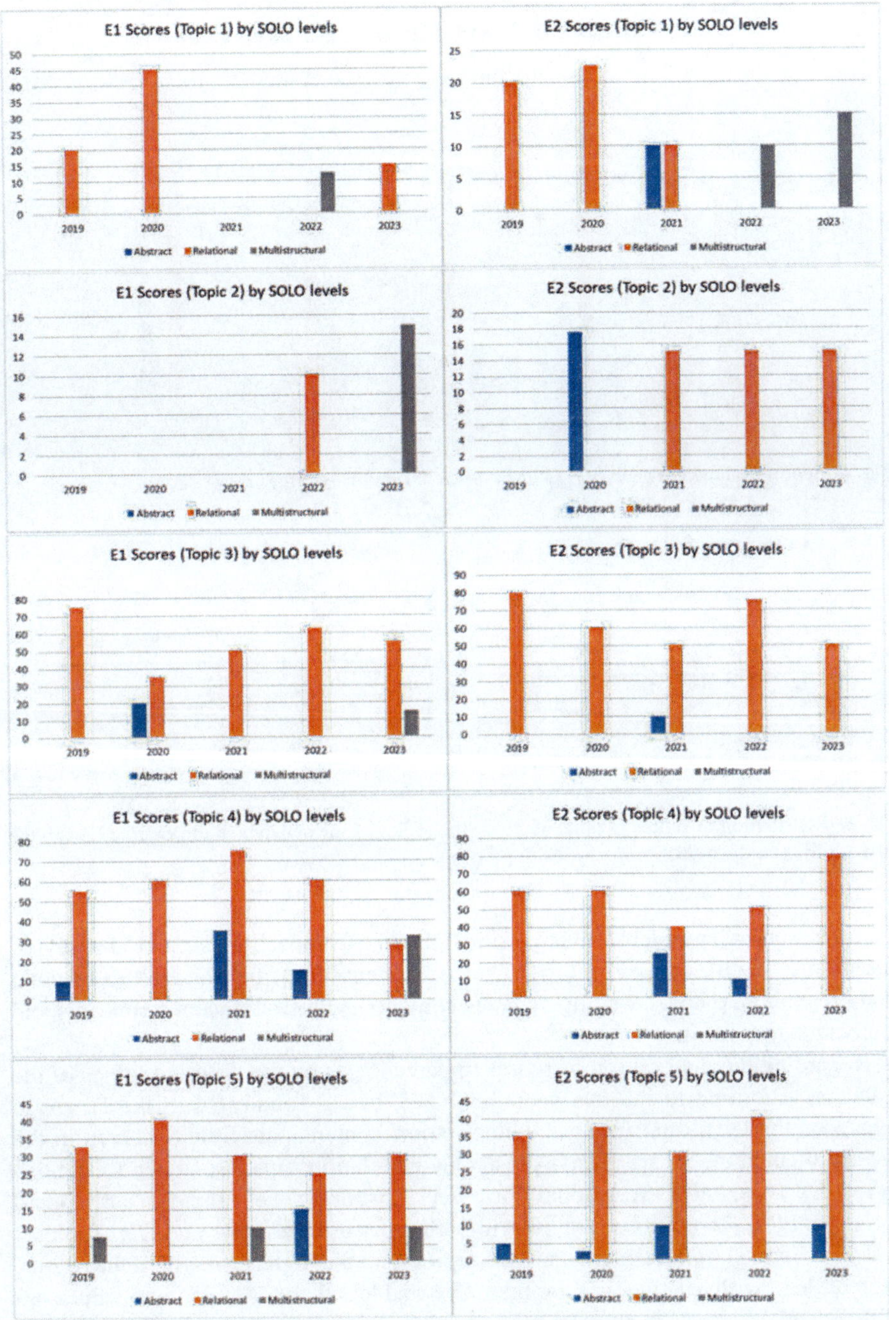

Fig. 3 Topic Scores assigned to each SOLO level on the E1 (on the left) and E2 (on the right) exams

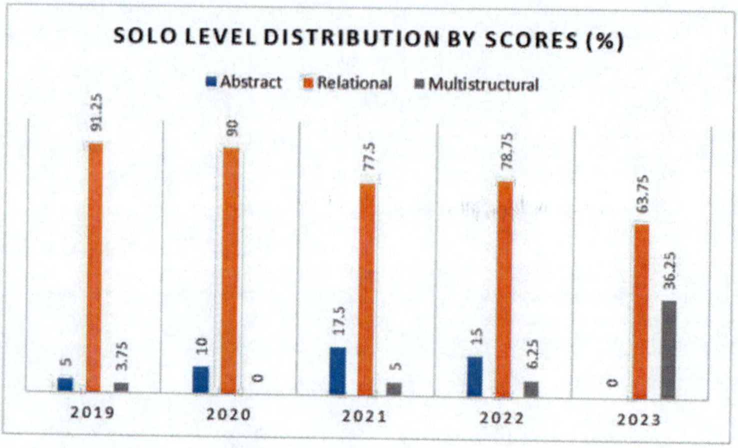

Fig. 4 Scores assigned to each SOLO level on the E1 and E2 exams

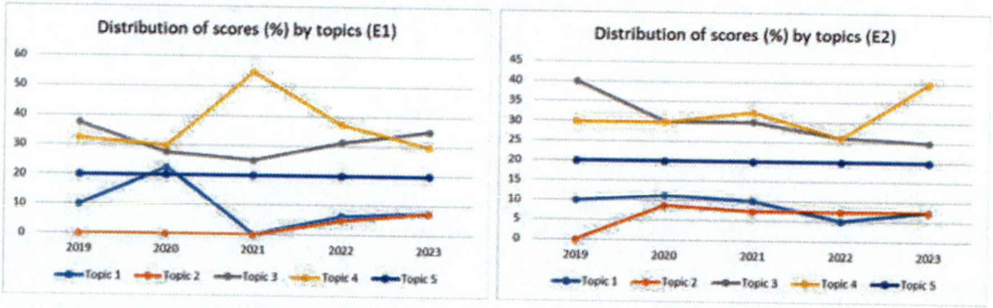

Fig. 5 Distribution of scores in percentage by the topics of the syllabus in the exams: E1 (on the left) and E2 (on the right)

qualities, such as, problem solving, critical thinking, innovation, structural analysis, attention to detail, and so on [18]. That is, students must be prepared to respond to more elaborate items, linking syllabus contents, with simultaneous concepts and innovative responses.

Regarding the distribution of scores in percentages by the different topics of the syllabus, according to the five years analyzed, it was found that in the E1 exam (Fig. 5 on the left), topic 1 and 2, being review contents, are little observed in the context of an item, but are always present in the relationships that students establish to resolve issues at a "Relational" or even "Abstract" level. Topic 5 has a fixed weight of 20% (20 points out of 100) in both types of exams (E1 and E2). Topics 3 and 4 (in both E1 and E2 exams) are the topics with the highest weight in the exams, each of them with a weight greater than 25% and less than 55% (25 to 55 values out of 100) and, together, they make up an average of about 68% in the E1 exam and 62% in the E2 exam. These topics are of great importance to engineering students and are planned to be taught in about half of the course's teaching hours.

Table 3 Index associated to SOLO level

c_i	SOLO level	index(c_i)
c_1	Abstract	20
c_2	Relational	16
c_3	Multistructural	12
c_4	Unistructural	8
c_5	Prestructural	4

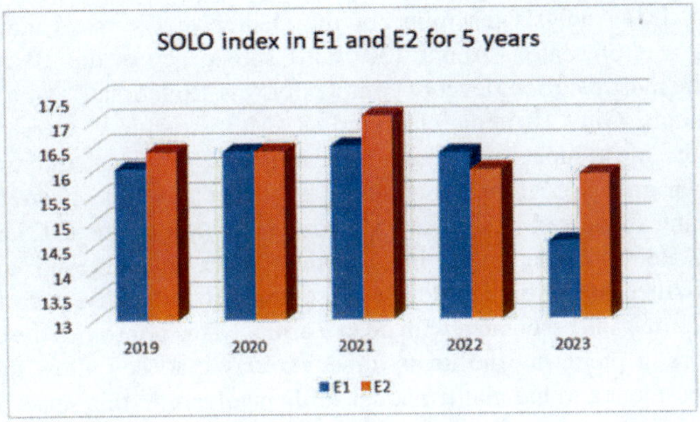

Fig. 6 SOLO index in exams E1 and E2 from 2019 to 2023

In order to be able to compare the type and "quality" of the exams proposed by the professor over the five years, the SOLO index (IS) defined in [7] was calculated using Eq. 1.

$$IS = \frac{\sum_{i=1}^{5} level(c_i) \times index(c_i)}{200}. \tag{1}$$

where $index(c_i)$, $i = 1, \ldots 5$ is obtained by Table 3 and $level(c_i)$ represents the exam score associated with the SOLO level c_i.

In Eq. 1, if all scores are from "Abstract" level items, the index will be 20 (maximum). If all items are of the "Relational" level, the final index will be 16, and so on. Figure 6 shows the SOLO index of each exam (E1 and E2) from 2019 to 2023. It can be seen that the index is always higher than 16, with the exception of the exam E1 of 2023 with an index of 14.55. Thus, all exams have a difficulty/quality between equal to or greater than the "Relational" level. E2 exams have a higher index than E1 exams in 2019 (IS(E1)=16.05, IS(E2)=16.4), 2021 (IS(E1)=16.5, IS(E2)=17.1) and in 2023 (IS(E1)=14.55, IS(E2)=15.9). Nevertheless, in 2020 they have the same index (16.4) and in 2022 the index of the E2 exam is lower than that of the E1 (IS(E1)=16.35, IS(E2)=16). In 2021, the E2 exam was the most difficult (IS(E2)=17.1). Therefore, it is not possible to conclude in which types of exam the difficulty/quality is greater because of the great variability that exists.

4 Conclusions

The SOLO Taxonomy is an important tool in educational evaluation, which helps teachers in their pedagogical practice. In this paper, the SOLO Taxonomy was applied in the evaluation of exams in order to assess the "quality" of the exams and allow the teacher to lead students to reach the deepest levels of learning. An analysis carried out over five years from 2019 to 2023 of two exams per year of the Mathematical Analysis discipline of the Degree in Electrical and Computer Engineering was presented. About 130 items spread across the 10 exams were analyzed. The most recurrent level of complexity was "Relational", which prevailed in the 10 exams. Only 10 items classified as "Multistructural" were found, with no item in the most superficial levels ("Prestructural" and Unistructural). At the SOLO level of greater cognitive complexity "Abstract", 14 items were found, with only one being displayed per exam, with the exception of the 2021 exams that introduced 2 items. These results indicate that the exams surveyed are made up of items classified at the highest levels of cognitive complexity, according to the SOLO Taxonomy, and associated with deep learning. This reinforces the effort made by the authors in preparing the items in such a way that they allow the student's learning assessment and the mathematical skills involved. In this sense, knowledge about the SOLO Taxonomy (cognitive complexity) allowed the authors to examine their pedagogical practice in relation to assessment and consequently prepare the assessment consciously, focusing on a deep learning, since it is a matter of assessing engineering students who attend higher education.

As future work, the authors intend to apply the SOLO Taxonomy to study students' answers to the items portrayed in the exams. The student will be able to answer in a more or less complex way than the SOLO level presented in the item. Thus, it is not possible to guarantee that the student answers with the same level of complexity required in the item. Therefore, by separating the level of complexity of the item and the student's response, the SOLO Taxonomy provides relevant information to the teacher, who will be able to identify whether the student is beyond or below the complexity of the item raised.

References

1. Brown, S., Knight, P.: Assessing Learners in Higher Education. Psychology Press (1994). ISBN: 9780749411138
2. Brown, S., Race, P., Rust, C.: Using and experiencing assessment. In: Assessment for Learning in Higher Education, pp. 75–85. Routledge (2012). ISBN: 9780203062074
3. Van de Watering, G., Gijbels, D., Dochy, F., Van der Rijt, J.: Students' assessment preferences, perceptions of assessment and their relationships to study results. Higher Educ. **56**, 645–658 (2008). https://doi.org/10.1007/s10734-008-9116-6
4. Biggs, J.B., Collis, K.F.: Evaluating the Quality of Learning: The Solo Taxonomy New York. Academic Press (1982). https://doi.org/10.1016/C2013-0-10375-3

5. de Almeida Santos Pereira, V.C.: Aplicação da taxonomia solo nos exames nacionais de matemática do 12º ano. Seven Editora (2023). http://sevenpublicacoes.com.br/index.php/editora/article/view/396
6. Pegg, J., Tall, D.: The fundamental cycle of concept construction underlying various theoretical frameworks. In: Sriraman, B., English, L. (eds.) Theories of Mathematics Education: Seeking New Frontiers, pp. 173–192. Springer, New York (2010). https://doi.org/10.1007/BF02655855
7. Pereira, V.C.A.S.: Aplicação da Taxonomia Solo na Análise da Qualidade da Avaliação. Validação do Método Analítico por Aplicação aos Exames Nacionais de Matemática A entre 2006 e 2014, vol. I, UBI, Covilhã (2019). http://hdl.handle.net/10400.6/7091
8. Filipe, M.A.E.R.: A Taxonomia SOLO nos Exames Nacionais de Matemática – 9º Ano. Universidade Nova de Lisboa. Lisboa - Portugal, p. 189 (2011). http://hdl.handle.net/10362/8173
9. Malik, S.I., Tawafak, R.M., Shakir, M.: Aligning and assessing teaching approaches with solo taxonomy in a computer programming course. Int. J. Inf. Commun. Technol. Educ. **17**(4), 1–15 (2021). https://doi.org/10.4018/IJICTE.20211001.oa
10. Mol, S.M., Matos, D.A.S.: Uma análise sobre a taxonomia solo: aplicações na avaliação educacional. Estudos Em Avaliação Educacional **30**(75), 722–747 (2019). https://doi.org/10.18222/eae.v30i75.6593
11. Malasari, P.N., Sari, A., Haqiqi, A.K., Masrukhin, M., Qomariyah, S: Using the SOLO taxonomy to investigate students' mathematical creative thinking proficiency. In: AIP Conference Proceedings, vol. 2595, No. 1. AIP Publishing (2023). https://doi.org/10.1063/5.0123686
12. Liu, S., Zhang, C.: Research on mathematics test questions of high school entrance examination based on the SOLO taxonomy theory: take the 2019–2021 Jinan High School entrance examination questions as an example, 2022. In: 12th International Conference on Information Technology in Medicine and Education (ITME), Xiamen, 2022, pp. 151–155. https://doi.org/10.1109/ITME56794.2022.00041
13. Aydin, U., Birgili, B.: Assessing Mathematical Higher-Order Thinking Skills: An Analysis of Turkish University Entrance Examinations. Educational Assessment, pp. 1–20 (2023). https://doi.org/10.1080/10627197.2023.2202311
14. Ceia, M.: A taxonomia SOLO e os níveis de van Hiele (2002). https://www.researchgate.net/publication/237709095
15. Castro, F.E.V., Fisler, K.: Designing a multi-faceted SOLO taxonomy to track program design skills through an entire course. In: Proceedings of the 17th Koli Calling International Conference on Computing Education Research, pp. 10–19 (2017). https://doi.org/10.1145/3141880.3141891
16. Caridade, C.M.R., Pereira, V.: SOLO taxonomy to assessment of skills and competencies required to students: comparison between online versus face-to-face exam. In: Mesquita, A., Abreu, A., Carvalho, J.V., Santana, C., de Mello, C.H.P. (eds.) Perspectives and Trends in Education and Technology. ICITED 2023. Smart Innovation, Systems and Technologies, vol. 366. Springer, Singapore (2023). https://doi.org/10.1007/978-981-99-5414-8_52
17. Silwana, A., Subanji, S., Manyunu, M., Rashahan, A.A.: Students' responses leveling in solving mathematical problem based on SOLO taxonomy viewed from multiple intelligences. Indones. J. Learn. Adv. Educ. **3**(1), 1–16 (2021). https://doi.org/10.23917/ijolae.v3i1.10528
18. Alpers, B.A., Demlova, M., Fant, C.H., Gustafsson, T., Lawson, D., Mustoe, L., Velichova, D.: A framework for mathematics curricula in engineering education: a report of the mathematics working group (2013). ISBN: 978-2-87352-007-6

Is Collaborative Learning a Voluntary Process?

Deolinda M. L. D. Rasteiro and Cristina M. R. Caridade

Abstract Mathematics learning process in Engineering courses is heavily studied, and the integration of Information and communication technologies (ICT) has grown significantly in recent years. Due to the pandemic, teaching processes were adjusted, and assessments were conducted considering the prevailing context we all encountered. Thus, students need innovative and stimulating teaching and learning practices that quickly motivate and involve them in the teaching/learning processes. In the last two years, ICT tools and digital platforms were extensively used, prompting questions about their optimal and effective utilization. Simultaneously, face-to-face group work and involvement with colleagues' needs lost some space for achievement and effectiveness. At the study's outset, the concern was about a potential preference for individual work and reduced solidarity among colleagues, except for possible friendships from previous school groups. Collaborative learning (CL) has been proposed as one of the solutions, with the Padlet platform (Padlet, 2023. Retrieved from http://www.padlet.com) being a popular tool for facilitating such interactions. In this paper, the authors propose two collaborative learning platforms using Padlet, one for Statistical Methods in Informatics Engineering and other for Mathematical Analysis I, with the aim of creating a space for virtual interaction among students. The study enrols 533 students from Informatics Engineering, many of whom are student workers, and from Mechanical Engineering 20 students, who were divided into small groups to foster teamwork and cooperation. The paper discusses the results such as students' participation, collaboration with colleagues, and course' performance, highlighting when it is, or it is not fruitful to use this type of activities.

D. M. L. D. Rasteiro (✉)
Polytechnic Institute of Coimbra, ISEC, Coimbra, Portugal

Applied Biomechanics Laboratory, ISEC, Coimbra, Portugal

C. M. R. Caridade
Polytechnic Institute of Coimbra, ISEC, Coimbra, Portugal

Applied Biomechanics Laboratory, ISEC, Coimbra, Portugal

Centre for Research in Geo-Space Science (CICGE), Porto, Portugal

V. Gayoso Martínez et al. (eds.), *Mathematical Methods for Engineering Applications*, Springer Proceedings in Mathematics & Statistics 439,
https://doi.org/10.1007/978-3-031-49218-1_9

1 Contextualization

The engagement of students within the educational institution has been identified as a primary determinant in diminishing risky behaviors, reducing dropout rates, and enhancing academic achievements [1]. Solidarity is rooted in the collective involvement of all individuals concerned, with the aim of fostering growth for everyone, emphasizing equality, cooperation, communal efforts, and respect for each individual's rights. To foster solidarity, it is essential to adopt a social behavior that is founded on collaboration and community-oriented thinking, rather than competition. Social behavior that embodies solidarity occurs when all individuals, either directly or indirectly, contribute to a shared endeavor, with dialogue serving as a fundamental catalyst. In essence, solidarity encompasses empathy, collaboration, unity, equality, teamwork, humility, respect, dialogue, generosity, interaction, and democracy. In order to foster solidarity through educational activities, it is imperative to propose group work, where cooperation plays a vital role in constructing knowledge collectively [2]. Collaborative activities permit the consideration of each person's life experiences, facilitating dialogue, knowledge exchange, empathy, and a deeper understanding of reality, among other benefits. Although group work is not commonly practiced by most teachers, modern teaching and learning technologies, such as collaborative methodologies, have made it possible to utilize various online platforms to facilitate group work [3]. One such platform is Padlet [4–7]. It allows students to share their resolutions of proposed exercises, correct or comment on colleagues' work, and access all work developed by their peers. In the case of Informatics Engineering graduation, the professors corrected the students' resolutions regularly, evaluating their involvement in the learning process and their collaboration and solidarity with colleagues.

The article explores the implementation of group work through the Padlet platform as a means to foster principles like solidarity, tolerance, and acceptance of diversity in two specific academic courses: Statistical Methods of Informatics Engineering and Mathematical Analysis I of Electrotechnical and Computer Engineering. Furthermore, the article presents the outcomes of this experiment, including students' level of engagement, their collaboration with peers, and their overall performance in the course. Noteworthy evidence is provided to ascertain the usefulness of this type of activity, as well as the potential factors that may influence its effectiveness.

2 Experience Methodology

2.1 Statistical Methods

A semester-long initiative was implemented for students enrolled in the Statistical Methods course of the Informatics Engineering graduation. The primary aim of this initiative was to establish a platform for collaborative learning, enabling students to interact and engage with each other within the context of the course.

A total of 533 students were registered for the Statistical Methods course, out of which 85 attended classes outside of regular hours. All students received an email invitation and had the opportunity to register for the Padlet project. The invitation to participate was accessible on the Moodle platform for a two-week period at the beginning of the semester. Among the 533 students, 137 expressed their intention to participate, including 28 students from the after-work course. Consequently, approximately 24.3% of the regular students and 33% of the after-work students actively took part in this initiative (approximately 25.7% of the total students enrolled in the course). However, despite the interest expressed by 137 students, only 50 of them actively posted and interacted with their peers.

The objective of the initiative was to enhance students' comprehension of the topics covered in class, including the provided references, and encourage them to review exercises completed by their classmates. To facilitate this process, the Padlet platform was utilized, allowing registered students to share their solutions to previous exam questions. Students were required to provide their identification (name and student number) while participating in the initiative. Furthermore, the teachers encouraged students to publish their own resolutions and provide constructive feedback on the solutions shared by their peers.

Student scores were calculated based on their level of participation across different content areas. The student who had the highest number of correct participations in the widest range of content areas received the highest rating. For instance, if there were X distinct content areas, a student who correctly solved 1 exercise from content A and 1 exercise from content B would receive a bonus value of $(\frac{2}{X}) \times 2$ in the final evaluation. Similarly, a student who correctly solved 2 exercises from content A and 0 exercises from content B would receive a bonus value of $(\frac{1}{X}) \times 2$ in the final evaluation. Engaging in exercises from the same content area, although not contributing to the final bonus values, still benefited the student by reinforcing their personal study and contributing to collaborative learning within the group.

The materials published by the students were reviewed and corrected by the professors, who provided feedback on whether the solutions required revision or if they were correct, as shown in Fig. 1. All participating students had access to the entire body of work developed throughout the initiative. Upon completion of the initiative, students had the potential to earn a maximum of 2 additional points to be added to their final grade.

The evaluation of students' involvement, collaboration, and solidarity, along with the results obtained, will be discussed and presented in the results and discussion section.

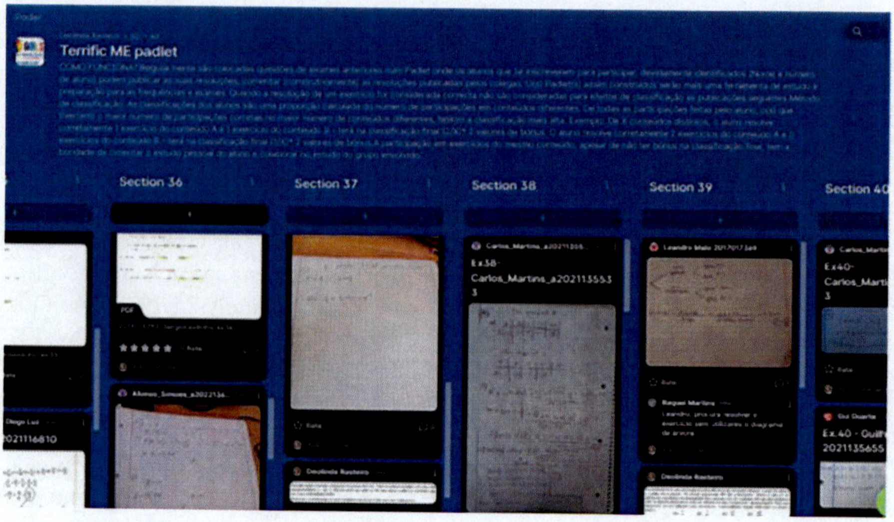

Fig. 1 Padlet activity capture

2.2 *Mathematical Analysis I*

The course on Mathematical Analysis I is offered in the first year of the Electrotechnical and Computer Engineering program, specifically targeting students who are new to higher education. Many of these students have weak foundational knowledge in mathematics and encounter significant difficulties. To address this, the Coimbra Engineering Institute offers the same course in the second semester in a progressive manner, allowing a larger number of students to acquire the necessary basic skills. The described experience takes place in the second semester and involves only 20 students who did not achieve a passing grade in the first semester but intend to continue studying this subject in the second semester. Since the students have already attended this course, the classes are conducted in a collaborative environment where the professor plays a guiding role.

The laboratory classes, lasting 1 hour and 30 minutes, are divided into three parts. In the first part, students spend 15 minutes solving an evaluation exercise based on the topics covered in the previous week's class. In the second part, they work in pairs to independently solve one or two exercises related to the current week's content. Initially, they access the proposed exercises for their respective groups through a QR code, consult the course materials on Moodle, and use available technologies such as GeoGebra, Wolfram Alpha, or calculators to solve the exercises. At the end of this hour-long segment, the solutions are shared on the Padlet platform in the corresponding group column. Students may use digital tablets, mathematical writing applications like Math Type, or even pen and paper.

Group work and solidarity play a crucial role in establishing strong connections, cooperation, and support between the members of each group, thereby fostering

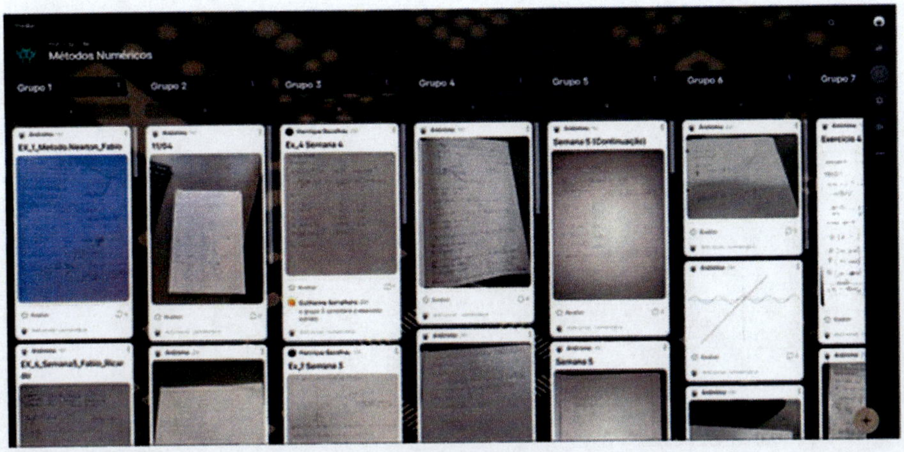

Fig. 2 Capture of part of the group work developed in Padlet

and enhancing the entire teaching-learning process. It is important to define roles within the group, such as identifying who will write the exercise solutions, who will upload them to the Padlet, and who will act as the group leader. Group members are encouraged to choose their partners based on affinity. The third part of the class, lasting 15 minutes, involves mixing the groups, and each group is assigned the task of correcting one or two exercises from another group. The groups that finish first will be the first to exchange assignments. In this segment, solidarity is displayed through a competitive yet productive atmosphere among the working groups.

Group work serves as a crucial factor for members to reach their full potential and identify errors or minor inaccuracies in other groups' work. In Fig. 2 above, an example of the work conducted by the students within the classroom can be observed.

3 Findings and Discussion

3.1 Statistical Methods

Padlet Posts and Satisfaction Questionnaire

The preliminary findings of the activity revealed a less-than-satisfactory rate of student participation, with only 25.7% of the students responding to the invitation, despite the opportunity to earn 2 bonus points. During the observation period (from 18/03/2023 to 28/04/2023), a total of 448 posts were generated, and their daily distribution can be observed in Fig. 3.

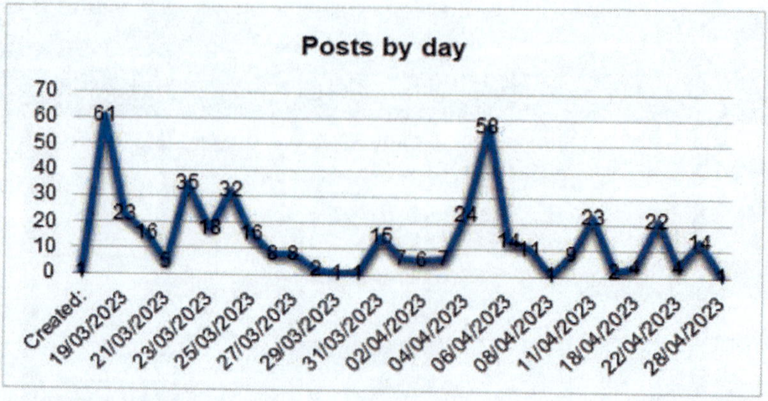

Fig. 3 Distribution of posts by day

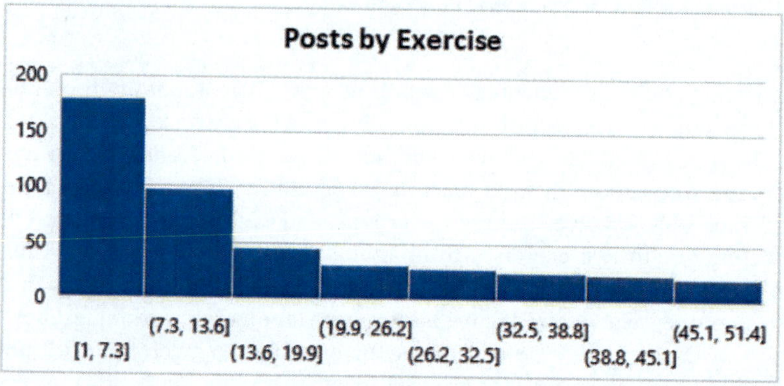

Fig. 4 Posts by Exercise

The distribution of posts by Exercise entrance is depicted on Fig. 4, above. As we may observe, in the beginning there was a higher response rate that may be attributed to two distinct reasons: one is natural curiosity, the other is because initial exercises were simpler that the following ones.

Given the stabilization of the response rate throughout the observation period and the consistent improvement in the difficulty of exercises, the authors are inclined to attribute the initial response index to curiosity.

In order to obtain feedback from the students involved, the teachers developed a satisfaction questionnaire using Google Forms (Fig. 5). The purpose of this questionnaire was to evaluate the students' satisfaction levels and their perceived usefulness of the Padlet activity.

From the data collected and resumed we may conclude that 61.9% of the engaged students had never used Padlet. Since the activity engagement in the beginning was only approximately 25.7%, the authors questioned if the students were only involved

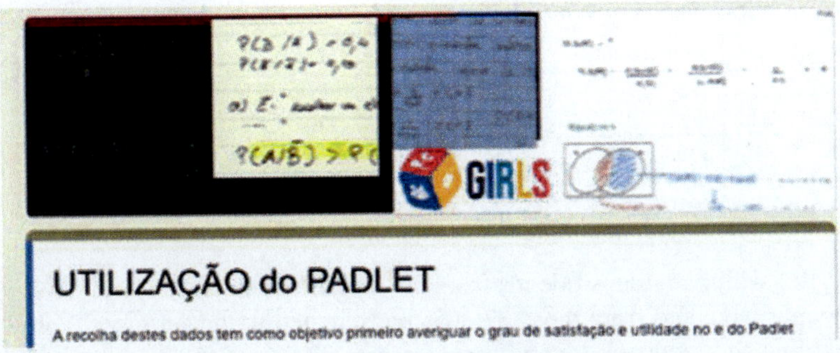

Fig. 5 Satisfaction questionnaire

Fig. 6 Activity influence in the assessment

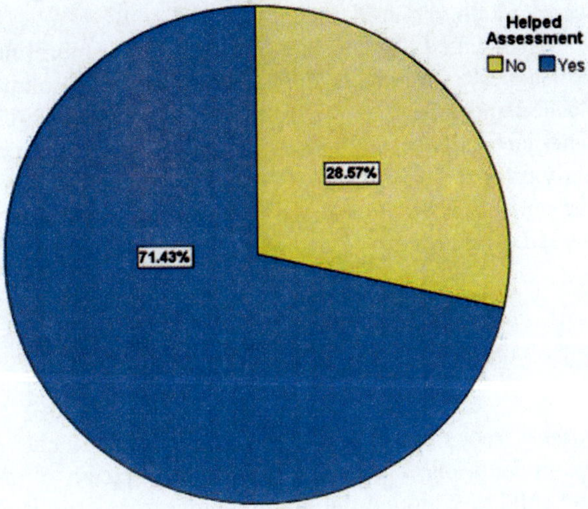

because of the 2-points bonus proposed. Besides the 2-points bonus proposed, 90.5% of the students agreed that this activity helped them to study. When asked "Do you consider that, in addition to the bonus, which you may have in your mark, this activity helped you to obtain a better classification in the written assessment?", the obtained answers indicate, as shown in Fig. 6, that 71.43% of the students consider the activity has helped them to prepare to the written assessment.

Padlet Posts and Written Assessment

To corroborate these received answers, we compared the number of students approved in the written assessment to the number of those students that participated in the activity. In fact, in case of after work students, from the 44.71% of the ones that were assessed, 60.53% approved, and from those 60.87% participated in the

Table 1 Statistics from the written assessment

Student type	Students total	Assessed	Approved	In activity	Approval rate
After work students	85	38	23	14	60.53%
Regular students	533	234	164	62	70.09%

activity. Regarding regular students, from the 43.90% of the ones that were assessed, 70.09% approved, and from those 37.80% participated in the activity (Table 1).

Regarding the feedback given by professors, 90.5% of the students considered that it was enough, and 85.7% of them wished that colleagues gave more feedback. As for the materials posted by their fellow colleagues, 95.2% considered that they were of help to complement their own study.

When asked "Would you like there to be more interaction with and from your colleagues? (Comments and questions to your resolutions)?", 66.7% of the students wished more interaction from their colleagues. Finally, to be able to conclude whether students' solidarity was only towards the colleagues participating in the activity or in general, it was asked to students "Do you agree with the sharing of the published resolutions with all your colleagues enrolled in the course?", 95.2% agreed.

3.2 *Mathematical Analysis I*

After 4 weeks (week 2, 3, 4 and 5) of laboratory classes in Mathematical Analysis, the students answered a survey with the objective of measuring interest, motivation, and solidarity among the group members. 14 students answered the questionnaire, demonstrating that the group work was productive, that they had an important role in this group and that they think that their learning was effective, as can be seen in Table 2.

Cooperation, perfectionism. motivation, responsibility, and commitment were the words most identified by the students about the relationship between the elements of the group. This reinforces the solidary experience of these students in small groups. Finally, in an open response about what the students liked most about the laboratory classes, 50% of the students chose group work (Fig. 7).

Regarding effective learning, it can be said that of the 20 students enrolled in the UC, only 18 carried out the assessments during the laboratory classes, obtaining the classifications shown in Fig. 8. It should be noted that in week 3 only 13 students were evaluated and that the assigned classifications correspond to a range of 0–10 points in each week.

Table 2 Identification of group work and the relationship between its elements

Questions	% Yes
Your contribution was important for the group as an element?	92.9%
The dynamics of group work contributed to the development of the proposed exercise?	100%
Are you confident in what you learned in the classroom?	78.6%
Did you enjoy using Padlet?	92.9%
Padlet promotes greater participation and collaboration among group members?	92.9%
The teacher/student mediation was essential for overcoming the different stages of work design?	100%
The mediation between the different elements of the group was essential for overcoming the different stages of work design?	92.9%
Do you like that innovative learning activities are applied in the classroom?	100%
Do you consider that this work methodology can contribute to the development of skills inherent to academic (engineering) and professional choices?	100%

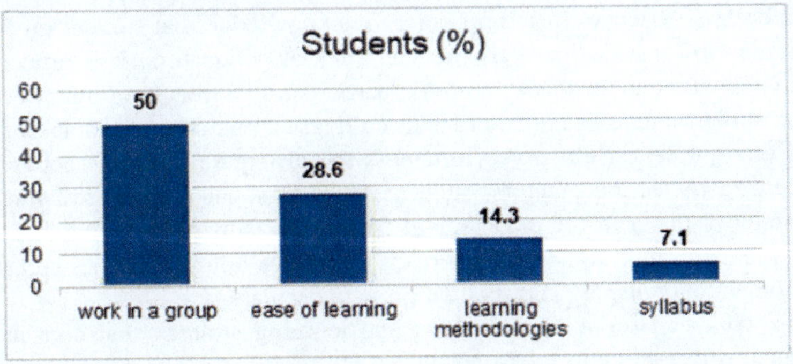

Fig. 7 What do you like most about laboratory classes?

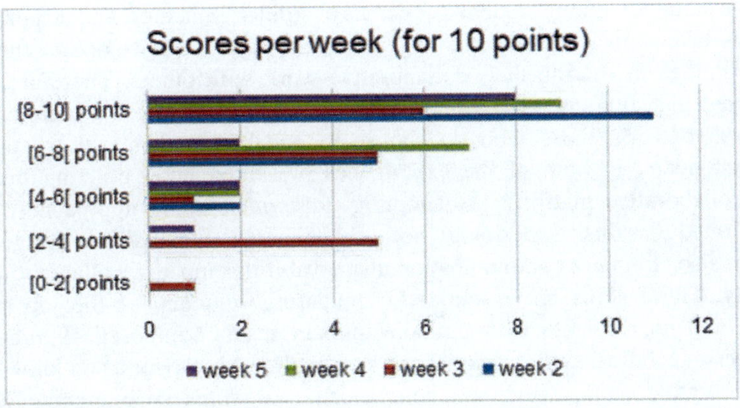

Fig. 8 Scores obtained in the 4-week assessments

4 Conclusions

With regards to the experience conducted in the Statistical Methods course, several conclusions can be drawn from this activity. The first conclusion pertains to the percentage of students who actively participated in the activity, which was found to be lower than anticipated by the professors. Additionally, it was surprising to observe that only approximately 51% of the students completed the written assessment. Students tend to prefer seeking clarification for content and exercise resolutions through individual means, such as office hours or email, where their interaction is limited to the professor alone. Consequently, individual study appears to be the preferred method for the majority of students.

Another noteworthy conclusion is that despite the widespread use of information and communication technologies (ICT), Padlet, a well-known collaborative platform, was unfamiliar to approximately 60% of the students. Almost all students agreed that this activity facilitated their studying process, and the feedback provided by the professors was deemed sufficient. However, a small percentage (14.3%) of the students expressed a desire for more feedback from their peers.

From both experiences, it is important to acknowledge that student preferences and attitudes towards online activities can vary. As educators, it is crucial to be responsive to student feedback, monitor participation levels, and adapt activities based on student engagement in order to cultivate a positive and inclusive online learning environment. Students' willingness to participate in online activities can be influenced by various factors, such as their personal preferences, the nature of the activity, the level of engagement fostered by instructors, and the overall course structure. While some students readily embrace online activities and actively engage, others may display less inclination to participate.

Group work can serve as a teaching and learning strategy that contributes to pedagogical differentiation by fostering an environment where students integrate into groups with the aim of both learning and teaching other members. This approach also facilitates the development of various skills. Through group work and reflective practices among students, concepts of solidarity, justice, and responsibility gradually take shape. The students engaged in the Padlet activity do not encounter any difficulties in sharing their collaborative work with others. Therefore, it can be inferred that although a small percentage of students express a preference for collaborative work, those who do choose to collaborate demonstrate solidarity with their peers. In both of the experiences presented here, utilizing the same Padlet collaborative platform but adopting different learning approaches, it was demonstrated that this methodology not only enhances students' knowledge-based skills but also cultivates communication, responsibility, empathy, collaboration, and solidarity. These skills are essential for preparing students for the job market, where companies operate with multidisciplinary teams composed of individuals with diverse technical skills, experiences, personalities, behaviors, and knowledge.

Collaborative work, combined with a sense of voluntarism, recognizes the importance of collective efforts and encourages students to actively contribute to group projects and initiatives. However, it is equally crucial to acknowledge the value of individual work, as it allows students to develop their independent thinking and problem-solving skills. Furthermore, fostering positive societal relations between students promotes a supportive and inclusive learning environment, where mutual respect, empathy, and cooperation thrive, enhancing the overall educational experience.

It is not uncommon for some students to prioritize their individual grades over collaborative values and societal relations. This behavior may stem from various factors such as a competitive academic environment, external pressure to achieve high grades, or personal motivations. While it is essential to acknowledge and respect individual aspirations, it is also important to emphasize the long-term benefits of collaboration, voluntarism, and positive societal interactions.

Educational institutions can play a vital role in promoting a balanced perspective by fostering a supportive learning environment that highlights the value of collaboration and societal relations alongside individual achievement. By incorporating group projects, encouraging peer feedback and mentorship, and fostering a culture of inclusion and cooperation, students can develop a broader understanding of the benefits of working together and appreciate the value of collective efforts.

Additionally, educators can provide guidance and support to students, emphasizing the importance of both individual growth and collaboration in achieving academic success and preparing for future professional and societal challenges. By reinforcing the idea that collaborative skills and positive relationships can enhance not only individual achievements but also contribute to a more harmonious and effective collective learning experience, students can be encouraged to find a balance between their personal goals and the broader values of collaboration and societal relations. Therefore, in the future, the authors intended to continue implementing this practice in their classrooms, as it has proven beneficial in fostering solidarity, tolerance, and acceptance of differences among students, as well as facilitating the development of cooperative group work.

References

1. Cyrre, M.R.L.: Um novo olhar para a sala de aula: privilegiando o trabalho em grupo. Ciências e Letras. Revista da Faculdade Porto-Alegrense de Educação, Ciências e Letras, Porto Alegre, (30), pp. 101–111 (2001)
2. Frankenberger, F., Anastacio, M., Tortato, U.: Education for solidarity: A case study at PUCPR. In: Handbook of Lifelong Learning for Sustainable Development, pp. 183–195 (2018). https://doi.org/10.1007/978-3-319-63534-7_13
3. Figueiredo, E.B., Zuchi, I.S., Azevedo, E.B., Moro, G.: A collaborative work experience in basic mathematics subjects in engineering courses. J. Eng. Educ. **33**(1), 13–23 (2014). https://doi.org/10.15552/2236-0158/abenge.v33n1p13-23
4. Fisher, C.D.: Padlet: An online tool for learner engagement and collaboration. Acad. Manag. Learn. Educ. **16**(1), 163–165 (2017). https://doi.org/10.5465/amle.2017.0055

5. Mehta, K.J., Miletich, I., Detyna, M.: Content-specific differences in Padlet perception for collaborative learning amongst undergraduate students. Res. Learn. Technol. **29** (2021). https://doi.org/10.25304/rlt.v29.2551
6. Zhi, Q., Su, M.: Enhance collaborative learning by visualizing process of knowledge building with Padlet. In: 2015 International Conference of Educational Innovation through Technology (EITT), Wuhan, China, 2015, pp. 221–225 (2015). https://doi.org/10.1109/EITT.2015.54
7. Padlet (2023). Retrieved from http://www.padlet.com. Accessed February 2023

Elliptic Biquaternionic Sequence with Vietoris' Numbers as Its Components

Regina de Almeida and Paula Catarino

Abstract In this study, we introduce an elliptic biquaternionic sequence with Vietoris' numbers as its components and discuss some of its properties. Also, the generating function and some identities in terms of elliptic biquaternionic sequence with Vietoris' numbers are given. Furthermore, the construction of this elliptic biquaternion sequence is presented using matrices that generate the quaternionic sequence where the components are Vietoris' number, and also by applying the determinant to a special kind of matrices.

1 Preliminaries

In order to introduce a new elliptic biquaternion sequence involving Vietoris' numbers that will appear in the next section, we begin this section by presenting some basic notions and properties of quaternion algebra, (for more detail see for example [6] and [7]).

The set of quaternions forms a four-dimensional associative and noncommutative algebra over the set of real numbers. Let $\{1, \mathbf{i}, \mathbf{j}, \mathbf{k}\}$ be the standard basis in $\mathbb{R}^4$ satisfying the following multiplication rules:

$$\mathbf{i}^2 = \mathbf{j}^2 = \mathbf{k}^2 = -1, \quad \mathbf{ij} = -\mathbf{ji} = \mathbf{k}, \quad \mathbf{jk} = -\mathbf{kj} = \mathbf{i}, \quad \mathbf{ki} = -\mathbf{ik} = \mathbf{j}.$$

Let $\mathbb{H}$ be the skew field of real quaternions defined by

$$\mathbb{H} = \{q = q_0 + q_1\mathbf{i} + q_2\mathbf{j} + q_3\mathbf{k}, \quad q_s \in \mathbb{R}, \ s = 0, 1, 2, 3\},$$

R. de Almeida (✉) · P. Catarino
Department of Mathematics, University of Trás-os-Montes e Alto Douro, Vila Real, Portugal

CMAT - Centre of Mathematics, Polo CMAT-UTAD, Vila Real, Portugal
e-mail: ralmeida@utad.pt; pcatarin@utad.pt

V. Gayoso Martínez et al. (eds.), *Mathematical Methods for Engineering Applications*, Springer Proceedings in Mathematics & Statistics 439,
https://doi.org/10.1007/978-3-031-49218-1_10

and for $q \in \mathbb{H}$ given by $q = q_0 + q_1\mathbf{i} + q_2\mathbf{j} + q_3\mathbf{k}$, the scalar part of the quaternion q is denoted by $\text{Sc}(q) = q_0$ and $\text{Vec}(q) = q_1\mathbf{i} + q_2\mathbf{j} + q_3\mathbf{k}$ is the vector part.

We recall that the prime involution in $\mathbb{H}$ is the mapping $q \to q'$ defined by

$$q' = q_0 - q_1\mathbf{i} - q_2\mathbf{j} + q_3\mathbf{k}$$

and the reversion in $\mathbb{H}$ is the mapping $q \to q*$ defined by

$$q^* = q_0 + q_1\mathbf{i} + q_2\mathbf{j} - q_3\mathbf{k}.$$

The conjugation in $\mathbb{H}$ is the mapping $q \to \overline{q}$ defined by

$$\overline{q} = (q')^* = (q^*)' = q_0 - q_1\mathbf{i} - q_2\mathbf{j} - q_3\mathbf{k}.$$

These involutions satisfy the following product rules

$$(q\,p)' = q'\,p', \quad (q\,p)^* = p^*\,q^*, \quad \overline{q\,p} = \overline{p}\,\overline{q}, \quad \forall q, p \in \mathbb{H}.$$

Furthermore, one has

$$q\mathbf{k} = \mathbf{k}q', \quad \forall q \in \mathbb{H}.$$

One also has the relation between the scalar part and the vector part as follows

$$q + \overline{q} = 2\text{Sc}(q), \qquad q - \overline{q} = 2\text{Vec}(q). \tag{1}$$

The modulus or absolute value of the quaternion, is given by

$$\|q\|^2 = q\overline{q} = \overline{q}q, \qquad \forall q \in \mathbb{H}. \tag{2}$$

Let $q, p \in \mathbb{H}$ represented by $q = q_0+q_1\mathbf{i}+q_2\mathbf{j}+q_3\mathbf{k}$ and $p = p_0+p_1\mathbf{i}+p_2\mathbf{j}+p_3\mathbf{k}$. The inner product is given by

$$\langle p, q\rangle = \frac{\overline{p}\,q + \overline{q}\,p}{2} = \frac{p\,\overline{q} + q\,\overline{p}}{2} = p_0q_0 + p_1q_1 + p_2q_2 + p_3q_3.$$

A generalization of the real quaternion algebra is the complex quaternion algebra, which can be represented by

$$\mathbb{H}_{\mathbb{C}} = \{z = z_0 + z_1\mathbf{i} + z_2\mathbf{j} + z_3\mathbf{k}, \quad z_s \in \mathbb{C}, \; s = 0, 1, 2, 3\}.$$

Moreover, a generalized complex numbers or binary numbers are introduced in [15] as

$$z = x + Iy, \; x, y \in \mathbb{R}, \; I^2 = iq + p, \; p, \; q \in \mathbb{R}.$$

In [8], A. A. Harkin and J. B. Harkin studied the geometry of an one parameter family of generalized complex number systems, in which $I^2 = p$, $(-\infty < p < +\infty)$ and $q = 0$. This family of complex number systems is called by p-complex plane

$$\{x + Iy : x, y \in \mathbb{R},\ I^2 = p,\ p \in \mathbb{R}\}. \tag{3}$$

Note that in the special case of $p < 0$, (3) corresponds to the system of elliptic numbers. Some properties of these type of numbers can be seen in [10] and [11].

Throughout this paper we will consider the system of elliptic numbers denoted by

$$\mathbb{C}_p = \{x + Iy :\ x, y \in \mathbb{R}, I^2 = p,\ p < 0\},$$

and denote the set of elliptic biquaternions, using the elliptic numbers, by

$$\mathbb{H}_{\mathbb{C}_p} = \left\{A_0 + A_1\mathbf{i} + A_2\mathbf{j} + A_3\mathbf{k}, \quad A_s \in \mathbb{C}_p,\ s = 0, 1, 2, 3\right\},$$

where I commutes with the quaternionic units [11]

$$\mathbf{i}I = I\mathbf{i}, \quad \mathbf{j}I = I\mathbf{j}, \quad \mathbf{k}I = I\mathbf{k}.$$

The structure of this paper is as follows: Sect. 2 is dedicated to the sequence of quaternions with Vietoris' numbers and some of its properties which are fundamental to the work that follows; in Sect. 3, a new biquaternion sequence where Vietoris' numbers are involved is introduced, as well as some of its properties, in particular, the generating function are stated; in Sect. 4, elliptic biquaternion matrix representations of Vietoris' elliptic biquaternions are obtained.

2 A Quaternion Sequence with Vietoris' Numbers $\{Q_n\}_{n\geqslant 0}$ and Basic Properties

In this section we present the Vietoris' sequence as well as the quaternionic sequence with Vietoris' numbers as its components. Some properties of the later sequence will also be introduced.

Special integers sequences have been the center of attention for many researchers, in this work we will focus our attention to a sequence of rational numbers denoted by *Vietoris' sequence*, $\{v_n\}_{n\geqslant 0}$, which is defined by

$$v_n = \frac{1}{2^n}\binom{n}{\lfloor \frac{n}{2} \rfloor}, \qquad n \geqslant 0,$$

where $\lfloor\cdot\rfloor$ is the floor function. The first Vietoris' numbers of the sequence are

$$1, \frac{1}{2}, \frac{1}{2}, \frac{3}{8}, \frac{3}{8}, \frac{5}{16}, \frac{5}{16}, \frac{35}{128}, \frac{35}{128}, \frac{63}{256}, \frac{63}{256}, \frac{231}{1024}, \frac{231}{1024}, \ldots,$$

which is related with the sequence A283208 in the OEIS in [13]. It is also known that the recurrence relation for this sequence is given by the following identity:

$$v_n = \begin{cases} \frac{n}{n+1} v_{n-1}, & n \text{ odd} \\ v_{n-1}, & n \text{ even} \\ 1, & n = 0 \end{cases}.$$

Although it is an immediate consequence, we want to emphasize the recurrence relation for $\{v_{2n}\}_{n\geqslant 0}$:

$$v_{2n+2} = d(2n)v_{2n}, \qquad n \geqslant 0, \tag{4}$$

where

$$d(k) = \frac{k+1}{k+2}, \qquad k \geqslant 0. \tag{5}$$

In 1958, L. Vietoris published a result in [14] where this sequence appeared. This sequence has applications in harmonic analysis, as can be seen in the work of R. Askey and J. Steinig, in [1], as well as, in the theory of stable holomorphic functions presented in the paper [12] by St. Ruscheweyh and L. Salinas. Some properties of Vietoris' number sequence and some techniques, using special types of matrices that generate this number sequence were presented in [5]. In [4], we started to investigate a quaternion sequence whose components are Vietoris' numbers, some of its properties are presented in this section.

A quaternion sequence with Vietoris' numbers can be represented by

$$Q_n = v_n + v_{n+1}\mathbf{i} + v_{n+2}\mathbf{j} + v_{n+3}\mathbf{k} \tag{6}$$

which can be rewritten in terms of one element of the Vietoris' sequence by:

$$Q_n = v_{n+\frac{1-(-1)^n}{2}} \left(a_{n+\frac{1-(-1)^n}{2}} + \frac{1-(-1)^n}{2} \left(b_{n+1} - a_{n+\frac{1-(-1)^n}{2}} \right) \right) \tag{7}$$

where

$$\begin{aligned} a_n &= 1 + d(n)(\mathbf{i}+\mathbf{j}) + d(n)d(n+2)\mathbf{k}, & n &= 2k \\ b_{n+1} &= 1 + \mathbf{i} + d(n+1)(\mathbf{j}+\mathbf{k}), & n &= 2k+1 \end{aligned} . \tag{8}$$

Some recurrence relations for the quaternion sequence with Vietoris' numbers, $\{Q_n\}_{n\geq 0}$, from [4] are presented next, and it will be used throughout the paper.

Proposition 1 ([4]) *Consider Q_s defined in (6) and $d(\cdot)$ defined in (5). Then*

$$Q_{2n+2} = Q_{2n+1}\phi_R(2n+2),$$

where

$$\phi_R(2n+2) = \frac{\phi_0(2n+2) + \phi_1(2n+2)\mathbf{i} + \phi_2(2n+2)\mathbf{j} + \phi_3(2n+2)\mathbf{k}}{2 + 2d^2(2n+2)}$$

with

$$\begin{aligned}
&\phi_0(2n+2) = 1 + d(2n+2) + d^2(2n+2)[1 + d(2n+4)];\\
&\phi_1(2n+2) = d(2n+2) - 1 + d^2(2n+2)[1 - d(2n+4)];\\
&\phi_2(2n+2) = d(2n+2)[d(2n+4) - d(2n+2)];\\
&\phi_3(2n+2) = d(2n+2)[d(2n+4) + d(2n+2)] - 2d(2n+4).
\end{aligned}$$

Proposition 2 ([4]) *Consider Q_s defined in (6), and $d(\cdot)$ defined in (5). Then*

$$Q_{2n+1} = Q_{2n}\psi_R(2n),$$

where

$$\psi_R(2n+2) = \frac{d(2n)\left(\psi_0(2n) + \psi_1(2n)\mathbf{i} + \psi_2(2n)\mathbf{j} + \psi_3(2n)\mathbf{k}\right)}{1 + d^2(2n)(2 + d^2(2n+2))}$$

with

$$\begin{aligned}
&\psi_0(2n) = 1 + d(2n) + d(2n)d(2n+2)[1 + d(2n+2)];\\
&\psi_1(2n) = 1 - d(2n) + d(2n)d(2n+2)[d(2n+2) - 1];\\
&\psi_2(2n) = d(2n+2) - d(2n);\\
&\psi_3(2n) = d(2n+2) + d(2n) - 2d(2n+2)d(2n).
\end{aligned}$$

Furthermore, one can deduce the next result.

Proposition 3 *Let Q_s be defined as in (6). Then*

$$Q_s P_1(s-1) = 2Q_{s-1}P_1^2(s-1),$$

where

$$P_1(s) = \begin{cases} \frac{1}{2}\psi_R(s), & s = 2k \\ \frac{1}{2}\phi_R(s+1), & s = 2k+1 \end{cases}$$

where $\phi_R(\cdot)$, $\psi_R(\cdot)$ are defined in Propositions 1 and 2.

Proof If $s = 2k$, then using Proposition 1, one gets

$$\begin{aligned} Q_{2k} P_1(2k-1) &= Q_{2k} \tfrac{1}{2}\phi_R(2k) \\ &= Q_{2k-1}\phi_R(2k)\tfrac{1}{2}\phi_R(2k) \\ &= 2Q_{2k-1} P_1^2(2k-1). \end{aligned}$$

For the case $s = 2k+1$, then using Proposition 2, one gets

$$\begin{aligned} Q_{2k+1} P_1(2k) &= Q_{2k+1} \tfrac{1}{2}\psi_R(2k) \\ &= Q_{2k}\psi_R(2k)\tfrac{1}{2}\psi_R(2k) \\ &= 2Q_{2k} P_1^2(2k). \end{aligned}$$

□

The next result, which the proof can be seen in [4], is a recurrence relation given by a two consecutive terms.

Theorem 1 ([4]) *Let Q_s be defined as in (6). Then*

$$Q_{s+1} = Q_s P_1(s) + Q_{s-1} P_0(s-1),$$

where $P_1(\cdot)$ is defined in Proposition 3 and

$$P_0(s-1) = \begin{cases} \frac{1}{2}\kappa_R(s), & s = 2k \\ \frac{1}{2}\theta_R(s-1), & s = 2k+1 \end{cases}$$

where

$$\theta_R(2n) = \frac{d(2n)\left(\theta_0(2n) + \theta_1(2n)\mathbf{i} + \theta_2(2n)\mathbf{j} + \theta_3(2n)\mathbf{k}\right)}{1 + d^2(2n)(2 + d^2(2n+2))}$$

with

$$\theta_0(2n) = 1 + d(2n)d(2n+2)(2 + d(2n+4)d(2n+2));$$
$$\theta_1(2n) = d(2n+2) - d(2n) + d(2n)d(2n+2)(d(2n+2) - d(2n+4));$$
$$\theta_2(2n) = d(2n+2) - d(2n) + d(2n)d(2n+2)(d(2n+4) - d(2n+2));$$
$$\theta_3(2n) = d(2n+2)(d(2n+4) - d(2n)),$$

and

$$\kappa_R(2n) = \frac{d(2n)\,(\kappa_0(2n) + \kappa_1(2n)\mathbf{j})}{1 + d^2(2n)}$$

with

$$\kappa_0(2n) = 1 + d(2n)d(2n+2); \qquad \kappa_1(2n) = d(2n+2) - d(2n).$$

3 The Elliptic Biquaternionic Sequence with Vietoris' Numbers and Some Properties

In this section, using (6), an elliptic biquaternionic sequence is introduced, as well as some identities and its generating function.

For a non negative integer n and for a negative real number p, let us consider the Vietoris' elliptic biquaternion of order n given by

$$V_n^p = Q_n + I\,Q_{n+1}, \qquad I^2 = p, \qquad p < 0,$$

where, for $m \in \mathbb{N}_0$,

$$Q_m = v_m + v_{m+1}\mathbf{i} + v_{m+2}\mathbf{j} + v_{m+3}\mathbf{k}.$$

Other way to write the Vietoris' elliptic biquaternion of order n is as follows:

$$V_n^p = (v_n + I v_{n+1}) + (v_{n+1} + I v_{n+2})\mathbf{i} + (v_{n+2} + I v_{n+3})\mathbf{j} + (v_{n+3} + I v_{n+4})\mathbf{k}.$$

Let us denote $A_i^p = v_i + I v_{i+1},\ i = 0, 1, \ldots$, hence

$$V_n^p = A_n^p + A_{n+1}^p\,\mathbf{i} + A_{n+2}^p\,\mathbf{j} + A_{n+3}^p\,\mathbf{k}.$$

Thus, any Vietoris' elliptic biquaternion includes a scalar part $Sc(V_n^p)$ and a vectorial part $\mathcal{V}ec(V_n^p)$ presented by:

$$Sc(V_n^p) = A_n^p = v_n + I v_{n+1}, \quad \mathcal{V}ec(V_n^p) = A_{n+1}^p\mathbf{i} + A_{n+2}^p\mathbf{j} + A_{n+3}^p\mathbf{k}.$$

The Hamilton, complex and total conjugates of Vietoris' elliptic biquaternion are, respectively,

$$\overline{V_n^p} = A_n^p - A_{n+1}^p\mathbf{i} - A_{n+2}^p\mathbf{j} - A_{n+3}^p\mathbf{k},$$

$$(V_n^p)^* = (A_n^p)^* + (A_{n+1}^p)^*\mathbf{i} + (A_{n+2}^p)^*\mathbf{j} + (A_{n+3}^p)^*\mathbf{k},$$

$$(V_n^p)^+ = (A_n^p)^* - (A_{n+1}^p)^*\mathbf{i} - (A_{n+2}^p)^*\mathbf{j} - (A_{n+3}^p)^*\mathbf{k},$$

where $(A_k^p)^* = v_k - I v_{k+1}$.

From here it can be stated the following properties that are easily proved.

Proposition 4 *Let V_n^p the Vietoris' elliptic biquaternion of order n. Then the following properties hold:*

(i) $V_n^p + \overline{V_n^p} = 2A_n^p$;
(ii) $V_n^p + (V_n^p)^* = 2Q_n$;
(iii) $V_n^p - (V_n^p)^* = 2I\, Q_{n+1}$;
(iv) $(A_n^p)^2 = (v_n^2 + p v_{n+1}^2) + I(2v_n v_{n+1})$.

Proof Let us start by proving (i)

$$\begin{aligned} V_n^p + \overline{V_n^p} &= A_n^p + A_{n+1}^p\mathbf{i} + A_{n+2}^p\mathbf{j} + A_{n+3}^p\mathbf{k} + A_n^p - A_{n+1}^p\mathbf{i} - A_{n+2}^p\mathbf{j} - A_{n+3}^p\mathbf{k} \\ &= 2A_n^p. \end{aligned}$$

The proof of (ii), is as follows

$$\begin{aligned} & V_n^p + (V_n^p)^* \\ &= A_n^p + A_{n+1}^p\mathbf{i} + A_{n+2}^p\mathbf{j} + A_{n+3}^p\mathbf{k} + (A_n^p)^* + (A_{n+1}^p)^*\mathbf{i} + (A_{n+2}^p)^*\mathbf{j} + (A_{n+3}^p)^*\mathbf{k} \\ &= (v_n + I v_{n+1}) + (v_{n+1} + I v_{n+2})\mathbf{i} + (v_{n+2} + I v_{n+3})\mathbf{j} + (v_{n+3} + I v_{n+4})\mathbf{k} \\ &+ (v_n - I v_{n+1}) + (v_{n+1} - I v_{n+2})\mathbf{i} + (v_{n+2} - I v_{n+3})\mathbf{j} + (v_{n+3} - I v_{n+4})\mathbf{k} \\ &= 2(v_n + v_{n+1}\mathbf{i} + v_{n+2}\mathbf{j} + v_{n+3}\mathbf{k}) \\ &= 2Q_n. \end{aligned}$$

By a straightforward calculation, one gets (iii)

$$\begin{aligned} & V_n^p - (V_n^p)^* \\ &= A_n^p + A_{n+1}^p\mathbf{i} + A_{n+2}^p\mathbf{j} + A_{n+3}^p\mathbf{k} - \big((A_n^p)^* + (A_{n+1}^p)^*\mathbf{i} \\ &\quad + (A_{n+2}^p)^*\mathbf{j} + (A_{n+3}^p)^*\mathbf{k}\big) \end{aligned}$$

$$
\begin{aligned}
&= (v_n + I v_{n+1}) + (v_{n+1} + I v_{n+2})\mathbf{i} + (v_{n+2} + I v_{n+3})\mathbf{j} + (v_{n+3} + I v_{n+4})\mathbf{k} \\
&- ((v_n - I v_{n+1}) + (v_{n+1} - I v_{n+2})\mathbf{i} + (v_{n+2} - I v_{n+3})\mathbf{j} + (v_{n+3} - I v_{n+4})\mathbf{k}) \\
&= 2I (v_{n+1} + v_{n+2}\mathbf{i} + v_{n+3}\mathbf{j} + v_{n+4}\mathbf{k}) \\
&= 2I Q_{n+1}.
\end{aligned}
$$

To prove (iv), one uses the fact that $I^2 = p$,

$$
\begin{aligned}
(A_n^p)^2 &= (v_n + I v_{n+1})^2 \\
&= v_n^2 + I^2 v_{n+1}^2 + 2I v_n v_{n+1} \\
&= (v_n^2 + p v_{n+1}^2) + I (2 v_n v_{n+1}).
\end{aligned}
$$

□

In the next result, one obtains a relation between the difference of two consecutive Vietoris' elliptic biquaternion elements with the real Vietoris' numbers.

Proposition 5 *Let V_n^p the Vietoris' elliptic biquaternion of order $n \in \mathbb{N}_0$. Then the following property holds:*

$$
V_n^p - V_{n+1}^p = \begin{cases} \frac{1}{n+3} v_{n+1}\mathbf{i} + \frac{1}{n+5} v_{n+3}\mathbf{k} + I\left(\frac{1}{n+3} v_{n+1} + \frac{1}{n+5} v_{n+3}\mathbf{j}\right), & n = 2k+1 \\ \left(\frac{1}{n+2} v_n + \frac{1}{n+4} v_{n+2}\mathbf{j}\right) + I\left(\frac{1}{n+4} v_{n+2}\mathbf{i} + \frac{1}{n+6} v_{n+4}\mathbf{k}\right), & n = 2k \end{cases}.
$$

Proof Let $k \in \mathbb{N}_0$ for the case $n = 2k$, using $v_{2k+1} = v_{2k+2}$ and (4), one gets

$$
\begin{aligned}
V_{2k}^p - V_{2k+1}^p &= (Q_{2k} - Q_{2k+1}) + I (Q_{2k+1} - Q_{2k+2}) \\
&= v_{2k} - v_{2k+1} + (v_{2k+1} - v_{2k+2})\mathbf{i} + (v_{2k+2} - v_{2k+3})\mathbf{j} \\
&\quad + (v_{2k+3} - v_{2k+4})\mathbf{k} + I\big(v_{2k+1} - v_{2k+2} + (v_{2k+2} - v_{2k+3})\mathbf{i} \\
&\quad + (v_{2k+3} - v_{2k+4})\mathbf{j} + (v_{2k+4} - v_{2k+5})\mathbf{k}\big) \\
&= v_{2k} - v_{2k+2} + (v_{2k+2} - v_{2k+4})\mathbf{j} \\
&\quad + I\left((v_{2k+2} - v_{2k+4})\mathbf{i} + (v_{2k+4} - v_{2k+6})\mathbf{k}\right) \\
&= \left(\frac{1}{2k+2} v_{2k} + \frac{1}{2k+4} v_{2k+2}\mathbf{j}\right) + I\left(\frac{1}{2k+4} v_{2k+2}\mathbf{i} + \frac{1}{2k+6} v_{2k+4}\mathbf{k}\right).
\end{aligned}
$$

Substituting $n = 2k$, one has

$$
V_n^p - V_{n+1}^p = \left(\frac{1}{n+2} v_n + \frac{1}{n+4} v_{n+2}\mathbf{j}\right) + I\left(\frac{1}{n+4} v_{n+2}\mathbf{i} + \frac{1}{n+6} v_{n+4}\mathbf{k}\right).
$$

Consider the case $n = 2k + 1$, using $v_{2k+1} = v_{2k+2}$ and (4), one obtains

$$\begin{aligned}V^p_{2k+1} - V^p_{2k+2} &= (Q_{2k+1} - Q_{2k+2}) + I\,(Q_{2k+2} - Q_{2k+3})\\ &= v_{2k+1} - v_{2k+2} + (v_{2k+2} - v_{2k+3})\mathbf{i} + (v_{2k+3} - v_{2k+4})\mathbf{j}\\ &\quad +(v_{2k+4} - v_{2k+5})\mathbf{k} + I\big(v_{2k+2} - v_{2k+3} + (v_{2k+3} - v_{2k+4})\mathbf{i}\\ &\quad +(v_{2k+4} - v_{2k+5})\mathbf{j} + (v_{2k+5} - v_{2k+6})\mathbf{k}\big)\\ &= (v_{2k+2} - v_{2k+4})\mathbf{i} + (v_{2k+4} - v_{2k+6})\mathbf{k}\\ &\quad +I\,(v_{2k+2} - v_{2k+4} + (v_{2k+4} - v_{2k+6})\mathbf{j})\\ &= \tfrac{1}{2k+4}v_{2k+2}\mathbf{i} + \tfrac{1}{2k+6}v_{2k+4}\mathbf{k} + I\left(\tfrac{1}{2k+4}v_{2k+2} + \tfrac{1}{2k+6}v_{2k+4}\mathbf{j}\right).\end{aligned}$$

Substituting $n = 2k + 1$, one gets

$$V^p_n - V^p_{n+1} = \frac{1}{n+3}v_{n+1}\mathbf{i} + \frac{1}{n+5}v_{n+3}\mathbf{k} + I\left(\frac{1}{n+3}v_{n+1} + \frac{1}{n+5}v_{n+3}\mathbf{j}\right).$$

□

The seminorm of Vietoris' elliptic biquaternion V^p_n can be given as

$$\mathcal{N}(V^p_n) = V^p_n\overline{V^p_n} = (A^p_n)^2 + (A^p_{n+1})^2 + (A^p_{n+2})^2 + (A^p_{n+3})^2,$$

from this the following results hold:

Proposition 6 *Let $n \in \mathbb{N}_0$. The seminorm of the Vietoris' elliptic biquaternion V^p_n is given by*

$$\mathcal{N}(V^p_n) = \|Q_n\|^2 + p\|Q_{n+1}\|^2 + 2I\,\langle Q_n, Q_{n+1}\rangle\,.$$

Proof Using Proposition 4 (iv), one obtains

$$\begin{aligned}\mathcal{N}(V^p_n) &= (A^p_n)^2 + (A^p_{n+1})^2 + (A^p_{n+2})^2 + (A^p_{n+3})^2\\ &= v_n^2 + pv_{n+1}^2 + 2I\,v_nv_{n+1} + v_{n+1}^2 + pv_{n+2}^2 + 2I\,v_{n+1}v_{n+2}\\ &+ v_{n+2}^2 + pv_{n+3}^2 + 2I\,v_{n+2}v_{n+3} + v_{n+3}^2 + pv_{n+4}^2 + 2I\,v_{n+3}v_{n+4}\\ &= \|Q_n\|^2 + p\|Q_{n+1}\|^2 + 2I\,\langle Q_n, Q_{n+1}\rangle\,.\end{aligned}$$

□

Furthermore, one can rewrite the seminorm of V^p_n in terms of its Vietoris' components as follows:

Proposition 7 *Let $m, n \in \mathbb{N}_0$, a_m and b_m defined in (8). The seminorm of the Vietoris' elliptic biquaternion V^p_n, is given by*

$$\mathcal{N}(V_n^p) = v^2_{n+\frac{1-(-1)^n}{2}} \left(\frac{1+(-1)^n}{2} \|a_{n+\frac{1-(-1)^n}{2}}\|^2 + \frac{1-(-1)^n}{2} \|b_{n+1}\|^2 \right)$$
$$+ p v^2_{n+1+\frac{1+(-1)^n}{2}} \left(\frac{1-(-1)^n}{2} \|a_{n+1+\frac{1+(-1)^n}{2}}\|^2 + \frac{1+(-1)^n}{2} \|b_{n+2}\| \right)$$
$$+ 2I v_{n+\frac{1-(-1)^n}{2}} v_{n+1+\frac{1+(-1)^n}{2}} \left(\frac{1+(-1)^n}{2} \left\langle a_{n+\frac{1-(-1)^n}{2}}, b_{n+2} \right\rangle \right.$$
$$\left. + \frac{1-(-1)^n}{2} \left\langle b_{n+1}, a_{n+1+\frac{1+(-1)^n}{2}} \right\rangle \right).$$

Proof Using Proposition 6 and (7), one gets

$$\mathcal{N}(V_n^p) = \|Q_n\|^2 + p\|Q_{n+1}\|^2 + 2I \langle Q_n, Q_{n+1} \rangle$$
$$= v^2_{n+\frac{1-(-1)^n}{2}} \left(\frac{1+(-1)^n}{2} \|a_{n+\frac{1-(-1)^n}{2}}\|^2 + \frac{1-(-1)^n}{2} \|b_{n+1}\|^2 \right)$$
$$+ p v^2_{n+1+\frac{1+(-1)^n}{2}} \left(\frac{1-(-1)^n}{2} \|a_{n+1+\frac{1+(-1)^n}{2}}\|^2 + \frac{1+(-1)^n}{2} \|b_{n+2}\| \right)$$
$$+ 2I v_{n+\frac{1-(-1)^n}{2}} v_{n+1+\frac{1+(-1)^n}{2}} \left(\frac{1+(-1)^n}{2} \left\langle a_{n+\frac{1-(-1)^n}{2}}, b_{n+2} \right\rangle \right.$$
$$\left. + \frac{1-(-1)^n}{2} \left\langle b_{n+1}, a_{n+1+\frac{1+(-1)^n}{2}} \right\rangle \right).$$

□

Some identities regarding the elements of the sequence $\{V_n^p\}_{n \geqslant 0}$ are presented below.

Proposition 8 *Let* $m, n \in \mathbb{N}_0$ *and* V_n^p *be an elliptic biquaternion with Vietoris' numbers. Then*

$$V_n^p = c_1(n) v_{n+\frac{1-(-1)^n}{2}} + I\, c_2(n) v_{n+1+\frac{1+(-1)^n}{2}},$$

where

$$c_k(n) = a_{n+(k-1)+\frac{1+(-1)^{n+k}}{2}} + \frac{1+(-1)^{n+k}}{2} \left(b_{n+k} - a_{n+(k-1)+\frac{1+(-1)^{n+k}}{2}} \right),$$

for a_m *and* b_m *defined in (8), and* $k = 1, 2$.

Proof Using (7), one gets

$$
\begin{aligned}
V_n^p &= Q_n + I Q_{n+1} \\
&= v_{n+\frac{1-(-1)^n}{2}} \left(a_{n+\frac{1-(-1)^n}{2}} + \frac{1-(-1)^n}{2} \left(b_{n+1} - a_{n+\frac{1-(-1)^n}{2}} \right) \right) \\
&+ I v_{n+1+\frac{1+(-1)^n}{2}} \left(a_{n+1+\frac{1+(-1)^n}{2}} + \frac{1+(-1)^n}{2} \left(b_{n+2} - a_{n+1+\frac{1+(-1)^n}{2}} \right) \right) \\
&= c_1(n) v_{n+\frac{1-(-1)^n}{2}} + I c_2(n) v_{n+1+\frac{1+(-1)^n}{2}},
\end{aligned}
$$

where

$$
\begin{aligned}
c_1(n) &= a_{n+\frac{1-(-1)^n}{2}} + \frac{1-(-1)^n}{2} \left(b_{n+1} - a_{n+\frac{1-(-1)^n}{2}} \right) \\
c_2(n) &= a_{n+1+\frac{1+(-1)^n}{2}} + \frac{1+(-1)^n}{2} \left(b_{n+2} - a_{n+1+\frac{1+(-1)^n}{2}} \right).
\end{aligned}
$$

□

Notice that, one can rewrite the elements os the sequence $\{V_n^p\}_{n \geqslant 0}$ in terms of the even index of the Vietoris' sequence $\{v_n\}_{n \geqslant 0}$.

Corollary 1 *Let $k \in \mathbb{N}_0$, $d(\cdot)$ defined in (5), and a_m and b_m defined in (8). Then*

$$
V_n^p = \begin{cases} (a_{2k} + I b_{2k+2} d(2k))\, v_{2k}, & n = 2k \\ (b_{2k+2} + I a_{2k+2})\, v_{2k+2}, & n = 2k+1 \end{cases}.
$$

Proof Using Proposition 8, one gets

$$
V_n^p = \begin{cases} c_1(2k) v_{2k} + I\, c_2(2k) v_{2k+2}, & n = 2k \\ c_1(2k+1) v_{2k+2} + I\, c_2(2k+1) v_{2k+2}, & n = 2k+1 \end{cases}. \tag{9}
$$

Substituting in (9) the expressions of c_1 and c_2 given by

$$
c_1(n) = \begin{cases} a_{2k}, & n = 2k \\ b_{2k+2}, & n = 2k+1 \end{cases}, \qquad c_2(n) = \begin{cases} b_{2k+2}, & n = 2k \\ a_{2k+2}, & n = 2k+1 \end{cases},
$$

one obtains

$$
V_n^p = \begin{cases} a_{2k} v_{2k} + I\, b_{2k+2} v_{2k+2}, & n = 2k \\ b_{2k+2} v_{2k+2} + I\, a_{2k+2} v_{2k+2}, & n = 2k+1 \end{cases}.
$$

Using the recurrence relation of the Vietoris' number sequence (4), one has

$$V_n^p = \begin{cases} (a_{2k} + I b_{2k+2} d(2k))\, v_{2k}, & n = 2k \\ \\ (b_{2k+2} + I a_{2k+2})\, v_{2k+2}, & n = 2k+1 \end{cases}.$$

□

Proposition 9 *Let V_n^p be an elliptic biquaternion with Vietoris' numbers. Then the following properties hold:*

(i) $V_n^p = \sum_{k=0}^{1} V_{n-2+k}^p P_k(n-2+k) + I \sum_{k=0}^{1} Q_{n+k-1} \left(P_k(n-1+k) - P_k(n-2+k)\right);$

(ii) $V_n^p = V_{n-1}^p P_1(n-1) + V_{n-2}^p P_0(n-2) + I\left(Q_{n+1} - Q_{n-1}\left(2P_1^2(n-1) + P_0(n-2)\right)\right),$

where $P_1(\cdot)$ and $P_0(\cdot)$ are defined in Theorem 1.

Proof Using Theorem 1, one gets (i)

$$\begin{aligned} V_n^p &= Q_n + I Q_{n+1} \\ &= Q_{n-1} P_1(n-1) + Q_{n-2} P_0(n-2) + I\left(Q_n P_1(n) + Q_{n-1} P_0(n-1)\right) \\ &= Q_{n-1} P_1(n-1) + I Q_n P_1(n-1) - I Q_n P_1(n-1) + I Q_n P_1(n) \\ &+ Q_{n-2} P_0(n-2) + I Q_{n-1} P_0(n-2) - I Q_{n-1} P_0(n-2) \\ &+ I Q_{n-1} P_0(n-1) \\ &= V_{n-1}^p P_1(n-1) + V_{n-2}^p P_0(n-2) \\ &+ I\left(Q_n (P_1(n) - P_1(n-1)) + Q_{n-1}\left(P_0(n-1) - P_0(n-2)\right)\right). \end{aligned}$$

To prove (ii) one uses (i) and Proposition 3

$$\begin{aligned} V_n^p &= V_{n-1}^p P_1(n-1) + V_{n-2}^p P_0(n-2) \\ &+ I\left(Q_n P_1(n) + Q_{n-1} P_0(n-1)\right) - I\left(Q_n P_1(n-1) + Q_{n-1} P_0(n-2)\right) \\ &= V_{n-1}^p P_1(n-1) + V_{n-2}^p P_0(n-2) + I Q_{n+1} \\ &- I\left(Q_n P_1(n-1) + Q_{n-1} P_0(n-2)\right) \\ &= V_{n-1}^p P_1(n-1) + V_{n-2}^p P_0(n-2) \\ &+ I\left(Q_{n+1} - Q_{n-1}\left(2P_1^2(n-1) + P_0(n-2)\right)\right). \end{aligned}$$

□

We dedicate the last part of this section to the generating function of elliptic biquaternion with Vietoris' numbers. One introduces the generating function for the sequence of $\{V_s^p\}_{s \geqslant 0}$, by using the generating function of the Vietoris' sequence $\{v_s\}_{n \geqslant 0}$ (see [2])

$$g(t) = \frac{\sqrt{1+t} - \sqrt{1-t}}{t\sqrt{1-t}} = \sum_{n=0}^{+\infty} v_n t^n, \qquad 0 < |t| < 1. \tag{10}$$

In [4] the generating function of the quaternionic sequences with Vietoris' components $\{Q_n\}_{n\geqslant 0}$ is given by

$$G(t) = \frac{1}{t^3}\left(g(t)(t^3 + t^2\mathbf{i} + t\mathbf{j} + \mathbf{k}) - H(t)\right), \qquad 0 < |t| < 1, \tag{11}$$

where $H(t) = \frac{1}{2}\left(t^2(2\mathbf{i} + \mathbf{j} + \mathbf{k}) + t(2\mathbf{j} + \mathbf{k}) + 2\mathbf{k}\right)$.

Using (10) and (11) one constructs the generating function for the sequence $\{V_n^P\}_{n\geqslant 0}$, as presented in the next result.

Theorem 2 *The generating function for* $\{V_n^P\}_{n\geqslant 0}$ *is*

$$P(t) = \left(1 + \frac{1}{t}I\right)G(t) - \frac{1}{t}IQ_0, \qquad 0 < |t| < 1,$$

where $G(t)$ *is given in (11) and* $Q_0 = 1 + \frac{1}{2}(\mathbf{i} + \mathbf{j}) + \frac{3}{8}\mathbf{k}$.

Proof Let

$$P(t) = \sum_{n=0}^{+\infty} V_n^P t^n$$

be the generating function for $\{V_s^P\}_{s\geqslant 0}$. Using the generating function $G(t)$ of $\{Q_n\}_{n\geqslant 0}$ presented in (11), one gets

$$\begin{aligned} P(t) &= \sum_{n=0}^{+\infty}(Q_n + IQ_{n+1})t^n \\ &= \sum_{n=0}^{+\infty} Q_n t^n + I\sum_{n=0}^{+\infty} Q_{n+1}t^n \\ &= G(t) + I\tfrac{1}{t}\sum_{n=0}^{+\infty} Q_{n+1}t^{n+1} \\ &= G(t) + I\tfrac{1}{t}\left(\sum_{n=0}^{+\infty} Q_n t^n - Q_0\right) \\ &= \left(1 + \tfrac{1}{t}I\right)G(t) - \tfrac{1}{t}IQ_0. \end{aligned}$$

□

4 Matrix Representations of Elliptic Biquaternionic Sequence with Vietoris' Numbers

We start this section presenting some techniques, using matrices, that generate the quaternionic sequence where the components are Vietoris' numbers and construct the elliptic biquaternion sequence. Furthermore, using determinant of a special kind of matrices one obtains a quaternionic component or an element of the elliptic biquaternion with Vietoris' number sequence.

We start by considering the following 3-by-3 matrix, whose entries are quaternionic numbers

$$T(n) = \begin{bmatrix} 0 & 1 & 0 \\ 0 & 0 & 1 \\ 0 & P_0(n+1) & P_1(n+2) \end{bmatrix}, \qquad n \in \mathbb{N}_0, \tag{12}$$

where $P_0(\cdot)$ and $P_1(\cdot)$ are defined in Theorem 1.

Using this kind of matrix we have a triple of quaternionic Vietoris' numbers sequences.

Theorem 3 *Let $n \in \mathbb{N}_0$, $T(n)$ be the matrix defined in (12) and $X_n = \begin{bmatrix} Q_n \\ Q_{n+1} \\ Q_{n+2} \end{bmatrix}$. Then*

$$T(n)X_n = X_{n+1}.$$

Proof Consider $n = 0$, and so

$$T(0)X_0 = \begin{bmatrix} 0 & 1 & 0 \\ 0 & 0 & 1 \\ 0 & P_0(1) & P_1(2) \end{bmatrix} \begin{bmatrix} Q_0 \\ Q_1 \\ Q_2 \end{bmatrix} = \begin{bmatrix} Q_1 \\ Q_2 \\ Q_1 P_0(1) + Q_2 P_1(2) \end{bmatrix} = \begin{bmatrix} Q_1 \\ Q_2 \\ Q_3 \end{bmatrix} = X_1,$$

the third entry in the vector is due to Theorem 1.

Considering $n \geqslant 1$ and using Theorem 1, one obtains

$$\begin{aligned} T(n)X_n &= \begin{bmatrix} 0 & 1 & 0 \\ 0 & 0 & 1 \\ 0 & P_0(n+1) & P_1(n+2) \end{bmatrix} \begin{bmatrix} Q_n \\ Q_{n+1} \\ Q_{n+2} \end{bmatrix} \\ &= \begin{bmatrix} Q_{n+1} \\ Q_{n+2} \\ Q_{n+1} P_0(n+1) + Q_{n+2} P_1(n+2) \end{bmatrix} = \begin{bmatrix} Q_{n+1} \\ Q_{n+2} \\ Q_{n+3} \end{bmatrix} = X_{n+1}. \end{aligned}$$

□

Applying successively Theorem 3, we obtain the triple X_n in terms of the triple X_0.

Corollary 2 *Let $n \in \mathbb{N}_0$, $T(n)$ be defined in (12) and $X_n = \begin{bmatrix} Q_n \\ Q_{n+1} \\ Q_{n+2} \end{bmatrix}$. Then*

$$X_n = \prod_{k=0}^{n-1} T(k) X_0.$$

In the next result one gets a 2×1 column vector of elliptic biquaternionic sequence with Vietoris' numbers in terms of a triple column vector of quaternionic Vietoris' numbers.

Proposition 10 *Consider $n \in \mathbb{N}_0$, the triple column vector of the quaternionic Vietoris' numbers sequence $X_n = \begin{bmatrix} Q_n \\ Q_{n+1} \\ Q_{n+2} \end{bmatrix}$ and the matrix*

$$W(n) = \begin{bmatrix} 0 & 1 & I \\ 0 & IP_0(n+1) & 1 + IP_1(n+2) \end{bmatrix}, \tag{13}$$

where $P_0(\cdot)$ and $P_1(\cdot)$ are defined in Theorem 1. Then

$$W(n)X_n = \begin{bmatrix} V_{n+1}^p \\ V_{n+2}^p \end{bmatrix}.$$

Proof Consider the matrix $A = \begin{bmatrix} 1 & I & 0 \\ 0 & 1 & I \end{bmatrix}$ and the matrix $T(n)$ defined in (12), then $W(n)$ is the product of A and $T(n)$:

$$W(n) = \begin{bmatrix} 1 & I & 0 \\ 0 & 1 & I \end{bmatrix} \begin{bmatrix} 0 & 1 & 0 \\ 0 & 0 & 1 \\ 0 & P_0(n+1) & P_1(n+2) \end{bmatrix}.$$

Using Theorem 3, one has

$$\begin{aligned} W(n)X_n &= AT(n)X_n = AX_{n+1} \\ &= \begin{bmatrix} 1 & I & 0 \\ 0 & 1 & I \end{bmatrix} \begin{bmatrix} Q_{n+1} \\ Q_{n+2} \\ Q_{n+3} \end{bmatrix} = \begin{bmatrix} Q_{n+1} + IQ_{n+2} \\ Q_{n+2} + IQ_{n+3} \end{bmatrix} = \begin{bmatrix} V_{n+1}^p \\ V_{n+2}^p \end{bmatrix}. \end{aligned}$$

□

In the following result one uses determinant of a special kind of tridiagonal matrices to obtain a quaternion component of the elliptic biquaternion with Vietoris' number sequence.

Using a similar technique as given in [3] and [9]. Let us consider a square matrix with biquaternion number, $X = \begin{bmatrix} x_{ij} \end{bmatrix}_{n \times n}$. Due to the noncommutativity we calculate the determinant of a matrix with biquaternion number using the Laplace expansion starting always with all entries of the last column, i.e., the determinant is given by

$$\det X = \sum_{i=1}^{n} c_{in} x_{in},$$

with

$$c_{in} = (-1)^{i+n} \det Y_{in},$$

where $\det Y_{in}$ is the i, n minor of X.

One defines the following tridiagonal matrix

$$Z_{n+1} = \begin{bmatrix} 2IQ_1 & 2IQ_2 & 0 & 0 & 0 & \dots & 0 \\ -1 & 0 & P_0(1) & 0 & 0 & \dots & 0 \\ 0 & -1 & P_1(2) & P_0(2) & 0 & \dots & 0 \\ 0 & 0 & -1 & P_1(3) & P_0(3) & \dots & 0 \\ \vdots & \ddots & \ddots & \ddots & \ddots & \ddots & \\ 0 & 0 & 0 & 0 & -1 & P_1(n-1) & P_0(n-1) \\ 0 & 0 & 0 & 0 & 0 & -1 & P_1(n) \end{bmatrix}_{(n+1)\times(n+1)} \quad (14)$$

where $P_0(\cdot)$ and $P_1(\cdot)$ are defined in Theorem 1.

Using the matriz Z_{n+1}, one gets the following result.

Proposition 11 *Consider the matrix Z_{n+1} defined in (14). Then*

$$\det Z_{n+1} = V_n^p - (V_n^p)^*.$$

Proof Using Proposition 4 (iii), for the case $n = 0$, one gets

$$\det Z_0 = 2IQ_1 = V_0^p - (V_0^p)^*,$$

and also for $n = 1$ one obtains

$$\det Z_2 = \begin{bmatrix} 2IQ_1 & 2IQ_2 \\ -1 & 0 \end{bmatrix} = 2IQ_2 = V_1^p - (V_1^p)^*.$$

Furthermore, for the case of $n = 2$, and using Theorem 1, one gets

$$Z_3 = \begin{bmatrix} 2IQ_1 & 2IQ_2 & 0 \\ -1 & 0 & P_0(1) \\ 0 & -1 & P_1(2) \end{bmatrix}$$

and then

$$\begin{aligned} \det Z_3 &= \det Z_2 P_1(2) + 2IQ_1 P_0(1) = 2IQ_2 P_1(2) + 2IQ_1 P_0(1) \\ &= 2IQ_3 = V_2^p - (V_2^p)^*. \end{aligned}$$

Supposing that $\det Z_n = V_{n-1}^p - (V_{n-1}^p)^*$ and using Theorem 1, one gets

$$\begin{aligned} \det Z_{n+1} &= \det Z_n P_1(n) + \det Z_{n-1} P_0(n-1) \\ &= \left(V_{n-1}^p - (V_{n-1}^p)^*\right) P_1(n) + \left(V_{n-2}^p - (V_{n-2}^p)^*\right) P_0(n-1) \\ &= 2IQ_n P_1(n) + 2IQ_{n-1} P_0(n-1) \\ &= 2IQ_{n+1} \\ &= V_n^p - (V_n^p)^*. \end{aligned}$$

□

In what follows, considering two different sequence of 2-by-2 matrices, $\{A_n\}_{n\geqslant 0}$ and $\{B_n\}_{n\geqslant 1}$, it is also possible to obtain the elliptic biquaternion Viectoris' numbers sequence, in the following way

$$\{\det A_n\}_{n\geqslant 0} = \left\{V_{2n}^p\right\}_{n\geqslant 0}, \qquad \{\det B_n\}_{n\geqslant 1} = \left\{V_{2n-1}^p\right\}_{n\geqslant 1}.$$

To do so, let us define the sequence of 2-by-2 matrices, $\{A_n\}_{n\geqslant 0}$ by

$$A_0 = \begin{bmatrix} 1 & 0 \\ 0 & \left(a_0 + I\frac{b_2}{2}\right) \end{bmatrix}, \quad A_1 = \begin{bmatrix} \frac{1}{2} & 0 \\ -\frac{1}{4} & v_0\left(a_2 + I\frac{3b_4}{4}\right) \end{bmatrix},$$

$$A_{k+1} = \begin{bmatrix} \frac{1}{2} & v_{2k-2}\left(a_{2k+2} + Ib_{2k+4}d(2k+2)\right) \\ \frac{1-2k}{4(k+1)} & v_{2k}\left(a_{2k+2} + Ib_{2k+4}d(2k+2)\right) \end{bmatrix}, \quad k \in \mathbb{N}. \tag{15}$$

Proposition 12 *Consider the sequence of matrices* $\{A_k\}_{k\geqslant 0}$ *defined in (15). Then*

$$\det A_k = V_{2k}^p.$$

Proof Let $k \in \mathbb{N}_0$, using Corollary 1 one gets

$$\det A_0 = a_0 + I\frac{b_2}{2} = V_0^p,$$

and

$$\det A_1 = v_0\left(a_2 + I\frac{3b_4}{4}\right)\frac{1}{2} = v_2\left(a_2 + I\frac{3b_4}{4}\right) = V_2^p.$$

For the case $k \geqslant 2$, using the recurence relation of the Vietoris' numbers sequence, presented in [5]

$$v_{2k+2} = \frac{1}{2}v_{2k} + \frac{2k-1}{4(k+1)}v_{2k-2}, \quad k \in \mathbb{N}, \tag{16}$$

one has

$$\begin{aligned} \det A_{k+1} &= \left(a_{2k+2} + Ib_{2k+4}d(2k+2)\right)\left(\frac{1}{2}v_{2k} - \frac{1-2k}{4(k+1)}v_{2k-2}\right) \\ &= \left(a_{2k+2} + Ib_{2k+4}d(2k+2)\right)v_{2k+2} \\ &= V_{2k+2}^p. \end{aligned}$$

□

In order to obtain the odd index of the elliptic biquaternion with Vietoris' numbers sequence, $\left\{V_{2k-1}^p\right\}_{n\geqslant 1}$, we introduce the sequence of 2-by-2 matrices, $\{B_k\}_{k\geqslant 1}$, given by

$$B_1 = \begin{bmatrix} \frac{1}{2} & 0 \\ -\frac{1}{4} & v_0\left(b_2 + Ia_2\right) \end{bmatrix},$$

$$B_{k+1} = \begin{bmatrix} \frac{1}{2} & v_{2k-2}\left(b_{2k+2} + I a_{2k+2}\right) \\ \frac{1-2k}{4(k+1)} & v_{2k}\left(b_{2k+2} + I a_{2k+2}\right) \end{bmatrix}, \; k \in \mathbb{N}. \tag{17}$$

Proposition 13 *Consider the sequence of matrices* $\{B_{k+1}\}_{k \geqslant 0}$ *defined in (17). Then*

$$\det B_{k+1} = V_{2k+1}^{p}.$$

Proof Let $k \in \mathbb{N}_0$, using Corollary 1 one gets

$$\det B_1 = v_0\left(b_2 + I a_2\right)\frac{1}{2} = v_2\left(b_2 + I a_2\right) = V_1^{p}.$$

For the case $k \geqslant 1$, using the recurence relation of the Vietoris numbers sequence (16), (see [5]), one has

$$\begin{aligned} \det B_{k+1} &= \left(b_{2k+2} + I a_{2k+2}\right)\left(\frac{1}{2}v_{2k} - \frac{1-2k}{4(k+1)}v_{2k-2}\right) \\ &= \left(b_{2k+2} + I a_{2k+2}\right) v_{2k+2} \\ &= V_{2k+1}^{p}. \end{aligned}$$

□

In the next two results, by computing a determinant of certain type of matrices, one has any element of the elliptic biquaternion with Vietoris' numbers sequence $\{V_n^p\}_{n \geqslant 0}$.

Consider the sequence of 2-by-2 matrices, $\{R_n\}_{n \geqslant 0}$, given by

$$R_0 = \begin{bmatrix} Q_1 & 1 \\ -Q_0 & I \end{bmatrix}, \quad R_n = \begin{bmatrix} Q_n & I P_0(n-1) \\ -Q_{n-1} & I P_1(n) + 1 \end{bmatrix}, \; n \in \mathbb{N}. \tag{18}$$

The determinant of these matrices lead to the elliptic biquaternionic sequence with Vietoris' numbers, as given in the following result.

Proposition 14 *Consider the sequence of matrices* $\{R_n\}_{n \geqslant 0}$ *defined in (18). Then*

$$\det R_n = V_n^{p}.$$

Proof Let $n \in \mathbb{N}_0$, then

$$\det R_0 = Q_1 I + Q_0 = V_0^{p}.$$

For the case $n \geqslant 1$, using Theorem 1, one has

$$\det R_n = Q_n + I(Q_n P_1(n) + Q_{n-1}P_0(n-1)) = Q_n + IQ_{n+1} = V_n^p.$$

□

By Proposition 8, we know that

$$V_n^p = c_1(n)v_{n+\left(\frac{1-(-1)^n}{2}\right)} + Ic_2(n)v_{n+1+\left(\frac{1+(-1)^n}{2}\right)}$$

and so, in the easy way, we can consider the following square 2-by-2 matrix

$$M_n = \begin{bmatrix} c_1(n) & v_{n+1+\left(\frac{1+(-1)^n}{2}\right)} \\ -Ic_2(n) & v_{n+\left(\frac{1-(-1)^n}{2}\right)} \end{bmatrix},$$

whose entries of first column are $c_k(n),\ k = 1, 2$ defined in Proposition 8 and those of the second column are Vietoris' numbers. Using this matrix we have immediately the next result that gives us another way to calculate any elliptic biquaternion with Vietoris' numbers by computing a determinant.

Proposition 15 *For $n \in \mathbb{N}_0$ and V_n^p be the elliptic biquaternion with Vietoris' numbers of order n, then*

$$\det M_n = V_n^p.$$

Acknowledgments The authors thank the support given by the Portuguese Funds through FCT (Fundação para a Ciência e a Tecnologia), within the Projects UIDB/00013/2020 and UIDP/00013/2020. Also the second author thanks the support given by the Project UID/CED/00194/2020.

References

1. Askey, R., Steinig, J.: Some positive trigonometric sums. Trans. AMS **187**(1), 295–307 (1974). https://doi.org/10.2307/1997054
2. Cação, I., Falcão, M.I., Malonek, R.: On generalized Vietoris' number sequences. Discrete Appl. Math. **269**, 77–85 (2019). https://doi.org/10.1016/j.dam.2018.10.017
3. Cahill, N.D., D'Errico, J.R., Narayan, D.A., Narayan, J.Y.: Fibonacci determinants. College Math. J. **33**(3), 221–225 (2002). https://doi.org/10.1080/07468342.2002.11921945
4. Catarino, P., De Almeida, R.: On a quaternionic sequence with Vietoris' numbers. Filomat **35**(4), 1065–1086 (2021). https://doi.org/10.2298/FIL2104065C
5. Catarino, P., De Almeida, R.: A note on Vietoris' number sequence. Mediterr. J. Math. **19**(1), 41 (2022). https://doi.org/10.1007/s00009-021-01952-w
6. Conway, J.H., Smith, D.A.: On Quaternions and Octonions: Their Geometry, Arithmetic, and Symmetry. A. K. Peters, Natick (2003)

7. Gürlebeck, K., Habetha, K., Sprößig, W.: Holomorphic Functions in the Plane and n-Dimensional Space. Birkhäuser, Basel (2008)
8. Harkin, A.A., Harkin, J.B.: Geometry of generalized complex numbers. Math. Mag. **77**(2), 118–129 (2004). https://doi.org/10.1080/0025570X.2004.11953236
9. Kiliç, E., Taşci, D., Haukkanen, P.: On the generalized Lucas sequences by Hessenberg matrices. Ars Combin. **95**, 383–395 (2010)
10. Özen, E.K., Tosun, M.: Elliptic biquaternions algebra. In: 6th International Eurasian Conference on Mathematical Sciences and Applications (IECMSA-2017), AIP Conf. Proc. 1926, 020032-1020032-6 (2018). https://doi.org/10.1063/1.5020481
11. Özen, E.K., Tosun, M.: A note on elliptic biquaternions. In: 6th International Eurasian Conference on Mathematical Sciences and Applications (IECMSA-2017), AIP Conference Proceedings 1926, 020032 (2018). https://doi.org/10.1063/1.5020482
12. Ruscheweyh, St., Salinas, L.: Stable functions and Vietoris' theorem. J. Math. Anal. Appl. **291**, 596–604 (2004). https://doi.org/10.1016/j.jmaa.2003.11.035
13. Sloane, N.J.A., Plouffe, S.: The Encyclopedia of Integer Sequences. Academic Press, San Diego (1995)
14. Vietoris, L.: Über das Vorzeichen gewisser trigonometrischer Summen, Sitzungsber. Österr. Akad. Wiss **167**, 125–135 (1958)
15. Yaglom, I.M.: Complex Numbers in Geometry. Academic Press, New York (1968)

Teaching Mathematics in STEM Education

Ion Mierluş-Mazilu and Fatih Yilmaz

Abstract Teaching mathematics in STEM education is a vital component of developing students' analytical thinking, problem-solving skills, and logical reasoning abilities. To effectively teach mathematics in STEM, educators should employ strategies such as active learning, integration of technology, problem-based learning, differentiated instruction, cross-disciplinary connections, visualization, real-world applications, collaborative learning, formative assessment, and cultivating a growth mindset.

The integration of technology plays a significant role in enhancing mathematics instruction. Educators can utilize interactive software, graphing calculators, spreadsheets, online resources, virtual manipulatives, coding and programming, online collaborative platforms, data visualization tools, virtual reality, augmented reality, online tutorials, and videos to engage students and deepen their understanding of mathematical concepts.

By incorporating technology, teachers can provide dynamic visualizations, simulations, and interactive activities that enable students to explore multiple representations of mathematical ideas. Technology tools also support data analysis, graphing, problem-solving, and collaborative learning experiences. Furthermore, technology facilitates real-world connections and applications of mathematics, helping students recognize its relevance in various STEM fields.

Overall, the integration of technology in mathematics instruction empowers students to actively participate in their learning, visualize abstract concepts, solve complex problems, and develop essential mathematical and STEM skills. By incorporating these strategies, educators can create a dynamic and engaging mathematics learning environment that prepares students for future STEM pursuits.

I. Mierluş-Mazilu (✉)
Technical University of Civil Engineering Bucharest, Bucharest, Romania
e-mail: ion.mierlusmazilu@utcb.ro

F. Yilmaz
Ankara Haci Bayram Veli University, Ankara, Turkey
e-mail: fatih.yilmaz@hbv.edu.tr

V. Gayoso Martínez et al. (eds.), *Mathematical Methods for Engineering Applications*, Springer Proceedings in Mathematics & Statistics 439,
https://doi.org/10.1007/978-3-031-49218-1_11

1 Preliminaries

Teaching mathematics in STEM (Science, Technology, Engineering, and Mathematics) education is crucial for developing students' analytical thinking, problem-solving skills, and logical reasoning abilities [1].

For effective mathematics instruction in STEM education we have to take in account:

- Active Learning: Encourage active participation and engagement of students in the learning process. Use hands-on activities, group discussions, and real-life applications to make mathematics concepts more relatable and tangible.
- Integration of Technology: Incorporate technology tools and resources to enhance mathematics instruction. Interactive software, educational apps, graphing calculators, and online simulations can help students visualize abstract concepts, explore mathematical relationships, and solve complex problems.
- Problem-Based Learning: Emphasize problem-solving skills by presenting students with real-world challenges that require the application of mathematical concepts. Encourage them to analyze problems, develop strategies, and use critical thinking to arrive at solutions. This approach fosters a deeper understanding of mathematics and its relevance in practical scenarios.
- Differentiated Instruction: Recognize and accommodate the diverse learning needs of students. Provide multiple entry points and offer varying levels of challenge to cater to individual abilities. Differentiation can involve modifying the content, process, or product of the learning experience to suit students' interests, strengths, and readiness levels.
- Cross-disciplinary Connections: Highlight the connections between mathematics and other STEM fields. Explore how mathematics is used in science experiments, engineering designs, data analysis in technology, and statistical modeling in various domains. This interdisciplinary approach helps students understand the application of mathematics in different contexts and promotes a holistic understanding of STEM subjects.
- Visualization and Manipulatives: Utilize visual aids, manipulatives, and concrete materials to support students' comprehension of abstract mathematical concepts. Visual representations, such as graphs, diagrams, and models, can help students visualize relationships and patterns, facilitating a deeper understanding of mathematical principles.
- Real-World Applications: Emphasize the practical applications of mathematics in everyday life and STEM careers. Connect mathematical concepts to real-world scenarios to demonstrate their relevance and usefulness. By understanding how mathematics is used in fields like engineering, computer science, and physics, students are motivated to learn and appreciate its significance.
- Collaborative Learning: Promote collaborative learning environments where students can work together, share ideas, and solve problems collectively. Collaborative activities encourage communication, teamwork, and the exchange of

different perspectives, fostering a deeper understanding of mathematics through peer interaction.
- Formative Assessment: Regularly assess students' understanding throughout the learning process using formative assessment strategies. Provide timely feedback to students, highlighting areas of improvement and reinforcing correct mathematical reasoning. This iterative process supports student growth and helps identify misconceptions that need to be addressed.
- Cultivate a Growth Mindset: Foster a growth mindset in students by emphasizing that mathematical skills can be developed through effort, perseverance, and a positive attitude. Encourage students to embrace challenges, learn from mistakes, and view mathematics as a field that is open to exploration and continuous improvement.

By implementing these strategies, teachers can create an engaging and effective mathematics learning environment in STEM education, equipping students with essential mathematical skills and preparing them for future STEM pursuits.

2 Mathematics Instruction in STEM Education

2.1 Active Learning

The "Active Learning" focuses on the instructional approach that encourages students to actively participate in the learning process and take responsibility for their own learning. Active learning strategies engage students in activities that go beyond passive listening and promote higher-order thinking, problem-solving skills, and deeper understanding [2].

Implementing active learning strategies we need:

- Student Engagement: Active learning aims to actively engage students in the learning process. It involves creating a dynamic and interactive learning environment that captures students' interest and motivates them to participate actively. This can be achieved through various techniques, such as group discussions, hands-on activities, debates, case studies, simulations, and role-playing.
- Collaborative Learning: Active learning often involves collaborative learning experiences. Students work together in groups or pairs to solve problems, complete projects, or engage in discussions. Collaborative learning fosters teamwork, communication, and critical thinking skills. It encourages students to learn from each other, share ideas, and consider different perspectives.
- Problem-Based Learning: Active learning frequently incorporates problem-based learning (PBL) approaches. In PBL, students are presented with real-world problems or challenges that require them to apply their knowledge and skills to find solutions. This approach promotes critical thinking, analytical reasoning, and the application of concepts in authentic contexts.

- Experiential Learning: Active learning encourages experiential learning opportunities where students can directly interact with the subject matter. This can involve hands-on experiments, field trips, simulations, or role-playing activities. Experiential learning helps students develop a deeper understanding of concepts by experiencing them firsthand and making connections to real-life situations.
- Inquiry-Based Learning: Active learning often incorporates inquiry-based approaches, where students are encouraged to ask questions, investigate, and explore concepts independently. This fosters curiosity, critical thinking, and problem-solving skills. Students take an active role in constructing their knowledge through exploration and discovery.
- Technology Integration: Active learning can be enhanced through the integration of technology tools and resources. Technology provides opportunities for interactive learning, multimedia presentations, virtual simulations, and online collaboration. It allows students to engage with content in dynamic and interactive ways, promoting deeper understanding and engagement.
- Formative Assessment: Active learning is supported by ongoing formative assessment practices. Regular feedback and assessment help students monitor their progress, identify areas of improvement, and adjust their learning strategies accordingly. Formative assessment can take the form of quizzes, class discussions, group presentations, or self-reflection activities.
- Reflective Practices: Active learning encourages students to reflect on their learning experiences. Reflection helps students consolidate their understanding, connect new knowledge with prior knowledge, and identify areas where further learning is needed. Encouraging students to reflect on their learning process promotes metacognition and self-directed learning.

Implementing active learning strategies in the classroom creates an engaging and student-centered learning environment. It promotes higher-level thinking skills, problem-solving abilities, and a deeper understanding of the subject matter. Teachers can create a dynamic and engaging mathematics learning environment. Active learning promotes student engagement, critical thinking, and a deeper understanding of mathematical concepts, preparing students for success in STEM fields where analytical skills, problem-solving abilities, and logical reasoning are essential. Active learning prepares students to become lifelong learners who can apply their knowledge and skills in real-world contexts.

2.2 *Integration of Technology*

The integration of technology in mathematics instruction can greatly enhance students' learning experiences and provide them with new tools and resources to explore mathematical concepts [3].

When incorporating technology in teaching mathematics in STEM education we have to consider:

- Interactive Software and Applications: Utilize interactive software and applications specifically designed for mathematics education. These tools often provide dynamic visualizations, simulations, and interactive problem-solving activities. Examples include Geogebra, Desmos, Wolfram Alpha, and Mathway. These platforms allow students to experiment with mathematical concepts, visualize relationships, and explore multiple representations of mathematical ideas.
- Graphing Calculators and Spreadsheets: Graphing calculators and spreadsheet software are powerful tools for data analysis, graphing functions, and performing complex calculations. They enable students to visualize and manipulate mathematical data, experiment with mathematical models, and analyze real-world scenarios. Popular graphing calculators include TI-Nspire, Casio ClassPad, and HP Prime.
- Online Resources and Virtual Manipulatives: Take advantage of the vast array of online resources that provide virtual manipulatives and interactive simulations. Websites like National Library of Virtual Manipulatives (NLVM), Illuminations, and Mathigon offer digital tools that allow students to explore mathematical concepts using virtual manipulatives, which can be particularly useful for visualizing geometric shapes, fractions, and algebraic expressions.
- Coding and Programming: Integrate coding and programming activities to strengthen mathematical reasoning and computational thinking skills. Programming languages like Python, Scratch, and MATLAB provide opportunities for students to solve mathematical problems, create algorithms, and analyze data. Students can develop their own mathematical models and simulations, fostering a deeper understanding of mathematical concepts and their applications.
- Online Collaborative Platforms: Utilize online collaborative platforms to facilitate virtual teamwork and problem-solving. Tools like Google Docs, Padlet, and whiteboard collaboration platforms enable students to work together on mathematical tasks, share ideas, and engage in discussions. Collaborative platforms encourage active participation, peer learning, and the exchange of mathematical strategies.
- Data Visualization Tools: Introduce data visualization tools to help students analyze and interpret large datasets. Tools like Tableau, Excel, and Google Sheets allow students to create graphs, charts, and visual representations of data. By visualizing data, students can identify patterns, make predictions, and draw conclusions, enhancing their understanding of statistical concepts and data analysis.
- Virtual Reality (VR) and Augmented Reality (AR): Explore the use of virtual reality and augmented reality technologies to immerse students in mathematical experiences. VR and AR tools can provide three-dimensional representations of mathematical concepts, allowing students to interact with mathematical objects and explore geometric shapes in a more tangible and immersive manner.
- Online Tutorials and Videos: Leverage online tutorials and instructional videos to provide additional support and resources for students. Platforms like Khan Academy, Math Antics, and YouTube channels dedicated to mathematics offer step-by-step explanations, practice exercises, and demonstrations of mathemat-

ical concepts. These resources can be used to reinforce classroom learning and provide additional examples and practice opportunities.

When integrating technology, it's essential to ensure that the selected tools align with the learning objectives, cater to diverse student needs, and are accessible for all learners. Additionally, teachers should provide guidance and structure to help students effectively use technology as a tool for mathematical exploration and understanding.

2.3 Problem-Based Learning

The "Problem-Based Learning" focuses on an instructional approach that centers around presenting students with real-world problems or challenges that require the application of mathematical concepts and problem-solving skills. Problem-Based Learning (PBL) engages students in active learning, critical thinking, and collaborative problem-solving.

The are aspects to consider when implementing PBL in mathematics instruction in STEM education:

- Real-World Problems: Present students with authentic and meaningful problems that reflect real-life situations or scenarios. These problems should require the use of mathematical concepts and skills to find solutions or make informed decisions. By connecting mathematics to real-world contexts, students understand the practical relevance of the subject.
- Student-Centered Approach: Shift the focus from the teacher as the sole provider of knowledge to the students as active participants and problem-solvers. Encourage students to take ownership of their learning by investigating problems, asking questions, and seeking solutions independently or collaboratively. The teacher serves as a facilitator, guiding and supporting students throughout the process.
- Inquiry and Exploration: Encourage students to inquire, explore, and investigate the problem before attempting to find a solution. This involves identifying what is known, what is unknown, and what information or strategies may be needed to solve the problem. Students can brainstorm ideas, analyze patterns, and consider different approaches to tackle the problem.
- Collaboration and Communication: Foster collaborative learning environments where students work together in groups or pairs to solve problems. Collaboration encourages the sharing of ideas, perspectives, and strategies, as well as the development of communication and teamwork skills. Students can engage in discussions, explain their reasoning, and collectively construct solutions.
- Application of Mathematical Concepts: PBL provides opportunities for students to apply mathematical concepts and skills in authentic contexts. Students analyze the problem, identify relevant mathematical concepts, and decide how to apply them to find solutions. This application of knowledge helps students develop

a deeper understanding of mathematical concepts and their real-world implications.

- Critical Thinking and Problem-Solving Skills: PBL enhances students' critical thinking abilities and problem-solving skills. Students engage in higher-order thinking processes, such as analyzing, evaluating, and synthesizing information, to arrive at solutions. They learn to approach problems systematically, develop logical reasoning, and make informed decisions based on mathematical evidence.
- Reflection and Evaluation: Encourage students to reflect on their problem-solving processes and evaluate their solutions. Reflection helps students assess their approaches, identify strengths and weaknesses, and consider alternative strategies. It promotes metacognition and self-assessment, enabling students to become more effective problem solvers.
- Technology Integration: Integrate technology tools and resources to support PBL. Technology can facilitate data analysis, provide interactive simulations or visualizations, and allow students to explore mathematical concepts in dynamic ways. It can also assist in data collection, organization, and presentation, enhancing the problem-solving process.
- Authentic Assessment: Assess students' understanding and problem-solving abilities through authentic assessment methods. Instead of relying solely on traditional tests or quizzes, consider performance-based assessments, portfolios, presentations, or projects that require students to apply their mathematical knowledge and skills to real-world problems.

Problem-Based Learning in mathematics instruction promotes active engagement, critical thinking, collaboration, and the application of mathematical concepts in authentic contexts. By addressing real-world problems, students develop a deeper understanding of mathematics and its relevance in STEM fields. They become confident problem solvers, equipped with the skills necessary to tackle complex challenges in their academic and professional lives.

2.4 *Differentiated Instruction*

The "Differentiated Instruction" chapter focuses on the instructional approach that recognizes and addresses the diverse learning needs, interests, and abilities of students. Differentiated instruction aims to create a learning environment where all students can thrive and achieve their full potential.

Some key aspects to consider when implementing differentiated instruction:

- Student-Centered Approach: Differentiated instruction places students at the center of the learning process. It acknowledges that students have unique learning styles, strengths, and challenges. The teacher takes into account students' readiness levels, interests, and learning profiles when designing instruction to ensure it meets their individual needs.

- Flexible Grouping: Differentiated instruction often involves flexible grouping strategies. Students may be grouped based on their readiness levels, interests, or learning styles. Grouping options can include whole-class instruction, small groups, pairs, or individual work. Flexible grouping allows teachers to tailor instruction to specific student needs and provides opportunities for peer collaboration and support.
- Varied Instructional Strategies: Differentiated instruction employs a range of instructional strategies to address diverse learning needs. Teachers may use a combination of direct instruction, hands-on activities, multimedia resources, cooperative learning, technology integration, and inquiry-based approaches. Multiple instructional strategies provide students with various ways to engage with and understand the content.
- Multiple Learning Pathways: Differentiated instruction recognizes that students may have different entry points or pathways to learning. It offers multiple options for students to explore and demonstrate their understanding of concepts. This can include providing different levels of challenge, alternative assignments, choice boards, or independent research projects. Offering various learning pathways ensures that students can access and engage with the content at their own pace and level.
- Ongoing Assessment: Differentiated instruction relies on continuous assessment to inform instructional decisions. Teachers use pre-assessment, formative assessment, and ongoing observation to gather data on student progress and understanding. Assessment data helps identify students' strengths, areas for growth, and instructional next steps. It guides the adaptation of instruction to meet individual student needs.
- Individualized Support: Differentiated instruction provides individualized support to students who require additional assistance or extension. This can involve one-on-one conferences, targeted interventions, enrichment activities, or personalized learning plans. Individualized support ensures that students receive the necessary scaffolding or challenge to succeed academically.
- Culturally Responsive Teaching: Differentiated instruction acknowledges and values students' diverse cultural backgrounds and experiences. Teachers strive to incorporate culturally relevant materials, examples, and perspectives into instruction. By recognizing and affirming students' cultural identities, differentiated instruction fosters an inclusive and supportive learning environment.
- Collaboration and Professional Learning Communities: Differentiated instruction benefits from collaboration among teachers and the establishment of professional learning communities. Teachers can share strategies, resources, and insights to enhance their differentiation practices. Collaboration enables continuous growth and improvement in meeting the diverse needs of students.

Differentiated instruction recognizes the uniqueness of each student and aims to create an inclusive learning environment where all students can succeed. By tailoring instruction to meet individual needs, interests, and abilities, differentiated instruction promotes engagement, motivation, and academic achievement. It supports the development of students as independent learners who can take ownership of their learning and reach their full potential.

2.5 Cross-Disciplinary Connections

The "Cross-disciplinary Connections" has to focuses on the importance of integrating multiple disciplines and making connections across different subject areas in education. It highlights the benefits of cross-disciplinary learning, such as promoting holistic understanding, fostering creativity, and addressing real-world problems.

Exploring cross-disciplinary connections we will consider:

- Interdisciplinary Approach: Cross-disciplinary connections involve an interdisciplinary approach to education. It goes beyond teaching subjects in isolation and instead seeks to integrate knowledge, concepts, and skills from different disciplines. By breaking down the traditional silos of subject areas, students can see the interconnectedness of knowledge and develop a broader perspective on various topics.
- Integration of Content: Cross-disciplinary connections require the intentional integration of content from different subject areas. Teachers can identify and highlight the relationships between concepts, themes, or issues across disciplines. For example, exploring the connections between mathematics and science, language arts and social studies, or art and technology. This integration helps students see how knowledge from different disciplines can be applied in diverse contexts.
- Real-World Relevance: Cross-disciplinary connections emphasize the application of knowledge in real-world contexts. By exploring connections between different subject areas, students can understand how academic concepts are relevant and interconnected in practical situations. This approach enhances their ability to solve complex problems and make informed decisions in various domains.
- Critical Thinking and Problem-Solving: Cross-disciplinary connections promote critical thinking and problem-solving skills. When students encounter complex problems that require knowledge and skills from multiple disciplines, they develop the ability to analyze, synthesize, and evaluate information from different sources. They learn to think critically, consider multiple perspectives, and apply diverse strategies to find innovative solutions.
- Creativity and Innovation: Cross-disciplinary connections stimulate creativity and innovation. When students explore connections between different subject areas, they can draw upon diverse perspectives, approaches, and techniques. This interdisciplinary perspective encourages creative thinking and the generation of new ideas. It fosters innovation by encouraging students to transfer and apply knowledge from one discipline to another.

- Collaboration and Communication: Cross-disciplinary connections often involve collaborative learning experiences. Students work together in interdisciplinary teams, combining their knowledge and skills from different subject areas. Collaboration promotes communication, teamwork, and the sharing of diverse perspectives. Students learn to effectively communicate their ideas and listen to others, fostering a more comprehensive understanding of complex issues.
- Project-Based Learning: Cross-disciplinary connections align well with project-based learning (PBL) approaches. PBL engages students in authentic, interdisciplinary projects that require them to apply knowledge and skills from different subject areas to solve real-world problems. These projects encourage research, analysis, creativity, and collaboration, promoting deeper learning and the development of essential twenty-first-century skills.
- Teacher Collaboration and Professional Development: Implementing cross-disciplinary connections often involves collaboration among teachers from different subject areas. Educators can share their expertise, explore common themes, and develop interdisciplinary units or projects. Ongoing professional development opportunities can support teachers in expanding their knowledge, pedagogical strategies, and assessment practices related to cross-disciplinary learning.

By embracing cross-disciplinary connections, educators can provide students with a holistic, interconnected view of the world. Students develop critical thinking, problem-solving, and creative skills while understanding the relevance of academic content in real-world contexts. This approach prepares students for future challenges by fostering a deep understanding of how knowledge is connected and encouraging interdisciplinary thinking.

2.6 Visualization and Manipulatives

The "Visualization and Manipulatives" focuses on the use of visual aids and manipulatives as effective instructional tools in teaching and learning. It explores how visualization techniques and hands-on manipulatives can enhance understanding, engagement, and retention of concepts across various subjects.

Visualization and manipulatives aspects:

- Visualization as a Learning Tool: Visualization involves creating mental images or representations of concepts, processes, or relationships. It helps students make sense of abstract or complex ideas by providing a visual framework for understanding. Visualization techniques include diagrams, charts, graphs, concept maps, models, and virtual simulations. By engaging the visual senses, students can grasp and retain information more effectively.
- Enhancing Conceptual Understanding: Visualization aids in developing conceptual understanding by providing a visual representation of abstract ideas. For example, in mathematics, visualizing geometric shapes, functions, or data

patterns can deepen comprehension. In science, visualizing molecular structures, biological processes, or physical phenomena can aid in understanding complex concepts. Visualization helps students connect theoretical knowledge to concrete images, making concepts more accessible and memorable.

- Supporting Problem Solving: Visualization techniques play a crucial role in problem-solving. They help students analyze problems, identify relevant information, and develop strategies for solving them. Visual representations, such as diagrams or charts, assist in organizing data, recognizing patterns, and exploring different problem-solving approaches. Visualization encourages students to think critically and make connections between different elements of a problem.
- Hands-on Manipulatives: Manipulatives are physical objects that students can touch, move, and manipulate to explore mathematical or scientific concepts. They provide a concrete and tactile experience, facilitating active engagement and understanding. Examples of manipulatives include blocks, counters, geometric shapes, fraction strips, measurement tools, and science models. Manipulatives allow students to experiment, discover relationships, and develop spatial reasoning skills.
- Multisensory Learning: Visualization and manipulatives engage multiple senses, leading to multisensory learning experiences. Combining visual, tactile, and sometimes auditory elements enhances learning and memory retention. When students can see, touch, and interact with objects or visual representations, they create more connections in the brain, reinforcing understanding and recall.
- Differentiation and Accessibility: Visualization and manipulatives can support differentiated instruction by accommodating diverse learning styles and abilities. Visual aids benefit visual learners, while manipulatives cater to kinesthetic or hands-on learners. They provide alternative pathways for understanding and allow students to approach concepts from different angles. Moreover, visualization and manipulatives can be particularly valuable for students with learning difficulties or special needs, providing concrete support and facilitating access to abstract concepts.
- Technology and Virtual Visualization: Technology offers a wide range of digital tools and software for visualizing and manipulating concepts. Virtual simulations, interactive models, and digital manipulatives provide opportunities for dynamic and immersive learning experiences. These tools enable students to explore complex phenomena, conduct virtual experiments, and visualize concepts that are otherwise challenging to observe or manipulate.
- Effective Instructional Strategies: To maximize the benefits of visualization and manipulatives, effective instructional strategies are crucial. Teachers should provide clear explanations, guide students in interpreting visual representations, and encourage reflective thinking. They can facilitate class discussions, encourage students to create their own visualizations, and connect visuals to real-world applications. Regular formative assessment helps gauge students' understanding and adjust instruction accordingly.

Incorporating visualization techniques and manipulatives into instruction, educators can enhance students' understanding, engagement, and retention of concepts. Visualization aids in conceptual development, problem-solving, and the connection of abstract ideas to real-world contexts. Hands-on manipulatives provide concrete experiences and multisensory learning opportunities. By using these powerful tools, teachers can create dynamic and interactive learning environments that support diverse learners and foster deep understanding.

2.7 *Real-World Applications*

The "Real-World Applications" focuses on the importance of connecting learning to real-world contexts and applications. It explores how real-world applications enhance students' understanding, motivation, and transfer of knowledge.

Here are some key aspects to consider when exploring real-world applications:

- Contextualizing Learning: Real-world applications provide meaningful contexts for learning. By connecting concepts and skills to real-world situations, students can see the relevance and practicality of what they are learning. This approach helps students understand how academic content applies to their lives and the world around them.
- Authentic Problem-Solving: Real-world applications engage students in authentic problem-solving experiences. They are presented with challenges or tasks that simulate real-life situations, requiring them to apply knowledge, critical thinking, and problem-solving skills to find solutions. Authentic problems foster higher-order thinking and develop students' ability to analyze, evaluate, and synthesize information.
- Interdisciplinary Connections: Real-world applications often require the integration of knowledge and skills from multiple disciplines. They encourage students to make connections across subject areas and apply interdisciplinary approaches to problem-solving. This helps students develop a holistic understanding of complex issues and promotes the transfer of knowledge between different contexts.
- Promoting Transfer of Knowledge: Real-world applications facilitate the transfer of knowledge from the classroom to real-life situations. When students can apply what they have learned in meaningful contexts, they develop a deeper understanding of the subject matter and its practical applications. Transferability of knowledge enables students to adapt and apply what they have learned to new challenges and contexts.
- Engaging and Motivating: Real-world applications increase student engagement and motivation. When students see the relevance of their learning to real-life situations, they become more invested in the learning process. Real-world applications tap into students' curiosity and allow them to explore topics that are meaningful and interesting to them.

- Collaboration and Communication: Real-world applications often involve collaboration and communication skills. Students work together to tackle complex problems, share ideas, and communicate their solutions effectively. Collaboration promotes teamwork, cooperation, and the development of interpersonal skills necessary for success in real-world settings.
- Authentic Assessment: Real-world applications offer opportunities for authentic assessment. Instead of traditional tests or quizzes, students can be assessed based on their ability to apply their knowledge and skills to real-world problems or tasks. Authentic assessments may include projects, presentations, portfolios, or performance-based assessments, providing a more comprehensive view of students' abilities.
- Career and Life Skills Development: Real-world applications help students develop essential career and life skills. They encourage critical thinking, problem-solving, communication, collaboration, adaptability, and creativity–skills valued in various professional and personal contexts. Real-world applications prepare students for future careers by bridging the gap between academic knowledge and practical skills.

Educators can incorporate project-based learning, case studies, simulations, community connections, and authentic tasks into their instruction. By connecting learning to real-world contexts, educators can inspire students, deepen their understanding, and prepare them for success beyond the classroom.

2.8 Collaborative Learning

The "Collaborative Learning" focuses on the benefits and strategies of collaborative learning, where students work together in groups to achieve shared learning goals. Collaborative learning emphasizes active participation, cooperation, and the development of interpersonal skills.

Collaborative learning involve:

- Active Engagement: Collaborative learning promotes active engagement among students. Instead of passively receiving information, students actively participate in discussions, problem-solving, and knowledge construction. This active engagement enhances their understanding of concepts and improves retention.
- Social Interaction: Collaborative learning emphasizes social interaction among students. Working in groups allows students to share ideas, ask questions, and explain their thinking. This social interaction enhances learning by providing opportunities for peer-to-peer teaching, feedback, and clarification.
- Cooperation and Teamwork: Collaborative learning fosters cooperation and teamwork skills. Students learn how to work together, negotiate, and respect diverse perspectives. They develop skills such as active listening, effective communication, and conflict resolution, which are essential for successful collaboration.

- Diverse Perspectives: Collaborative learning brings together students with diverse backgrounds, experiences, and perspectives. This diversity enriches discussions and problem-solving processes. It allows students to gain new insights, challenge assumptions, and consider multiple viewpoints, fostering critical thinking and broadening their understanding of complex issues.
- Cognitive Development: Collaborative learning promotes cognitive development. Through interactions with peers, students engage in higher-order thinking processes such as analyzing, evaluating, and synthesizing information. They learn to question, reason, and construct knowledge collectively, leading to deeper understanding and improved problem-solving skills.
- Shared Responsibility: Collaborative learning encourages shared responsibility for learning. Each group member has a role and contributes to the group's success. This shared responsibility cultivates a sense of accountability and motivates students to actively participate and contribute their best effort.
- Positive Interdependence: Collaborative learning establishes a sense of positive interdependence among group members. Students realize that their success is linked to the success of the group. This interdependence fosters a supportive and cooperative learning environment, where students actively assist and encourage each other's learning.
- Scaffolded Learning: Collaborative learning provides opportunities for scaffolding. Students can support and scaffold each other's learning by providing explanations, sharing strategies, and offering constructive feedback. This peer support enhances learning outcomes and helps students develop metacognitive skills.
- Reflection and Metacognition: Collaborative learning encourages reflection and metacognition. Through group discussions and reflection activities, students can reflect on their own thinking processes, evaluate their learning strategies, and monitor their progress. This metacognitive awareness enhances their ability to regulate their learning and develop self-directed learning skills.
- Technology Integration: Collaborative learning can be facilitated through technology integration. Online collaboration tools, virtual platforms, and digital communication tools enable students to collaborate regardless of geographical limitations. Technology also provides opportunities for asynchronous collaboration, where students can work together at their own pace and contribute at their convenience.

Effective implementation of collaborative learning involves establishing clear group roles, providing guidelines for effective collaboration, and monitoring group dynamics. Teachers play a facilitative role, guiding and supporting students' collaborative efforts. Assessment in collaborative learning can include both individual and group assessments, focusing on both process and product.

By implementing collaborative learning strategies, educators can foster an interactive and inclusive learning environment that promotes active engagement, critical thinking, and interpersonal skills. Collaborative learning prepares students for future collaboration and cultivates essential skills for success in their personal and professional lives.

2.9 *Formative Assessment*

The "Formative Assessment" focuses on the importance and strategies of formative assessment in the teaching and learning process. Formative assessment involves gathering information about students' understanding and progress during instruction to inform instructional decisions.

Formative assessment is an ongoing, interactive process that provides feedback to both teachers and students. Its purpose is to monitor student learning, identify areas of strength and weakness, and guide instructional decisions. Unlike summative assessment, which evaluates learning at the end of a unit or course, formative assessment is used during instruction to enhance learning. Formative assessment can be viewed under the following:

- Assessment for Learning: Formative assessment is often referred to as assessment for learning. It is designed to support and improve learning by providing timely feedback and guiding instructional adjustments. It focuses on the learning process rather than just measuring outcomes, helping students become active participants in their own learning.
- Informing Instruction: Formative assessment informs instructional decisions by providing insights into student understanding and needs. Teachers can use the information gathered to adjust their teaching strategies, address misconceptions, provide targeted interventions, and differentiate instruction to meet individual student needs. This responsive approach enhances student learning and improves instructional effectiveness.
- Varied Assessment Strategies: Formative assessment employs a variety of strategies beyond traditional tests and quizzes. These strategies include observations, questioning techniques, class discussions, student self-assessments, exit tickets, concept maps, journals, and portfolios. Using diverse assessment methods allows for a comprehensive understanding of student progress and provides multiple opportunities for feedback.
- Timely and Specific Feedback: Formative assessment involves providing timely and specific feedback to students. Feedback should be constructive, actionable, and focused on helping students understand their strengths and areas for improvement. It should also provide guidance on how students can enhance their learning and offer strategies for further development.
- Student Involvement and Self-assessment: Formative assessment encourages student involvement in the assessment process. Students can engage in self-assessment by reflecting on their own learning, setting goals, and monitoring their progress. Peer-assessment activities, where students provide feedback to

their peers, also promote active learning and encourage students to develop critical thinking and evaluative skills.

- Data-Informed Instruction: Formative assessment provides valuable data that can guide instructional decision-making. By analyzing assessment results, teachers can identify patterns, trends, and areas where students may be struggling. This data helps teachers make informed choices about instructional strategies, interventions, and adjustments to ensure optimal learning outcomes.
- Cultivating a Positive Learning Culture: Formative assessment contributes to a positive learning culture by emphasizing growth, progress, and improvement. It shifts the focus from grades and scores to the learning process itself. By providing constructive feedback and recognizing effort and improvement, formative assessment encourages a growth mindset and fosters a supportive and motivating learning environment.
- Assessment as Learning: Formative assessment allows students to engage in assessment as learning. Through self-assessment and reflection activities, students develop metacognitive skills, become more aware of their learning strategies, and actively monitor their progress. This self-regulation promotes independent learning and empowers students to take ownership of their learning journey.
- Continuous Improvement: Formative assessment is an ongoing process that supports continuous improvement. It is not a one-time event but rather an integral part of the instructional cycle. Teachers can continuously assess, adjust, and refine their instructional practices based on the feedback and insights gained from formative assessment.

Implementing effective formative assessment practices involves clear learning targets, aligned assessment strategies, constructive feedback, and a supportive classroom culture. By integrating formative assessment into their teaching practices, educators can support student learning, guide instruction, and promote continuous improvement.

2.10 Cultivate a Growth Mindset

The "Cultivate a Growth Mindset" based on the concept of a growth mindset and its importance in promoting learning, resilience, and success. It explores strategies to foster a growth mindset among students and create a positive learning environment.

A growth mindset is the belief that abilities and intelligence can be developed through effort, learning, and persistence. It is in contrast to a fixed mindset, which assumes that abilities are fixed and cannot be changed. Embracing a growth mindset allows students to view challenges, setbacks, and mistakes as opportunities for growth and learning.

Here are some aspects to consider when exploring the chapter on cultivating a growth mindset:

- Importance of Effort and Persistence: Cultivating a growth mindset emphasizes the importance of effort and persistence in the learning process. Students are encouraged to see effort as a pathway to improvement and to persevere through challenges. This mindset shift helps students overcome obstacles, develop resilience, and maintain motivation even when faced with difficulties.
- Embracing Challenges: Developing a growth mindset involves embracing challenges as opportunities for learning and growth. Students are encouraged to step out of their comfort zones, take risks, and tackle challenging tasks. By reframing challenges as learning opportunities, students develop problem-solving skills, become more resilient, and build confidence in their ability to overcome obstacles.
- Emphasizing Learning and Progress: A growth mindset focuses on the process of learning rather than solely on outcomes. Students are encouraged to value the learning experience, celebrate progress, and view mistakes as opportunities to learn and improve. This shift in perspective fosters a love for learning, intrinsic motivation, and a willingness to take on new challenges.
- Developing a Positive Mindset Language: Language plays a crucial role in cultivating a growth mindset. Teachers can use positive and empowering language that promotes effort, resilience, and the belief in students' ability to improve. Encouraging phrases like "I believe in your potential", "Keep trying", or "Mistakes help us learn" reinforce the growth mindset and create a supportive classroom culture.
- Providing Constructive Feedback: Feedback is an essential tool for cultivating a growth mindset. Teachers can provide specific, constructive feedback that focuses on effort, progress, and areas for improvement. Feedback should be seen as an opportunity for growth rather than as a judgment of abilities. Encouraging students to see feedback as valuable information that helps them grow enhances their ability to self-assess and self-adjust their learning strategies.
- Modeling a Growth Mindset: Teachers play a significant role in modeling a growth mindset for their students. By sharing personal stories of challenges, failures, and successes, teachers demonstrate that growth and improvement are possible. Teachers can also openly discuss and address their own mistakes, demonstrating the importance of a growth mindset in their professional development.
- Creating a Safe and Supportive Learning Environment: Cultivating a growth mindset requires creating a safe and supportive learning environment. Students should feel comfortable taking risks, asking questions, and making mistakes without fear of judgment or failure. Encouraging collaboration, providing opportunities for reflection, and fostering a sense of belonging contribute to a positive learning environment that supports the growth mindset.
- Nurturing Student Self-efficacy: Developing a growth mindset goes hand in hand with nurturing student self-efficacy—the belief in one's ability to succeed. Teachers can foster self-efficacy by setting attainable goals, providing support and resources, and acknowledging and celebrating student achievements. Building

students' confidence in their abilities enhances their motivation and willingness to embrace challenges.

- Promoting a Love for Learning: Ultimately, cultivating a growth mindset aims to promote a love for learning. When students believe that they can develop their abilities through effort and learning, they become more engaged, curious, and motivated learners. By fostering a growth mindset, educators inspire students to become lifelong learners and pursue their goals with enthusiasm.

Incorporating growth mindset practices involves intentional language use, fostering a positive classroom culture, providing meaningful feedback, and modeling growth mindset behaviors. By nurturing a growth mindset, educators empower students to embrace challenges, persist through difficulties, and unlock their full potential.

By fostering a growth mindset in students, educators can empower them to embrace challenges, persevere through setbacks, and develop a love for learning. Cultivating a growth mindset helps students develop resilience, motivation, and a belief in their ability to achieve their goals.

3 Integration of Technology in the Teaching of Mathematics: A Case Study

This case study explores the integration of technology to enhance the teaching and learning of mathematics at the university level. The study focuses on a course titled "Advanced Calculus" offered to undergraduate mathematics majors. The goal was to investigate the impact of integrating technology tools on students' mathematical understanding and engagement. The case study involved the use of a graphing calculator application, collaborative online platforms, and dynamic geometry software.

At the university level, the integration of technology in mathematics education holds significant importance and offers numerous benefits to both students and educators. Here is an overview of the significance of technology integration in mathematics education at the university level:

- Enhancing Learning Opportunities: Technology provides students with access to a wide range of resources, tools, and interactive platforms that can enhance their learning experience. Through the integration of technology, students can engage in virtual simulations, explore complex mathematical concepts visually, and access online databases and interactive tutorials that provide additional practice and reinforcement.
- Promoting Active Learning: Technology integration encourages active learning in mathematics education. Students can actively participate in problem-solving activities, manipulate and visualize mathematical objects, and engage in collaborative projects facilitated by technology tools. This active engagement fosters

a deeper understanding of mathematical concepts and promotes critical thinking and problem-solving skills.

- Facilitating Personalized Learning: Technology allows for personalized learning experiences tailored to individual student needs. Adaptive learning systems can provide personalized feedback, adaptive assessments, and customized learning pathways based on students' strengths and weaknesses. This individualized approach helps students progress at their own pace, address their specific learning needs, and build a strong foundation in mathematics.
- Supporting Visualization and Conceptual Understanding: Technology tools enable students to visualize abstract mathematical concepts and processes. Graphing software, dynamic geometry tools, and 3D modeling applications allow students to explore mathematical relationships, patterns, and transformations visually. This visual representation enhances conceptual understanding and promotes the development of spatial reasoning skills.
- Facilitating Real-World Applications: Technology integration in mathematics education provides opportunities for students to apply mathematical concepts and skills to real-world scenarios. Through computer-based simulations, data analysis software, and mathematical modeling tools, students can explore practical applications of mathematics in various fields, such as engineering, finance, and computer science. This integration helps students see the relevance and utility of mathematics in real-life contexts.
- Encouraging Collaboration and Communication: Technology tools enable students to collaborate with peers, share ideas, and communicate their mathematical thinking effectively. Online platforms, discussion forums, and collaborative software allow students to work together on problem-solving tasks, engage in mathematical discourse, and receive feedback from their peers and instructors. This collaborative environment nurtures communication skills, fosters teamwork, and prepares students for real-world collaborative work settings.
- Promoting Data Analysis and Statistical Skills: Technology provides powerful tools for data analysis and statistical exploration. Students can utilize statistical software and spreadsheets to analyze and interpret data, conduct experiments, and make informed decisions based on mathematical models and statistical reasoning. The integration of technology enables students to develop data literacy and prepares them for careers that require analytical and quantitative skills.
- Preparing for Technological Advancements: In today's digital era, technological literacy is essential for success in various fields. Integrating technology in mathematics education equips students with the necessary skills to navigate and leverage technology effectively. By utilizing technology tools and applications, students develop digital literacy, problem-solving abilities, and adaptability to technological advancements in their future careers.

The integration of technology in mathematics education at the university level provides students with engaging and interactive learning experiences, supports deeper conceptual understanding, promotes real-world applications, and prepares them for the demands of the digital age. It empowers students to become active learners, critical thinkers, and proficient users of technology in their pursuit of mathematical knowledge and in their future professional endeavors.

The objective of the case study is to examine the impact of technology integration in the teaching of mathematics at the university level on various aspects, including:

- Learning Outcomes: Investigate the effect of technology integration on students' learning outcomes in terms of improved understanding, problem-solving skills, and achievement of learning objectives.
- Student Engagement: Explore how the integration of technology enhances student engagement and active participation in mathematical learning activities, leading to increased motivation and interest in the subject.
- Student Perceptions: Examine students' perceptions and attitudes towards mathematics as influenced by the use of technology, including their confidence in mathematical abilities, enjoyment of the subject, and perceived relevance of mathematics in their lives and future careers.
- Pedagogical Approaches: Assess the effectiveness of different pedagogical approaches facilitated by technology, such as flipped classrooms, online discussions, interactive simulations, or virtual laboratories, in promoting student learning and conceptual understanding.
- Collaborative Learning: Investigate the impact of technology-supported collaborative learning environments on students' collaboration skills, communication abilities, and teamwork in the context of mathematics education.
- Accessibility and Inclusivity: Explore how technology integration addresses the needs of diverse learners, including students with disabilities or different learning styles, by providing alternative representations, adaptive features, and personalized learning opportunities.

The case study aims to provide insights into the benefits, challenges, and effective strategies for integrating technology in mathematics education at the university level. By examining these various aspects, the objective is to contribute to the existing knowledge base and inform educators, administrators, and policymakers about the potential impact and implications of technology integration in mathematics teaching and learning.

The case study employed a mixed-methods approach, combining quantitative data collection and qualitative analysis. The participants included 120 undergraduate mathematics students enrolled in the Advanced Calculus course. Data collection methods included pre and post assessments, classroom observations, student surveys, and interviews with both students and the instructor.

Pre- and post-tests are administered to measure students' learning outcomes and assess the impact of technology integration on their mathematical understanding. The pre-test is conducted at the beginning of the course to establish a baseline of students' knowledge and skills. The post-test is administered at the end of the course to evaluate the effectiveness of technology integration in improving learning outcomes.

Surveys are used to gather quantitative data on students' perceptions, attitudes, and experiences related to technology integration in mathematics education. The surveys may include Likert-scale questions, multiple-choice questions, and open-ended questions. They assess students' engagement, confidence, motivation, and perceived usefulness of technology in their learning experience.

In this case study, the university setting involves a well-established mathematics department within a renowned research university. The mathematics department is known for its rigorous academic programs and commitment to excellence in teaching and research. It comprises a team of dedicated faculty members who are experts in their respective areas of mathematics.

The specific course under investigation is "Advanced Calculus" offered at the undergraduate level. This course is typically taken by mathematics majors or students pursuing a related field of study. It is designed to provide a deeper understanding of calculus, including topics such as limits, continuity, derivatives, integrals, and applications of calculus.

The target audience for this case study includes undergraduate students majoring in mathematics, as well as those pursuing related disciplines with a strong mathematical component, such as physics, engineering, or computer science. These students have a solid foundation in calculus and are motivated to deepen their understanding of advanced mathematical concepts and applications.

The university setting provides a rich academic environment with access to resources such as libraries, research facilities, and state-of-the-art technology infrastructure. The mathematics department offers a supportive learning environment with opportunities for students to engage in mathematical research, participate in seminars and workshops, and collaborate with faculty on various projects.

The students in this university setting are typically intellectually curious, self-motivated, and have a strong passion for mathematics. They are eager to explore advanced topics and develop their problem-solving and analytical skills. The department fosters a collaborative and interactive learning environment, encouraging students to engage in discussions, share ideas, and work on challenging mathematical problems.

Overall, the university setting provides a conducive environment for investigating the integration of technology in mathematics education, as it offers a diverse group of motivated students, experienced faculty, and ample resources to support innovative teaching and learning approaches.

The instructor integrated technology tools into various aspects of the course. The graphing calculator application was used for visualizing and analyzing functions, exploring graphical transformations, and solving equations. Collaborative online platforms were employed to facilitate group work, discussion forums, and online problem-solving activities. Dynamic geometry software allowed students to manipulate geometric figures and explore geometric concepts in a dynamic and interactive way.

The integration of technology tools in the Advanced Calculus course had several positive outcomes. Quantitative analysis of pre- and post-assessment scores showed a significant improvement in students' mathematical understanding and problem-solving skills (Fig. 1).

Fig. 1 Quantitative analysis of pre- and post-assessment scores

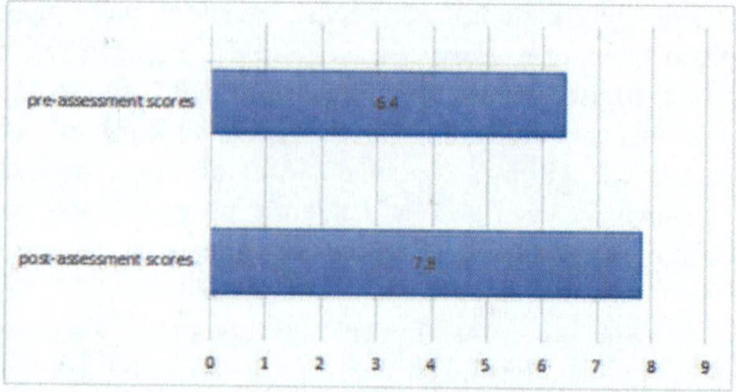

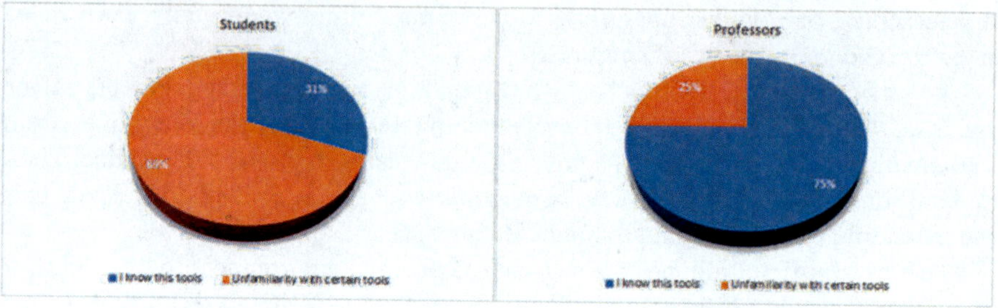

Fig. 2 Challenges in integrating technology in the mathematics classroom

Students reported increased engagement and motivation due to the interactive nature of the technology tools. Qualitative analysis of student interviews and classroom observations revealed that technology enhanced students' visualization abilities, deepened their conceptual understanding, and facilitated collaborative learning experiences.

The case study also identified some challenges in integrating technology in the mathematics classroom (Fig. 2).

These included technical issues, students' unfamiliarity with certain tools, and the need for additional time to incorporate technology effectively. Recommendations for future implementation include providing adequate training for both students and instructors, gradually introducing technology tools throughout the course, and addressing technical challenges proactively.

The case study demonstrated that the integration of technology in the teaching of mathematics at the university level can positively impact students' mathematical understanding, engagement, and collaboration. The findings support the use of graphing calculator applications, collaborative online platforms, and dynamic geometry software as effective tools for enhancing the learning experience in mathematics education.

This case study provides evidence of the benefits of integrating technology in university mathematics education and offers insights into successful implementation strategies. It highlights the importance of leveraging technology tools to promote active learning, visualization, and collaborative problem-solving in the study of mathematics.

4 Conclusions

Integration of technology enhances mathematics education: The integration of technology tools and platforms in mathematics education enhances the teaching and learning experience. Technology provides visualizations, simulations, and interactive activities that engage students, deepen their understanding of mathematical concepts, and promote problem-solving skills.

Active learning strategies improve learning outcomes: Implementing active learning strategies, such as problem-based learning, collaborative learning, and differentiated instruction, leads to improved learning outcomes in mathematics. These approaches encourage student engagement, critical thinking, and application of mathematical concepts in real-world contexts.

Cross-disciplinary connections enhance relevance: Making connections between mathematics and other STEM disciplines fosters interdisciplinary thinking and highlights the relevance of mathematics in various fields. Students gain a deeper understanding of the practical applications of mathematics and develop the skills needed to solve complex problems in STEM-related areas.

Visualization and manipulatives aid comprehension: Visualization tools and manipulatives play a vital role in mathematics education. They help students grasp abstract concepts, visualize geometric relationships, and develop spatial reasoning skills. The use of technology-based visualizations and manipulatives enhances student comprehension and facilitates deeper understanding.

Real-world applications promote relevance and engagement: Incorporating real-world applications and authentic problems in mathematics instruction increases student engagement and demonstrates the practical relevance of mathematical concepts. By solving real-world problems, students understand the importance of mathematics in addressing real-life challenges and develop transferable problem-solving skills.

Collaborative learning fosters critical thinking and communication: Collaborative learning environments encourage students to work together, share ideas, and engage in meaningful mathematical discourse. Collaboration enhances critical thinking skills, promotes different perspectives, and improves communication and argumentation skills, which are crucial for success in STEM fields.

Formative assessment guides instruction: Formative assessment techniques, such as quizzes, concept maps, and classroom observations, provide valuable feedback to both students and instructors. Assessing student understanding throughout the learning process allows for timely interventions, adjustments in instruction, and personalized support, leading to improved learning outcomes.

Cultivating a growth mindset promotes perseverance: Nurturing a growth mindset, emphasizing effort, resilience, and the belief that mathematical abilities can be developed, positively impacts students' attitudes and motivation towards mathematics. Students are more likely to persevere through challenges, embrace mistakes as learning opportunities, and develop a positive attitude towards mathematics and STEM education.

In conclusion, teaching mathematics in STEM education is enriched by integrating technology, implementing active learning strategies, promoting cross-disciplinary connections, utilizing visualization tools and manipulatives, incorporating real-world applications, fostering collaborative learning environments, employing formative assessment, and cultivating a growth mindset. These approaches enhance student engagement, understanding, and success in mathematics and contribute to their overall STEM education.

References

1. Burrill, G., Allison, J., Breaux, G., McClain, K.: Teaching Mathematics in the 21st Century: Methods and Activities for Grades 6-12. ASCD, Alexandria (2012)
2. English, L.D.: Promoting STEM learning with real-world problem solving. Math. Teacher **109**(6), 436–441 (2016)
3. Kilpatrick, J., Swafford, J., Findell, B.: Adding It Up: Helping Children Learn Mathematics. National Academies Press, Washington (2001)

Application of Discrete Wavelet Transform and Tree-Based Ensemble Machine Learning for Modeling of Particulate Matter Concentrations

Maya Stoimenova-Minova, Snezhana Gocheva-Ilieva, and Atanas Ivanov

Abstract The study of air pollution is an extremely important and urgent problem to be solved on a global and local scale. In this field, huge arrays of measurement data are accumulating, for the analysis of which various approaches based on mathematical, statistical, and machine learning (ML) methods are developed. In this paper, we investigate the application of different discrete wavelet transforms (DWT) families, coupled with state-of-the-art ML algorithms to predict concentrations of particulate matter PM10. Average daily data for this pollutant and several meteorological time series for a period of 630 days were used. A hybrid type models with wavelet decomposition of the initial time series and the application of predictive ensembles (Arcing, Arc-x4) were obtained. All models are cross-validated. The models are applied for short-term pollution forecasts.

1 Introduction

Among the most concerning problems of our time is environmental pollution, due to both natural and anthropogenic sources such as industrial enterprises, agriculture, households, transport and others. One of the most harmful air pollutants is particulate matter with a diameter of less than 10 microns (PM_{10}). Increased concentrations of PM_{10} are the cause of numerous lung and heart diseases, complications of the chronically ill, young children, the elderly and pregnant women [1–4]. To counteract the negative impact of air pollutants, including PM_{10}, many researchers study and analyze the current state and causes of harmful emissions. With vast arrays of observational data available, these studies also include methods for modelling and predicting future pollution. For this purpose, various statistical and machine learning (ML) methods and techniques are developed and applied.

M. Stoimenova-Minova (✉) · S. Gocheva-Ilieva · A. Ivanov
Faculty of Mathematics and Informatics, University of Plovdiv Paisii Hilendarski, Plovdiv, Bulgaria
e-mail: mstoimenova@uni-plovdiv.bg; snow@uni-plovdiv.bg; aivanov@uni-plovdiv.bg

V. Gayoso Martínez et al. (eds.), *Mathematical Methods for Engineering Applications*, Springer Proceedings in Mathematics & Statistics 439,
https://doi.org/10.1007/978-3-031-49218-1_12

In this study, we will look at the possibilities of forecasting time series using Wavelet transform (WT) coupled with ML methods. In the scientific literature, WT is one the preferred mathematical methods for studying and modelling environmental data [5–7]. For example, in [8] the authors show that hybrid models based on WT and Autoregressive integrated moving average (ARIMA) methods are more reliable in predicting future atmospheric air pollution with particulate matter compared to the classical Box-Jenkins ARIMA method [9]. Another study [5] applied WT to filter out noise effects that reduce the accuracy of simulation processes. A comparison is made between a hybrid wavelet—artificial neural network (ANN) and a classical ANN and it is found that the hybrid model outperforms the classical ANN. In another study, the WT is combined with a neural network (NN) ensemble, combining multilayer perceptron, radial basis function, Elman network, support vector machine, and a linear ARX used to predict the daily average concentration of PM_{10} [10]. Wavelet analysis was applied to identify the strongest harmonics in the variability of air pollutants' concentrations in the metropolitan area of São Paulo, Brazil in [11]. Authors of [12] construct four different topologies of deep artificial NN, for the prediction of $PM_{2.5}$ concentrations by using five levels of discrete wavelet transform (DWT). Also a hybrid NN composed of LSTM and 1D-CNN is built to improve the forecasts. Other results based on WT in environmental science can be found at [13–15].

In this paper, we propose a hybrid approach based on discrete wavelet transform (DWT) coupled with the boosting decision tree ensembles for modelling and forecasting PM_{10} time series, depending on a few meteorological factors and lagged variables. A combination of WT and the adaptive resampling and combining (Arcing) [16]), in its Arc-x4 variant, is considered. To the best of the authors' knowledge, this hybrid W-Arcing approach is applied for the first time in this study. We will perform the wavelet analysis with several different wavelet mother functions and statistical selection of threshold levels to improve the forecasting efficiency.

Statistical analyses are carried out using the software packages Wolfram Mathematica [17], Salford Predictive Modeler (SPM) [18] and IBM SPSS [19].

2 Study Area and Data

The studied area is the city of Plovdiv, the second largest city in Bulgaria. It is located in southern Bulgaria. The altitude is relatively low and light winds are observed. The climate in the city is transitional-continental. Fogs are characteristic during the cold months of the year.

The current analysis regarding the state of atmospheric air in the city of Plovdiv was conducted on the basis of real measurements from a mobile measuring station, which is located at one of the exits of the city. The station has analyzers for measuring basic and specific atmospheric pollutants such as particulate matter, nitrogen dioxide, carbon dioxide and sulphur dioxide. The data consists of average

Table 1 Basic descriptive statistics of the data used in this study

Variable	Mean	Median	Std. Dev.	Skewness	Kurtosis	Minimum	Maximum
PM10, μg/m^3	15.67	11.12	16.1	3.78	19.53	1.27	143.49
PM10w, μg/m^3	15.62	11.11	15.91	3.81	20.19	1.27	143.49
TEMP, °C	17.07	17.64	9.09	−0.09	−1.04	−3.03	35.29
HUM, %	57.81	52.61	18.83	0.72	−0.44	22.16	100
PRESS, mbar	996.25	995.78	6.49	0.18	0.23	975.46	1018.1
LUM, Luxe	3851.51	4128.61	1742.31	−0.42	−1.03	176	7652
trB, μg/m^3	15.63	11.23	15.71	3.92	21.52	−7.75	145.75
trD, μg/m^3	15.6	11.92	15.13	4.12	23.86	−12.17	142.69
trH, μg/m^3	15.51	11.92	15.19	4.24	24.29	−5.37	143.49

Note: The number of cases for PM10, TEMP, HUM, PRESS and LUM is N = 630; for other variables it is reduced to 620

day-night measurements for the period from 11th April 2021 to 31st December 2022, or N = 630 days. In the original database, 12.5% of measurements were missing, which were replaced by the method of linear interpolation. For the present study, the data from the pollutant PM_{10} (μg/m^3) were used, as well as the available data for the following four meteorological variables—TEMP (air temperature, °C), HUM (relative humidity, %), PRESS (air pressure, mbar), and LUM (luminosity, Luxe).

Table 1 shows the basic descriptive statistics of the used data (see the first 6 variables). For the studied period, maximum values of PM_{10} reaching up to 143.49 μg/m^3 were observed. The results clearly show that in the studied urban area the permissible safety daily standards of 50 μg/m^3, determined by the EU and national regulations are exceeded [20]. For the pollutant PM_{10} it is observed almost 4 times above the prescribed limit.

3 Proposed Hybrid Wavelet-Arcing Methodology

The developed methodology includes combining WT and Arcing algorithm.

3.1 Wavelet Transform

The foundations of wavelet transform (WT) theory were developed in the classic works of Mallat and Daubechies [21, 22], providing rich tools for time–frequency analysis. WT is considered as an analogue and a generalization of Fourier transform. For this purpose, some type of wavelet mother function $\psi(t)$ or "small wave" is selected, defined in a short interval. With a parameterization of this function, a given time series function or signal $f(t)$ is decomposed. As a result, $f(t)$ is presented as a linear sum of a set of components that represent trend, seasonality, amplitudes and other characteristics, as well as noise inherent in most empirical time series. In this way, local time-varying periods and intensities of the time series behaviour are taken into account [21–24]. Wavelet decomposition, however, is performed within the domain of definition of $f(t)$. To build models capable of predicting beyond this interval, various methods such as ARIMA, NN or other ML predictive methods and techniques are applied to the decomposed components obtained as a result of WT. In some cases, this hybrid approach improves prediction, but in others it may not be successful enough [25].

WT can be continuous or discrete type, denoted by CWT and DWT, respectively. Let some wavelet function $\psi(t)$ be selected. In the continuous case, it is parameterized in the form [23]

$$\psi_{a,b}(t) = \frac{1}{\sqrt{|a|}}\psi\left(\frac{t-b}{a}\right) \tag{1}$$

under the conditions

$$\int_0^{\infty} \frac{|\psi(t)|}{t}dt < \infty, \quad \int_{-\infty}^{\infty} |\psi(t)|^2 dt = 1. \tag{2}$$

Here "a" is a scaling parameter, "b" is a location (or position) parameter. Using (1) the WT is defined for any function $f(t)$, defined in the Hilbert space as:

$$W(a,b) = \int_t f(t)\frac{1}{\sqrt{|a|}}\psi\left(\frac{t-b}{a}\right)dt, \quad f(t) \in L^2(\mathbf{R}). \tag{3}$$

For each a and b the corresponding wavelet transform coefficient is calculated, which approximates $f(t)$ at location $t = a/b$. To switch to an orthonormal basis and reduce the number of coefficients, (3) is used with a special selection of a and b, by introducing two integer parameters j and k. For $a = 2^{-j}$, formula (3) gets the form:

$$W(j,k) = \int f(t)\psi_{j,k}(t)dt \tag{4}$$

where the wavelet coefficients are

$$\psi_{j,k}(t) = 2^{j/2}\psi(2^j t - k). \tag{5}$$

In the case of real-world problems, we need a discrete analogue of the CWT, or DWT. Depending on the type of known values of $f(t)$, an appropriate wavelet and a scaling functions are chosen. Using these functions and replacing the corresponding integrals by quadrature formulas, the DWT coefficients are calculated. The general form of decomposition is

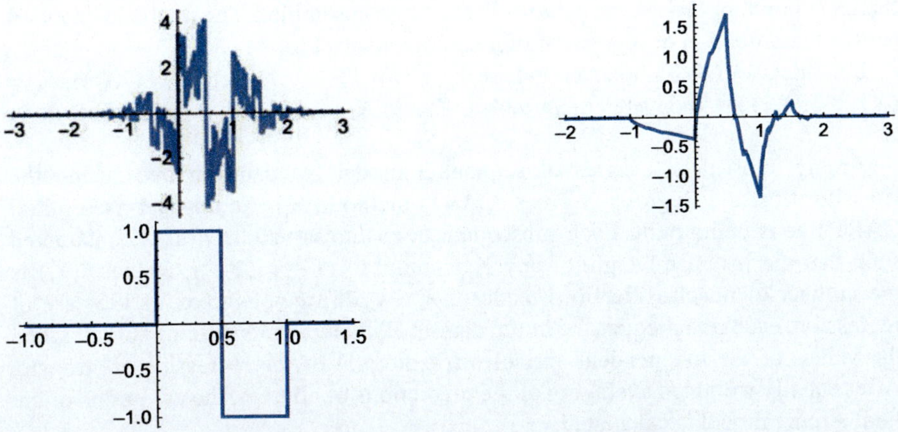

Fig. 1 Plots of the used wavelet functions: in the top left is the biorthogonal spline (B); in the top right—Daubechies (D); at the bottom—Haar (H)

$$f(t) = \sum_{k \in \mathbf{Z}} c_{o,k}\phi_{o,k} + \sum_{j=0}^{\infty} \sum_{k \in \mathbf{Z}} d_{j,k}\psi_{j,k}(t), \tag{6}$$

$$c_{j,k}(f) = \langle f, \phi_{j,k} \rangle = \int f(t)\phi_{j,k}(t)dt, \tag{7}$$

$$d_{j,k}(f) = \langle f, \psi_{j,k} \rangle = \int f(t)\psi_{j,k}(t)dt, \tag{8}$$

where $\phi_{j,k}(t)$ are the corresponding scaling functions of the wavelet.

The conditions, properties, types of CWT and DWT and their inverse transformations are presented in theoretical papers (see [23, 24]).

In this article, we will use three types of WT families: biorthogonal splines (B), Daubechies (D), and Haar (H) wavelets [23]. Figure 1 shows their wavelets.

After decomposing $f(t)$ the resulting components are transformed with a corresponding inverse DWT and can be modelled with predictive methods. A key element in DWT is the ability to select a threshold value so as to control the number of decomposition coefficients and more accurately filter the time series or signal. In our study, the thresholds are chosen to be equal to the standard deviation of the target function $f(t)$. We decompose the main part, called approximation (trend) and its residual, or details. The two parts are predicted separately with the Arcing method.

3.2 Arcing

Arcing stands for Adaptively Resample and Combine algorithm, a family of boosting ML algorithms proposed by Leo Breiman in [16]. In this study, we use the most popular Arc-x4 variant to construct tree ensembles. The method is applied for the classification or regression of a heterogeneous type of data.

Let a dataset (Y, X) be given, where $Y = (Y_1, Y_2, \ldots .Y_N)$, $Y_t = f(X(t))$, $X = (X_1, \ldots, X_p)$ are independent variables, Y, and X_k, $k = 1, 2, \ldots, p$ are vectors in R^N.

Arcing constructs a series of sequential models of decision trees using the Classification and Regression Trees (CART) method [26]. An unsupervised initial CART tree is being built. Each subsequent tree is cross-validated. It is constructed such that the instance weight of any Y_t is equal to $(1 + m(Y_t)^4)$, where $m(Y_t)$ is the number of misclassifications made on Y_t by all previous trees. In the case of regression, each tree predicts the initial dataset Y, including its missing values, when the values of the independent variables (predictors) of the regression are known. After equally weighted averaging of the predictions of all trees, the prediction of the final Arcing model is calculated.

Hyperparameters of the model are: number of trees in the ensemble, minimum number of instances (cases) in parent and child nodes of the tree, training method (we use standard 10-fold cross-validation), number of maximum randomly selected predictors at each node splitting of each tree and some others. As a result of varying the hyperparameters, a large number of ensemble models are obtained. We will select the best predictive model using the following statistics: Root mean squared error (RMSE), Mean absolute percentage error (MAPE) and coefficient of determination R^2, given by the expressions:

$$RMSE = \sqrt{\frac{1}{N}\sum_{t=1}^{N}(Y_t - P_t)^2}, \quad MAPE = \frac{1}{N}\sum_{t=1}^{N}\left|\frac{Y_t - P_t}{Y_t}\right|, \tag{9}$$

$$R^2 = \frac{\left\{\sum_{t=1}^{N}(Y_t - \overline{Y})(P_t - \overline{P})\right\}^2}{\sum_{t=1}^{N}(Y_t - \overline{Y})\sum_{t=1}^{N}(P_t - \overline{P})} \tag{10}$$

where Y_t and $\overline{Y}$ are the values and the mean of the dependent variable $Y = PM10$, respectively; P_t and $\overline{P}$ are the predicted values and their mean, respectively; and N is the sample size. A good predictive model should have a small errors for RMSE and MAPE , and a large R^2 , closed to 1.

3.3 *Description of the Hybrid Wavelet-Arcing (HWA) Approach*

The concentrations of the air pollutant PM_{10} will be modeled according to the following algorithm:

Step 1. We divide the values of the dependent variable $Y = \text{PM10}$ into two subsamples—learning and testing. The training sample, denoted PM10w, contains the first 620 cases. This data will be used to build the models, and the test sample contains the last 10 cases and will be used to validate the models in the forecasting process.

Step 2. We choose a threshold value equal to the standard deviation s of PM10w. We apply DWT to PM10w and extract Approximation part (trend) and its residuals. For this purpose, we use separately the wavelet functions of biorthogonal splines, Daubechies and Haar families. For each we obtain a trend denoted by trB, trD, and trH, respectively. We denote the corresponding residuals by Res_trW, where

$$Res\_trW = PM10w - trW, \quad W = B, D, H \tag{11}$$

Step 3. We build Arcing models of the three trends separately. As independent variables, we use the meteorological time series from Table 1 and several lagged variables of the corresponding trend. We select the best predictive Arcing model according to indicators (9)–(10). We denote it by A_Y.

Step 4. Analogously, we separately construct an arcing model of the residual of each trend and denote it by A_Res _trW,

Step 5. We calculate the predictions of the HWA model by the expression

$$Hyb\_W = A\_trW + A\_Res\_trW, \quad W = B, D, H \tag{12}$$

Step 6. We evaluate the accuracy and predictive ability of the model (12) using (9)–(10).

4 Results

In this section, we present the main results of the implementation of the algorithm described in Sect. 3.3.

4.1 *Application of DWT*

First, the variable $Y = \text{PM10w}$ is decomposed with a given selected wavelet function W. Then, the two components from DWT—the approximation part (trend) and the details (the residuals of the trend) are extracted. This is performed using a threshold equal to the standard deviation of PM10w, the value of which is $s = 15.9115$. For each of the three families of wavelet functions we obtained the trends illustrated in Fig. 2a–c together with the values of the time series PM10w. It is clear from Fig. 2 that the W-trends capture well the high values of PM10w and smooth out (filter) the low values.

Table 2 Basic performance statistics of the DWT trends

Model	RMSE	MAPE	R^2	D-W
trB	5.052	0.351	0.900	1.633
trD	5.052	0.351	0.900	1.633
trH	5.052	0.351	0.900	1.633

Note: The trends trB, trD and trH are compared to PM10w (N = 620)

The calculated corresponding statistical indicators from (9)–(10) are given in Table 2. In addition, the values of the Durbin-Watson (D-W) statistic for the time series residual are also given. In this case, all the values of this indicator are not close to 2, therefore, serial correlation is present in the errors of the model trend variables. This will require to examine and approximate their residuals as well, as provided in the steps in Sect. 3.3.

Their residuals were calculated using (11). To note that the amplitudes of the residuals are within approximate limits $(-s, s)$, where s is the standard deviation of the PM10w, used for the threshold value.

4.2 Building and Exploring Arcing Models

To build the arcing models of the three trends trB, trD, and trH the hyperparameters were varied as follows: number of trees in the ensemble of 30, 40, 50, ..., 80; minimum number of instances (cases) in parent and child nodes of the tree: 10 to 5, 5 to 5, and 5 to 3; learning method: standard 10-fold cross-validation; number of maximum randomly selected predictors at each node splitting of each tree—3.

After comparing the statistics of the obtained models according to criteria (9)–(10), one model was selected for each trend. Statistics were calculated for N = 630, i.e. they also include the estimated forecasts for the last 10 days. For models with close indicators, the simpler model was chosen. Best models were obtained for ensembles with 70 trees, minimum number of cases in parent to child node of the tree: 5 to 5. The main statistics of the selected models, as well as the predictors used for their construction, are given in the first three rows of Table 3. The order of the lagged variables is indicated by brackets <q>, where q is the lag number. From Table 3 it can be seen that the three trends have very similar statistics.

Following the algorithm of Sect. 3.3. we also build the arcing models of the residuals of each of the three trends in a similar way. For this purpose, we have used the same hyperparameter settings, except for their lagged variables. The statistics of the selected predictive Res models compared to the corresponding wave residuals are given in the middle part of Table 3.

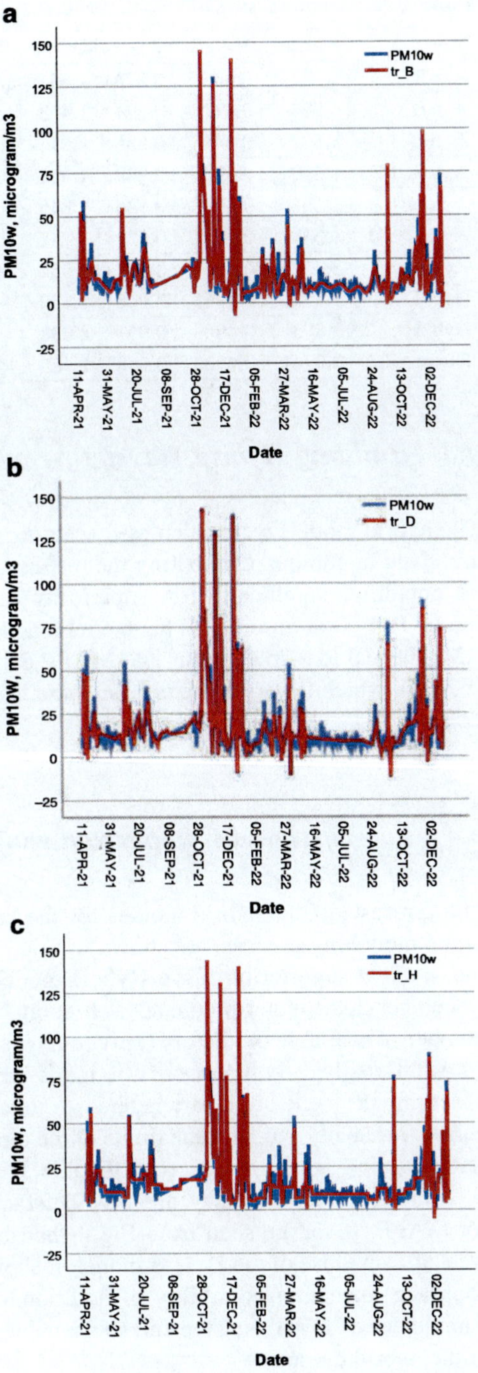

Fig. 2 Line plots of the identified trends (in red color) and the original variables PM10w (in blue color) from top to bottom: trB, trD, trH

Table 3 Summary statistics of the selected best Arcing and hybrid models of PM10

Model	RMSE	MAPE	R^2	D-W	Predictors
A_trB	5.861	0.655	0.874	1.682	trB <1,2,3,4>, Temp, Hum, Press, Lum
A_trD	5.828	0.671	0.874	1.463	trD <1,2,3,4>, Temp, Hum, Press, Lum
A_trH	5.582	0.641	0.885	1.615	trH <1,2,3,4>, Temp, Hum, Press, Lum
A_Res_trB	1.610	0.570	0.949	1.961	Res_trB <1,3,9>, Temp, Hum, Press, Lum
A_Res_trD	1.631	0.190	0.945	1.845	Res_trD <1,9>, Temp, Hum, Press, Lum
A_Res_trH	1.548	0.194	0.939	1.737	Res_trH <1,9>, Temp, Hum, Press, Lum
Hyb_B	3.752	0.500	0.954	2.054	–
Hyb_D	3.872	0.496	0.948	1.850	–
Hyb_H	3.803	0.496	0.951	2.038	–
DirectARC	3.9741	0.192	0.948	1.985	PM10w <1,2,3,4>, Temp, Hum, Press, Lum

4.3 Building Hybrid Wavelet-Arcing Models

The hybrid models were calculated according to formula (12). Their main statistics are given in Table 3. Comparing the results, we can conclude that the three models do not differ significantly. A superiority of Biorthogonal spline wavelet-Arcing model Hyb_B is observed. This model reaches coefficient of determination 95.4%, RMSE equal to 3.752 $\mu g/m^3$ and MAPE equal to 0.5. The Durbin-Watson statistic is 2.054, which is very close to 2, so there is no autocorrelation in the hybrid model errors.

4.4 Forecasts and Comparison with the Direct Arcing Model

The forecasts of the hybrid models for the last 10 days, which data were not used in the modelling process, are shown in Fig. 3, to the right of the vertical line. Here too, a slight superiority of the Hyb_B model is observed. The only outlier on 29 December 2022 is not predicted well enough. This can be explained by the small number of available predictors (only four) with which the study was conducted.

Additionally, a comparison was made with a direct arcing model of PM10w, observing the same basic parameters—70 trees, 10-fold cross validation, number of lagged variables, 5 to 5 for the limits of the terminal nodes of the trees. The obtained statistics are given in the last row of Table 3.

It can be seen that they are less satisfactory than the hybrid models, except for MAPE. It can be seen from Fig. 4 that the forecasts for the last days are also weaker than those of the Hyb_B model. It should be added here that in the proposed approach there is a possibility of additional adjustment and optimization of the parameters of DWT and the threshold value, which would improve the statistics of the hybrid models.

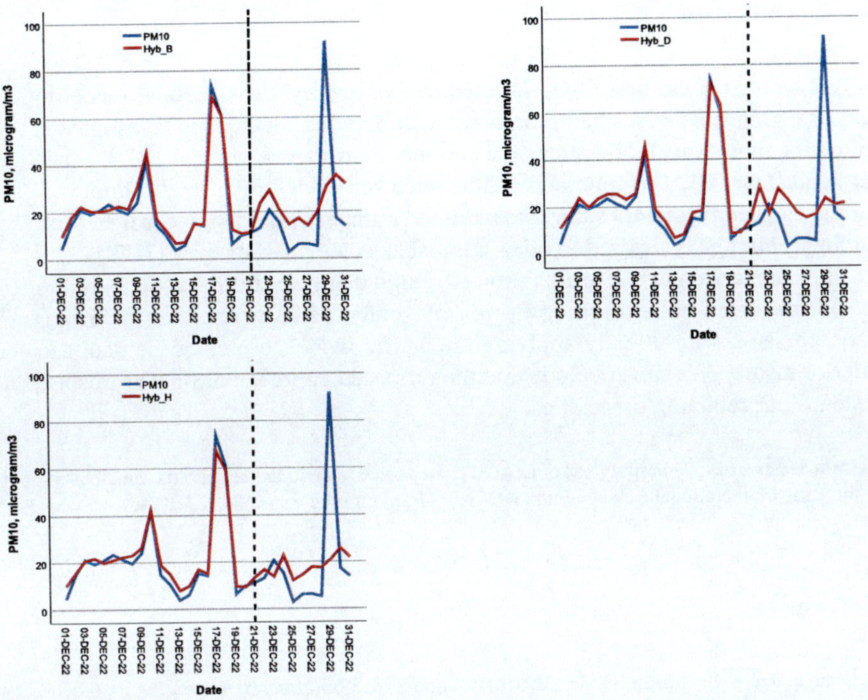

Fig. 3 PM_{10} predicted values from the hybrid models (in red color) for the last 30 days compared to actual data (in blue color). To the right of the vertical line are the forecasts for the last 10 days

Fig. 4 PM_{10} predicted values from the direct arcing model (in green color) for the last 30 days compared to actual data (in blue color). To the right of the vertical line are the forecasts for the last 10 days

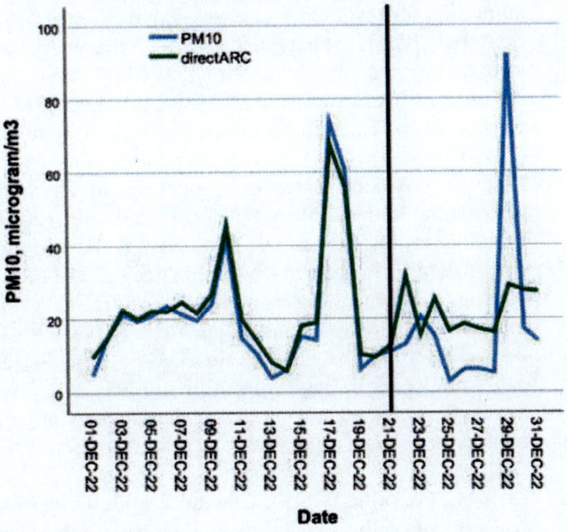

5 Conclusion

In this work, for the first time, the effectiveness of hybrid statistical modelling of fast-changing time series based on wavelet and arcing methods is investigated. The following three wavelet functions are applied: biorthogonal splines (B), Daubechies (D), and Haar (H). The approach is illustrated for modelling a small sample of data for the air pollutant PM_{10} subject to a limited number of meteorological factors. Our findings could be compared to other approaches (see for instance [27–29]).

The comparison of the results showed a high degree of fitting of the built models to the measured data, reaching up to 95% and RMSE, with value 3.752 $\mu g/m^3$. A comparison with directly applying the arcing model to predict the data showed inferior results. We can conclude that the proposed approach has great prospects as a means of predicting time series.

Acknowledgments This study has emanated from a research conducted with the financial support of the Bulgarian National Science Fund (BNSF), Grant number KP-06-IP-CHINA/1.

References

1. Hamanaka, R.B., Mutlu, G.M.: Particulate matter air pollution: effects on the cardiovascular system. Front. Endocrinol. **9**, art. no 680 (2018)
2. Davidson, C.I., Phalen, R.F., Solomon, P.A.: Airborne particulate matter and human health: a review. Aerosol Sci. Technol. **39**(8), 737–749 (2005)
3. Ibrahim, M.F., Hod, R., Tajudin, M.A.B.A., Mahiyuddin, W.R.W., Nawi, A.M., Sahani, M.: Children's exposure to air pollution in a natural gas industrial area and their risk of hospital admission for respiratory diseases. Environ. Res. **210**, 112966 (2022)
4. Shahrbaf, M.A., Akbarzadeh, M.A., Tabary, M., Khaheshi, I.: Air pollution and cardiac arrhythmias: a comprehensive review. Curr. Probl. Cardiol. **46**(3), 100649 (2021)
5. AlOmar, M.K., Hameed, M.M., AlSaadi, M.A.: Multi hours ahead prediction of surface ozone gas concentration: robust artificial intelligence approach. Atmos. Pollut. Res. **11**(9), 1572–1587 (2020)
6. Kisi, O.: Wavelet regression model as an alternative to neural networks for river stage forecasting. Water Resour. Manag. **25**, 579–600 (2011)
7. Mandrikova, O., Fetisova, N., Polozov, Y.: Hybrid model for time series of complex structure with ARIMA components. Mathematics **9**, 1122 (2021)
8. Aladağ, E.: Forecasting of particulate matter with a hybrid ARIMA model based on wavelet transformation and seasonal adjustment. Urban Climate **39**, 100930 (2021)
9. Box, G.E.P., Jenkins, G.M., Reinsel, G.S.: Time Series Analysis, Forecasting and Control, 3rd edn. Prentice-Hall, Englewood Cliffs (1994)
10. Siwek, K., Osowski, S.: Improving the accuracy of prediction of PM10 pollution by the wavelet transformation and an ensemble of neural predictors. Eng. Appl. Artif. Intell. **25**(6), 1246–1258 (2012)
11. Zeri, M., Carvalho, V.S.B., Cunha-Zeri, G., Oliveira-Júnior, J.F., Lyra, G.B., Freitas, E.D.D.: Assessment of the variability of pollutants concentration over the metropolitan area of São Paulo, Brazil, using the wavelet transform. Atmos. Sci. Lett. **17**(1), 87–95 (2016)
12. Galvão, S.L.J., Matos, J.C.O., Kitagawa, Y.K.L., Conterato, F.S., Moreira, D.M., Kumar, P., Nascimento, E.G.S.: Particulate matter forecasting using different deep neural network topologies and wavelets for feature augmentation. Atmosphere **13**(9), 1451 (2022)

13. Bai, Y., Li, Y., Wang, X., Xie, J., Li, C.: Air pollutants concentrations forecasting using back propagation neural network based on wavelet decomposition with meteorological conditions. Atmos. Pollut. Res. **7**(3), 557–566 (2016)
14. Ebrahimi-Khusfi, Z., Taghizadeh-Mehrjardi, R., Kazemi, M., Nafarzadegan, A.R.: Predicting the ground-level pollutants concentrations and identifying the influencing factors using machine learning, wavelet transformation, and remote sensing techniques. Atmos. Pollut. Res. **12**(5), 101064 (2021)
15. Liu, B., Yu, X., Chen, J., Wang, Q.: Air pollution concentration forecasting based on wavelet transform and combined weighting forecasting model. Atmos. Pollut. Res. **12**(8), 101144 (2021)
16. Breiman, L.: Arcing classifiers. Ann. Stat. **26**, 801–849 (1998)
17. Wolfram Mathematica, https://www.wolfram.com/mathematica/. Last Accessed 22 June 2023
18. Salford Predictive Modeler 8, https://www.salford-systems.com. Last Accessed 22 June 2023
19. SPSS IBM Statistics, https://www.ibm.com/products/spss-statistics. Last Accessed 22 June 2023
20. Directive 2008/50/EC of the European Parliament and of the council of 21 May 2008 on ambient air quality and cleaner air for Europe. Official Journal of the European Union, L **152/1** (2008)
21. Mallat, S.G.: A theory for multiresolution signal decomposition: the wavelet representation. IEEE Trans. Pattern Anal. Mach. Intell. **11**(7), 674–693 (1989)
22. Daubechies, I.: Orthonormal bases of compactly supported wavelets. Commun. Pure Appl. Math. **41**(7), 909–996 (1988)
23. Burrus, T.V., Burrus, C., Narasimhan, K., Guo, Y., Li, C.: Introduction to Wavelets and Wavelet Transforms: A Primer. Prentice Hall, Upper Saddle River (1998)
24. Walden, A.T., Percival, D.B.: Wavelet Methods for Time Series Analysis. Cambridge University Press, Cambridge (2000)
25. Schlüter, S., Deuschle, C.: Using wavelets for time series forecasting: does it pay off?. IWQW Discussion Papers 4, Friedrich-Alexander University Erlangen-Nuremberg, Institute for Economics (IWQW) (2010)
26. Breiman, L., Friedman, J.H., Olshen, R.A., Stone, C.J.: Classification and Regression Trees. Wadsworth Advanced Books and Software, Belmont (1984)
27. Tsvetanova, I., Zheleva, I., Filipova, M., Stefanova, A.: Statistical analysis of ambient air PM10 contamination during winter periods for Ruse region, Bulgaria. In: MATEC Web of Conferences, EDP Sciences, vol. 145, art. 01007 (2018)
28. Veleva, E., Zheleva, I.: Statistical modeling of particle mater air pollutants in the city of Ruse, Bulgaria. In: MATEC Web of Conferences, EDP Sciences, vol. 145, art. 01010 (2018)
29. Veleva, E., Georgiev I.R.: Seasonality of the levels of particulate matter PM10 air pollutant in the city of Ruse, Bulgaria. In: AIP Conference Proceedings, vol. 2302, art. 030006 (2020)

Fixed Point Theorems in Orthogonal F-Metric Spaces

Vildan Ozturk

Abstract F-metric spaces are defined as a generalization of metric spaces with properties of nonnegativity, self-distance, symmetry and generalized triangular inequality. Moreover, notion of orthogonal F-metric spaces are introduced given by establishing a binary relation on the set. In this work, we give some properties of orthogonal relation on a set and some topological properties of orthogonal F-metric spaces. We introduce some rational type contractions in orthogonal F-metric spaces and prove fixed point theorems for this type contraction. Our results generalize the Jaggi and Ciric type contractions.

1 Introduction and Preliminaries

Fixed point theory started with Brouwer fixed point theorem in 1912. Brouwer proved the existence of the fixed point of the continuous mappings defined from the closed unit ball of $\mathbf{R}^n$ to itself. Fixed point theory studies developed in three main branches as topological, metric and seperated spaces. Metric fixed point theory started with the Banach contraction principle in 1922 [2]. Banach fixed point theorem became a fundamental and rapidly developing topic in nonlinear analysis recently. It has useful applications in fields such as science, economics, engineering sciences as well as mathematics. In some cases where this principle is insufficient, some generalized metric spaces and generalized contraction principles were defined and many fixed point theorems were proved.

Jleli and Samet [10] presented the concept of F-metric space and proved Banach contraction principle. The authors defined the F-metric by making use of the F-functions in Wardowski's definition of F-contraction [21]. Later on Mitrovic et al. [16] obtained common fixed results for Banach, Jungck, Reich, and Berinde type

V. Ozturk (✉)
Department of Mathematics, Faculty of Polatli Science and Letters, University of Ankara Haci Bayram Veli, Ankara, Turkey
e-mail: vildan.ozturk@hbv.edu.tr

V. Gayoso Martínez et al. (eds.), *Mathematical Methods for Engineering Applications*, Springer Proceedings in Mathematics & Statistics 439,
https://doi.org/10.1007/978-3-031-49218-1_13

contractions in F-metric spaces and applied the results to dynamic programming. Zhou et al. [22] proved best proximity results in F-metric spaces. Lateefa et al. [14] proved fixed point theorems for Dass-Gupta type contaction. Jahangir et al. [9] gived nonlinear contraction principles in F-metric spaces. Faraji et al. [4] defined $(\alpha$-$\beta)$-admissible type contraction and proved fixed point theorems in F-metric spaces.

Recently, the concept of the orthogonal sets was introduced in [1, 6]. A generalization of Banach fixed point theorem was proved by Gordji and Habibi [5]. They worked on classical existence and uniqueness theorems of solutions to the first-order differential equation in orthogonal metric spaces. Ramezani and Baghani [17] gave the concept the strongly orthogonal sets and analyzed the existence of fixed points for generalized contractive operators in strongly metric spaces. Bilgili and Turkoglu [8] presented fixed point theorems on orthogonal metric spaces via altering distance functions. Beg et al. [3] and Mani et al. [15] introduced the concept of F-Suzuki contraction and gave fixed-point theorems in orthogonal b-metric spaces. Sawangsup et al. defined F-contraction in orthogonal metric spaces [18]. Senapati et al. and Bilgili proved some orthogonal fixed point theorems using w-distance function [7, 19]. Kanwal et al. [11] prove some fixed and periodic point theorems for orthogonal contraction in orthogonal F-metric spaces and apply the results for existence and uniqueness of the solution of nonlinear fractional differential equation. Lael [13] introduced common fixed point theorems in orthogonal modular metric spaces. Uddin et al. [20] gave the concept of orthogonal m-metric space.

Every metric space is orthogonal metric space and also F-metric space. But the converse of both cases may not be true. In this work we will define rational type contractions in orthogonal F-metric spaces and prove fixed point theorems. The results generalize the existing fixed point results in metric and orthogonal metric spaces in the literature.

Definition 1 [12] Let $Q \neq \emptyset$ be a set. $d : Q^2 \to [0, \infty)$ satisfies:

(w_1) $w(u,t) = 0 \Leftrightarrow u = t$
(w_2) $w(u,t) = w(t,u)$
(w_3) $w(u,t) \leq d(u,r) + d(r,t)$

for all $u, r, t \in Q$. Then (A, w) is called a metric space.

Definition 2 [12] Let $Q \neq \emptyset$ be a set and let $h : Q \to Q$ be a self mapping. If $h(s) = s$ for any $s \in Q$, then s is fixed point of h.

Theorem 1 ([2] Banach Contraction Principle) *Let (Q, w) be a complete metric space and let $h : Q \to Q$ be a self mapping. If there is a constant $0 < k < 1$ such that the inequality*

$$w(hs, hu) \leq kw(s, u)$$

is provided for each $s, u \in Q$. Then h has a unique fixed point.

In this work, we denote Ω the family of all functions $F : (0, \infty) \to (-\infty, +\infty)$ satisfying following properties;

(F_1) for all $s, u \in (0, +\infty)$ such that $s < u, \Rightarrow F(s) < F(u)$,
(F_2) For each sequence $\{s_n\}_{n \in \mathbf{N}}$, $\lim_{n \to \infty} s_n = 0 \Leftrightarrow \lim_{n \to \infty} F(s_n) = -\infty$;

Definition 3 [10] Let $Q \neq \emptyset$ be a set and let $(F, \alpha) \in \Omega \times [0, +\infty)$ be elements. If $d : Q^2 \to [0, \infty)$ satisfies:

(d$_1$) $d(s, u) = 0 \Leftrightarrow s = u$,
(d$_2$) $d(s, u) = d(u, s)$
(d$_3$) for $\forall n \geq 2$ $(n \in \mathbf{N})$ and $\forall \{s_n\} \subset Q$ such that

$$d(s_1, s_n) > 0 \Longrightarrow F(d(s_1, s_n)) \leq F\left(\sum_{i=1}^{n-1} d(s_i, s_{i+1})\right) + \alpha$$

for all $\varrho_1, \varrho_2 \in Q$. Then (Q, d) is called a F-metric space.

Example 3 Let $Q = \{1, 2, 3, \ldots\}$ and let $d : Q^2 \to [0, \infty)$ be a mapping defined by

$$d(s, u) = \left\{ \begin{array}{ll} (s-u)^2, & if\ s, u \in \{1, 2, \ldots 5\} \\ \\ |s-u|, & if\ s\ or\ u \notin \{1, 2, \ldots 5\} \end{array} \right| .$$

d satisfies (d1) and (d2).

$$d(1, 4) = 9 > 4 + 1 = d(1, 3) + d(3, 4).$$

Hence d is not a metric on Q. But d satisfies generalized triangular inequality with $F(t) = \ln t$ $(t > 0)$ and d is a F-metric on Q.

Definition 4 [10] Let $\{a_n\} \subset Q$ be a sequence. Then

(i) $\{a_n\}$ is called F-convergent if there is $a \in Q$ such that $d(a_n, a) \to 0$ as $n \to \infty$.
(ii) $\{a_n\}$ is callef F-Cauchy sequence if $d(a_n, a_m) \to 0$ as $n, m \to \infty$.
(iii) (Q, d) is called F-complete if each F-Cauchy sequence is F-convergent.

Definition 5 [6] Let $Q \neq \emptyset$ and let $\perp$ be a binary relation on $Q \times Q$. If

$$\exists a_0 \in Q; (\forall b \in Q, b \perp a_0) \vee (\forall b \in Q, a_0 \perp b)$$

is provided then $(Q, \perp)$ is called an orthogonal set (O-set). And a_0 is called an orthogonal element (O-element).

Definition 6 [6] Let $(Q, \perp)$ be an O-set. Any $a, b \in Q$ are called to be orthogonally related if $a \perp b$.

Definition 7 [6] A sequence $\{a_n\}$ is called orthogonal sequence (O-sequence)if

$$(\forall n \in \mathbf{N}; a_n \perp a_{n+1}) \vee (\forall n \in \mathbf{N}; a_{n+1} \perp a_n).$$

Definition 8 [6] A function $h : Q \to Q$ is called orthogonal preserving ($\perp_{preserving}$) if $ha \perp hb$ whenever $a \perp b$.

Definition 9 [11] Let $(Q, \perp)$ be an O-set and d be an F-metric on A. Then $(Q, \perp, d)$ is called an orthogonal F-metric space (shortly $F_\perp$-ms).

Example 4 Take $Q = [0, 1]$ and define $\perp$ on $Q \times Q$ by $a_1 \perp a_2 \Leftrightarrow a_1 a_2 < \frac{1}{6}$. Define

$$d : Q \times Q \to [0, \infty), \quad d(a_1, a_2) = \begin{cases} e^{|a_1 - a_2|}, & if\ a_1 \neq a_2 \\ 0, & if\ a_1 = a_2 \end{cases}.$$

Then d is a F-metric on Q with $F(a) = \frac{-1}{a}$ $(a > 0)$. Therefore $(Q, \perp, d)$ is $F_\perp$-ms.

Definition 10 [11] Let $(Q, \perp, d)$ be a $F_\perp$-ms

(i) A sequence $\{a_n\}$ in Q is said to be an orthogonal F-Cauchy sequence if

$$(\forall n \in \mathbf{N}; \ a_n \perp a_{n+1}) \vee (\forall n \in \mathbf{N}; \ a_{n+1} \perp a_n).$$

(ii) $(Q, \perp, d)$ is called orthogonal F-complete (complete $F_\perp$-ms), if every orthogonal F-Cauchy sequence is F-convergent.
(iii) A mapping $h : Q \to Q$ is said to be orthogonally continuous ($\perp_{continuous}$) if for each O-sequence $\{a_n\} \subseteq Q$ with $a_n \to a$ implies $h(a_n) \to h(a)$.

Example 5 Take $Q = [0, 1)$ and define $\perp$ on $Q \times Q$ by $a_1 \perp a_2 \Leftrightarrow a_1 a_2 \leq \max\{\frac{a_1}{6}, \frac{a_2}{6}\}$. Let define

$$d : Q \times Q \to [0, \infty), \quad d(a_1, a_2) = |a_1 - a_2|,$$

$$h : Q \to Q, \ h(a) = \begin{cases} \frac{a}{6}, & if\ a \in (0, \frac{1}{6}] \\ \frac{1}{36} & if\ a \notin (0, \frac{1}{6}] \end{cases}.$$

Then $(Q, \perp, d)$ is complete $F_\perp$-ms with $F(a) = \frac{-1}{a}$ $(a > 0)$. Also the mapping h is $\perp_{continuous}$ and $\perp_{preserving}$. Indeed, suppose that $a_1 \perp a_2$. Without lose of generality, $a_1 a_2 \leq \frac{a_1}{6}$ can be chosen. So $a_1 \geq 0$ and $0 \leq a_2 \leq \frac{1}{6}$. Thus there are two cases:

Case I. If $0 \leq a_1 \leq \frac{1}{6}$ and $0 \leq a_2 \leq \frac{1}{6}$, then $h(a_1) = \frac{a_1}{6}$ and $h(a_2) = \frac{a_2}{6}$.
Case II. If $a_1 \notin (0, \frac{1}{6}]$ and $0 \leq a_2 \leq \frac{1}{6}$, then $h(a_1) = \frac{1}{36}$ and $h(a_2) = \frac{a_2}{6}$.
These cases imply that $h(a_1) \perp h(a_2)$.

2 Fixed Point Results

Definition 11 Let $(Q, \perp, d)$ be a complete $F_{\perp}$-ms and let $h : Q \to Q$ a self mapping such that for all $a, b \in Q$,

$$d(ha, hb) \le \gamma \left(\frac{d(ha, a)\, d(hb, b)}{d(a, b) + 1} \right) + \eta d(a, b) \tag{1}$$

where $0 < \gamma, \eta < 1$ with $\gamma + \eta < 1$. Then h is called a Jaggi type contraction.

Theorem 2 *Let $(Q, \perp, d)$ be a complete $F_{\perp}$-ms and let $h : A \to A$ be a $\perp_{preserving}$ and $\perp_{continuous}$ Jaggi type contraction. Then h has a unique fixed point.*

Proof Let $\varepsilon > 0$ be fixed. From $(F, \alpha) \in \Omega \times [0, \infty)$, (d_3) and (F_2)

$$0 < s < u \Rightarrow F(t) < F(\varepsilon) - \alpha \tag{2}$$

Since $(Q, \perp)$ is an $\perp_{set}$,

$$\exists a_0 \in Q, (\forall b \in A,\ b \perp a_0) \vee (\forall b \in Q, a_0 \perp b).$$

And since h is a self mapping on Q, for any O—element $a_0 \in Q$, $a_1 \in Q$ can be chosen as $a_1 = ha_0$. Thus,

$$a_0 \perp ha_0 \vee ha_0 \perp a_0 \Rightarrow a_0 \perp a_1 \vee a_1 \perp a_0.$$

Then, if we continue in the same way,

$$a_1 = ha_0,\ a_2 = ha_1 = h^2 a_0 =, \ldots\ldots, a_n = ha_{n-1} = h^n a_0.$$

So that $\{h^n(a_0)\}$ is an iteration sequence. $\{h^n(a_0)\}$ is an O-sequence. If $a_n = a_{n+1}$ for any n, then we get $a_n = ha_n$ and so h has a fixed point. So we assume $a_n \neq a_{n+1}$ for each n. Since h is $\perp_{preserving}$, $\{a_n\}$ is an O-sequence. By (1)

$$\begin{aligned} d(a_n, a_{n+1}) = d(ha_{n-1}, ha_n) &\le \gamma \left(\frac{d(ha_{n-1}, a_{n-1})\, d(ha_n, a_n)}{d(a_{n-1}, a_n) + 1} \right) \\ &+\eta (a_{n-1}, a_n) \\ &= \gamma \left(\frac{d(a_n, a_{n-1})\, d(a_{n+1}, a_n)}{d(a_{n-1}, a_n) + 1} \right) + \eta (a_{n-1}, a_n) \\ &\le \gamma d(a_n, a_{n+1}) + \eta (a_{n-1}, a_n). \end{aligned}$$

Thus we have

$$d\left(a_n, a_{n+1}\right) \leq \frac{\eta}{1-\gamma}\left(a_{n-1}, a_n\right).$$

Then, if we continue in the same way,

$$d\left(a_n, a_{n+1}\right) \leq \left(\frac{\eta}{1-\gamma}\right)^n d\left(a_0, a_1\right)$$

Hence we have for $m, n \in \mathbf{N}$ with $m > n$

$$\begin{aligned}\sum_{i=n}^{m-1} d\left(a_i, a_{i+1}\right) &\leq \sum_{i=n}^{m-1}\left(\frac{\eta}{1-\gamma}\right)^i d\left(a_0, a_1\right)\\ &\leq \frac{c^n}{1-c} d\left(a_0, a_1\right)\end{aligned}$$

Since $c = \frac{\eta}{1-\gamma} < 1$, for all $\delta > 0$ there exists $n_0 \in \mathbf{N}$ such that for all $n \geq n_0$,

$$0 < \frac{c^n}{1-c} d\left(a_0, a_1\right) < \delta.$$

By (2) and (d3),

$$F\left(\sum_{i=n}^{m-1} d\left(a_i, a_{i+1}\right)\right) \leq F\left(\frac{c^n}{1-c} d\left(a_0, a_1\right)\right) < F(\varepsilon)-\alpha. \tag{3}$$

for all $m, n \in \mathbf{N}$ with $m > n \geq n_0$. Using (d3) and (3) we have

$$d\left(a_n, a_m\right) > 0 \Longrightarrow F\left(d\left(a_n, a_m\right)\right) \leq F\left(\sum_{i=n}^{m-1} d\left(a_i, a_{i+1}\right)\right)+\alpha < F(\varepsilon).$$

By (F1), we have

$$d\left(a_n, a_m\right) < \varepsilon.$$

Thus $\{a_n\}$ is an orthogonal F-Cauchy sequence.

Since A be a complete $-F_{\perp}$, there exists a $s \in A$ such that $\lim_{n\to\infty} a_n = s$. orthogonal continuity of h gives $\lim_{n\to\infty} ha_n = hs$. Hence

$$hs = \lim_{n\to\infty} ha_n = \lim_{n\to\infty} a_{n+1} = s.$$

So s is a fixed point of h. Now, we show uniqueness. Suppose that there exists $s, u \in Q$ such that $hs = s$ and $hu = u$ with $d(s, u) \neq 0$.

Case I.
If $s \perp u$, then by (1)

$$\begin{aligned} d(s,u) &\le \gamma \left(\frac{d(hs,s)\, d(hu,u)}{d(s,u)+1} \right) + \eta d(s,u) \\ &= \gamma \left(\frac{d(s,s)\, d(u,u)}{d(s,u)+1} \right) + \eta d(s,u) \\ &= \eta d(s,u) \end{aligned}$$

which is a contradiction with $\eta < 1$. Thus $s = u$.
Case II.
If not, for the choosen orthogonal element $a_0 \in Q$,

$$a_0 \perp s \text{ and } a_0 \perp u, \text{ or } s \perp a_0 \text{ and } u \perp a_0$$

and since h is $\perp_{preserving}$,

$$ha_0 \perp hs \text{ and } ha_0 \perp hu \text{ or } hs \perp ha_0 \text{ and } hu \perp ha_0$$

$$\vdots$$

$$a_n \perp s \text{ and } a_n \perp u \text{ or } s \perp a_n \text{ and } u \perp a_n.$$

By (1),

$$d(u,a_n) \le \gamma \left(\frac{d(hu,u)\, d(ha_n,a_n)}{d(u,a_n)+1} \right) + \eta d(u,a_n)$$

Taking limit as $n \to +\infty$, we get $\lim_{n\to+\infty} d(u,a_n) = 0$. By the uniqueness of limit $u = s$. Thus h has a unique fixed point.

Definition 12 Let $(Q, \perp, d)$ be a complete $F_\perp$-ms and let $h : Q \to Q$ a self mapping such that for all $a, b \in Q$

$$d(ha,hb) \le \gamma \max \left\{ d(a,b), \frac{d(ha,a)\, d(hb,b)}{d(a,b)+1}, d(ha,a), d(hb,b) \right\} \tag{4}$$

where $0 < \gamma < 1$. Then h is called a Ciric type rational contraction.

Theorem 3 *Let $(Q, \perp, d)$ be a complete $F_\perp$-ms and $h : Q \to Q$ be a $\perp_{preserving}$ and $\perp_{continuous}$ Ciric type rational contraction. Then h has a unique fixed point.*

Proof Let $\varepsilon > 0$ be fixed. From $(F, \alpha) \in \Omega \times [0, \infty)$, ($d_3$) and ($F_2$)

$$0 < s < u \Rightarrow F(t) < F(\varepsilon) - \alpha \tag{5}$$

Since $(Q, \perp)$ is an O-set,

$$\exists a_0 \in Q,\ (\forall b \in Q,\ b \perp a_0) \vee (\forall b \in Q, a_0 \perp b).$$

And since h is a self mapping on Q, for any O-element $a_0 \in Q$, $a_1 \in Q$ can be chosen as $a_1 = ha_0$. Thus,

$$a_0 \perp ha_0 \vee ha_0 \perp a_0 \Rightarrow a_0 \perp a_1 \vee a_1 \perp a_0.$$

Then, if we continue in the same way,

$$a_1 = ha_0,\ a_2 = ha_1 = h^2 a_0 =, \ldots\ldots., a_n = ha_{n-1} = h^n a_0.$$

So that $\{h^n(a_0)\}$ is an iteration sequence. $\{h^n(a_0)\}$is an O-sequence. If $a_n = a_{n+1}$ for any n, then we get $a_n = ha_n$ and so h has a fixed point. So we assume $a_n \neq a_{n+1}$ for each n. Since h is $\perp_{preserving}$, $\{a_n\}$ is an O-sequence . By (4)

$$\begin{aligned} d(a_n, a_{n+1}) &= d(ha_{n-1}, ha_n) \\ &\leq \gamma \max \left\{ \begin{array}{c} d(a_{n-1}, a_n), \frac{d(ha_{n-1},a_{n-1}) d(ha_n,a_n)}{d(a_{n-1},a_n)+1}, \\ d(ha_{n-1}, a_{n-1}), d(ha_n, a_n) \end{array} \right\} \\ &= \gamma \max \left\{ \begin{array}{c} d(a_{n-1}, a_n), \frac{d(a_n,a_{n-1}) d(a_{n+1},a_n)}{d(a_{n-1},a_n)+1}, \\ d(a_n, a_{n-1}), d(a_{n+1}, a_n) \end{array} \right\} \\ &= \gamma \max \{d(a_{n-1}, a_n), d(a_{n+1}, a_n)\} \end{aligned}$$

If $\max \left\{d(a_{n-1}, a_n), \frac{d(a_n,a_{n-1}) d(a_{n+1},a_n)}{d(a_{n-1},a_n)+1}, d(a_n, a_{n-1}), d(a_{n+1}, a_n)\right\} = d(a_{n+1}, a_n)$, then we have a contradiction with $0 < \gamma < 1$.

Thus we have

$$d(a_n, a_{n+1}) \leq \gamma (a_{n-1}, a_n).$$

Then, similar to Theorem 2, we can prove that $\{a_n\}$ is an orthogonal F-Cauchy sequence.

Since Q be a complete $F_\perp$-ms , there exists a $s \in Q$ such that $\lim_{n\to\infty} a_n = s$. Using orthogonal continuity of h, we get $\lim_{n\to\infty} ha_n = hs$. Hence

$$hs = \lim_{n\to\infty} ha_n = \lim_{n\to\infty} a_{n+1} = s.$$

So s is a fixed point of h. Now, we show uniqueness. Suppose that there exists $s, u \in Q$ such that $hs = s$ and $hu = u$ with $d(s, u) \neq 0$.

Case I.

If $s \perp u$, then by (4)

$$d(s,u) = d(hs,hu) \leq \gamma \max\left\{d(s,u), \frac{d(hs,s)\,d(hu,u)}{d(s,u)+1}, d(hs,s), d(hu,u)\right\}$$

$$= \gamma \max\left\{d(s,u), \frac{d(s,s)\,d(u,u)}{d(s,u)+1}, d(s,s), d(u,u)\right\}$$

$$= \gamma d(s,u).$$

which is a contradiction with $0 < \gamma < 1$. Thus $s = u$.

Case II.

If not, for the choosen orthogonal element $a_0 \in Q$,

$$a_0 \perp s \text{ and } a_0 \perp u, \text{ or } s \perp a_0 \text{ and } u \perp a_0$$

and since h is $\perp_{preserving}$,

$$ha_0 \perp hs \text{ and } ha_0 \perp hu \text{ or } hs \perp ha_0 \text{ and } hu \perp ha_0$$

$$\vdots$$

$$a_n \perp s \text{ and } a_n \perp u \text{ or } s \perp a_n \text{ and } u \perp a_n.$$

By (4)

$$d(u,a_n) \leq \gamma \max\left\{d(u,a_n), \frac{d(hu,u)\,d(ha_n,a_n)}{d(u,a_n)+1}, d(hu,u), d(ha_n,a_n)\right\}.$$

Taking limit as $n \to +\infty$, we get $\lim_{n\to+\infty} d(u,a_n) = 0$. By the uniqueness of limit $u = s$. Thus h has a unique fixed point.

Corollary 1 *Let* $(Q, \perp, d)$ *be a complete* $F_{\perp}$*−ms and let* $h : A \to A$ *be a* $\perp_{preserving}$ *and* $\perp_{continuous}$ *self mapping such that for all* $a, b \in Q$,

$$d(ha,hb) \leq \gamma \left(\frac{d(ha,a)\,d(hb,b)}{d(a,b)+1}\right)$$

where $0 < \gamma < 1$. Then h has a unique fixed point.

Corollary 2 *Let* $(Q, \perp, d)$ *be a complete* $F_{\perp}$*−ms and let* $h : Q \to Q$ *be a* $\perp_{preserving}$ *and* $\perp_{continuous}$ *self mapping such that for all* $a, b \in Q$.

$$d(ha,hb) \leq \gamma \max\left\{d(a,b), \frac{d(ha,a)+d(hb,b)}{2}, d(ha,a), d(hb,b)\right\}$$

where $0 < \gamma < 1$. *Then* h *has a unique fixed point.*

Corollary 3 *Let $(Q, \perp, d)$ be a complete $F_{\perp}-ms$ and let $h : Q \to Q$ be a $\perp_{preserving}$ and $\perp_{continuous}$ self mapping such that for all $a, b \in Q$,*

$$d(ha, hb) \leq \gamma d(a, b)$$

where $0 < \gamma < 1$. Then h has a unique fixed point.

Corollary 4 *Let $(Q, \perp, d)$ be a complete $F_{\perp}-ms$ and let $h : Q \to Q$ be a $\perp_{preserving}$ and $\perp_{continuous}$ self mapping such that for all $a, b \in Q$,*

$$\int_0^{d(ha,hb)} \Theta(t)\, dt \leq \gamma \int_0^{\frac{d(ha,a)d(hb,b)}{d(a,b)+1}} \Theta(t) dt + \eta \int_0^{d(a,b)} \Theta(t) dt$$

where $0 < \gamma, \eta < 1$ with $\gamma + \eta < 1$. If $\Theta : [0, \infty) \to [0, \infty)$ is Lebesque integrable and $\int_0^{\epsilon} \Theta(t)\, dt > 0$ for any $\epsilon > 0$. Then h has a unique fixed point.

Example 6 Take $Q = [0, 1)$ and define $\perp$ on $Q \times Q$ by $a \perp b \Leftrightarrow ab \leq \max\{\frac{a}{6}, \frac{b}{6}\}$. Define

$$d : Q \times Q \to [0, \infty),\ d(a, b) = |a - b|,$$

$$h : A \to A,\ h(a) = \begin{cases} \frac{a}{6}, & if\ a \in (0, \frac{1}{6}] \\ 0, & if\ a \notin (0, \frac{1}{6}] \end{cases}$$

Then $(Q, \perp, d)$ is orthogonal F-complete metric space with $F(a) = \frac{-1}{a}$ $(a > 0)$. The mapping h is $\perp_{continuous}$ and $\perp_{preserving}$. Therefore for $\eta = \frac{1}{2}$ and $\gamma < \frac{1}{2}$, if $a, b \in (0, \frac{1}{6}]$,

$$d(ha, hb) = \frac{1}{6}|a - b| \leq \frac{1}{2}|a - b| \leq \gamma \left(\frac{d(ha, a)\, d(hb, b)}{d(a, b) + 1} \right) + \eta |a - b|.$$

If $a, b \notin (0, \frac{1}{6}]$,

$$d(ha, hb) = 0 \leq \gamma \left(\frac{d(ha, a)\, d(hb, b)}{d(a, b) + 1} \right) + \eta |a - b|.$$

If $a \in (0, \frac{1}{6}]$, $b \notin (0, \frac{1}{6}]$,

$$d(ha, hb) = \left| \frac{a}{6} - 0 \right| \leq \frac{1}{2}|b - a| \leq \gamma \left(\frac{d(ha, a)\, d(hb, b)}{d(a, b) + 1} \right) + \eta |a - b|.$$

Then all conditions of Theorem 2 are satisfied. 0 is fixed point of h.

Notes and Comments The classical Banach contraction principle and most of generalisations give unique fixed point results for self mappings. In this work we will proved fixed point theorems for rational type contractions and integral type contractions in orthogonal F-metric spaces which was defined a orthogonal relation.

In future studies, weak contractive conditions can be defined and both fixed point theorems and common fixed point theorems can be proved in orthogonal spaces. In addition, applications can be given to Fredholm and Volterra integral equations and differential equations. There will be useful applications, especially in mathematics and engineering.

References

1. Baghani, H., Gordji, M.E., Ramezani, M.: Orthogonal sets: the axiom of choice and proof of a fixed point theorem. J. Fixed Point Theory Appl. **18**(3), 465–477 (2016)
2. Banach, S.: Surles operations dans les ensembles abstraits et leur application aux equationsitegrales. Fund. Math. **3**, 133–181 (1922)
3. I. Beg, G. Mani, A.J. Gnanaprakasam, Fixed point of orthogonal F-Suzuki contraction mapping on O-complete b-metric spaces with applications. J. Funct. Spaces **2021**, 6692112 (2021)
4. Faraji, H., Mirkov, N., Mitrović, Z.D., Ramaswamy, R., Abdelnaby, O.A.A., Radenović, S.: Some new results for (alpha-beta)-admissible mappings in F-metric spaces with applications to integral equations. Symmetry **14**(11), 2429 (2022)
5. Gordji, M.E., Habibi, H.: Fixed point theory in generalized orthogonal metric space. J. Linear Topol. Algebra **6**(3), 251–260 (2017)
6. Gordji, M.E., Ramezani, M., De La Sen, M., Cho, Y.J.: On orthogonal sets and Banach fixed point theorem. Fixed Point Theory **18**(2), 569–578 (2017)
7. Gungor, N.B.: Extensions of orthogonal p-contraction on orthogonal metric spaces. Symmetry **14**(4), 746 (2022)
8. Gungor, N.B., Turkoglu, D.: Fixed point theorems on orthogonal metric spaces via altering distance functions. AIP Conf. Proc. **2183**, 040011 (2019)
9. Jahangir, F., Haghmaram, P., Nourouzi, K.: A note on F-metric spaces. J. Fixed Point Theory Appl. **23**, 1–14 (2021)
10. Jleli, M., Samet, B.: On a new generalization of metric spaces. J. Fixed Point Theory Appl. **20**, 128 (2018)
11. Kanwal, T., Hussain, A., Baghani, H., De la Sen, M.: New fixed point theorems in orthogonal F-metric spaces with application to fractional differential equation. Symmetry **12**, 832 (2020)
12. Kreyszig, E.: Introductry Functional Analysis with Application. Wiley, New York (1978)
13. Lael, F.: Common fixed point of generalized weakly contractive mappings on orthogonal modular spaces with applications. Int. J. Nonlinear Anal. Appl. **12**(2), 1121–1140 (2021)
14. Lateefa, D., Ahmad, J.: Dass and Gupta's fixed point theorem in F-metric spaces. J. Nonlinear Sci. Appl. **12**, 405–411 (2019)
15. Mani, G., Gnanaprakasam, A.J., Mishra, L.N., Mishra, V.N.: Fixed point theorems for orthogonal F-Suzuki contraction mappings on O-complete metric space with an applications. Malaya J. Mat. **9**(1), 369–377 (2021)
16. Mitrovic, Z.D., Aydi, H., Hussain, N., Mukheimer, A.: Reich, Jungck, and Berinde common fixed point results on F-metric spaces and an application. Mathematics **7**, 387 (2019)
17. Ramezani, M., Baghani, H.: Contractive gauge functions in strongly orthogonal metric spaces. Int. J. Nonlinear Anal. Appl. **8**(2) 23–28 (2017)

18. Sawangsup, K., Sintunavarat, W., Cho, Y. J.: Fixed point theorems for orthogonal F-contraction mappings on O-complete metric spaces. J. Fixed Point Theory Appl. **22**, 10 (2020)
19. Senapati, T., Dey, L.K., Damjanovic, B., Chanda, A.: New fixed results in orthogonal metric spaces with an application. Kragujevac J. Math. **42**, 505–516 (2018)
20. Uddin, F., Park, C., Javed, K., Arshad, M., Lee, J.R.: Orthogonal m-metric spaces and an application to solve integral equations. Adv. Differ. Equ. **2021**, 159 (2021)
21. Wardowski, D.: Fixed points of a new type of contractive mappings in complete metric spaces. Fixed Point Theory Appl. **2012**, 94 (2012)
22. Zhou, M., Saleem, N., Ali, B., Misha, M.M., Hierro, A.F.R.L.: Common best proximity points and completeness of F-metric spaces. Mathematics **11**, 81 (2023)

New Trends on Malware Propagation: From IoT Environments to Drone Swarms

A. Martín del Rey

Abstract In this work a brief review of theoretical techniques to design mathematical models for malware propagation is introduced. It is shown that the classical methodology, that considers the basic assumptions of the Mathematical Epidemiology and which is wide employed for designing such mathematical models, is not suitable for the special case of malware spreading on IoT networks. Specifically, some recommendations for modeling are introduced and the individual-based paradigm is proposed as the fundamental framework.

1 Introduction

Traditionally, mathematical epidemiology is the scientific discipline devoted to the design and analysis of mathematical models to simulate biological agents propagation. In this sense, the fundamental questions it tries to answer are the following:

(1) Does an outbreak cause an epidemic?
(2) If an outbreak causes an epidemic, what is the growth rate of the number of infected during the rise of the epidemic process?
(3) What is the total amount of people infected during the epidemic?

The first mathematical attempt to model an epidemic disease propagation was made by D. Bernoulli at 1760 on the inoculation against smallpox that in this century was endemic. We had to wait until the first third of the twentieth century to obtain the main results that represented a significant advance in the development of modern mathematical epidemiology: W.H. Hamer described in 1906 the propagation process in terms of a mass action law for the infection rate, that is, the incidence should depend on the number of both susceptible, and infectious individuals [6]. R. Ross

A. Martín del Rey (✉)
Universidad de Salamanca, Department of Applied Mathematics, IUFFyM, Salamanca, Spain
e-mail: delrey@usal.es

V. Gayoso Martínez et al. (eds.), *Mathematical Methods for Engineering Applications*, Springer Proceedings in Mathematics & Statistics 439,
https://doi.org/10.1007/978-3-031-49218-1_14

proposed in 1911 a compartmental model for malaria spreading considering the population of humans and mosquitoes, and the so-called *mosquito theorem* was stated: there exists a critical density of mosquitoes below which the infection process decreases [15]. Moreover, some years later R. Ross and H. Hudson introduced for the first time a very general equation to characterize the incidence of an epidemic process [16]:

$$\iota(t) = \frac{S}{N}\int_0^{\infty} cF(t,\tau)\,d\tau, \tag{1}$$

where c stands for the infectivity, τ is the time elapsed since an individual was infected, $F(t,\tau)$ is the number of infected individuals at time t which were infected at $t-\tau$, and N is the total population size. The breakthrough of mathematical epidemiology was established by Kermack and McKendrick with their famous compartmental model (described in Sect. 2) and the associated *threshold theorem* that stated that an outbreak grows if the population density, N, is greater than a certain threshold:

$$N > \frac{1}{\int_0^{\infty}\phi(\tau)B(\tau)d\tau}, \tag{2}$$

where $B(t)$ represents the probability that a newly infected individual is still infected at infection-age τ, and $\phi(\tau)$ stands for the infectivity at τ. The idea underlying in all these results regarding the existence of a threshold epidemiological coefficient was formally introduced by G. McDonald in 1952 when he was studying the propagation of malaria, and using the name *basic reproduction rate* [11]. Subsequently, H. Hethcote and K. Dietz defined the basic reproduction rate as a mathematical concept given by the inverse of the proportion of susceptible in an endemic steady state [3, 8]. The next great milestone in mathematical epidemiology was reached at 1990 when O. Dieckmann, J.A.P. Heesterbeek and J.A.J. Metz proposed the Next Generation Method for computing in an explicit way the basic reproductive number, $\mathcal{R}_0$ [2].

Obviously this mathematical framework is based on the assumption that biological agents (virus, bacteria, fungus, etc.) are the propagable entities considered. Nowadays, and mainly due to the global deployed of different types of communication networks, other entities of different nature are susceptible to spread: rumors, fanatic ideologies, fake-news, viral marketing and, of course, malicious code (malware). Note that the first time that techniques of the classical mathematical epidemiology were proposed to simulate the propagation of ideas was in 1964 in a work by W. Goffman and V.A. Newill [5]. Of special interest, due to its impact in our technological society, is the mathematical study of the propagation process of malware. In this sense, the first mathematical model for malware propagation was proposed in 1989 by W. Gleissner [4]. Since then, several models have appeared in the scientific literature and the great majority considers the classical approach to biological diffusion processes ignoring the particular characteristics of propagation and infection processes that exhibit the different specimens of malware [12]. This is particularly serious in the case of considering new paradigms and technological environments such as Internet of Things (IoT), Industry 4.0, etc. It is obvious that the behavior of malware on IoT networks or on drone swarms is very different from that which biological agents propagation exhibits on a human population, and this should be reflected in the design of mathematical models. In fact, two types of specimens of malware exist and could be considered for designing ad hoc propagation models: (1) those that exhibit an indiscriminate propagation on very large networks such as Internet; and (2) those specimens that can be classified as "smart" or "intelligent" malware (advanced malware, APT malware) which are used on sophisticated cyber-attacks against specific networks with high security levels (critical infrastructures, etc.).

Obviously the first type of malware (which was the first to appear) has a behavior more similar to that of biological agents than the second type, but this does not mean that classical epidemiological models can be used without an adequate review. Consequently the main goal of this work is to highlight this fact and to propose a novel and formal framework for designing efficient mathematical models for malware propagation.

The rest of the work is organized as follows: In Sect. 2 a brief review of the design of models for malware propagation based on the classical mathematical epidemiology is introduced; moreover, a critical analysis is presented. Section 3 is devoted to the study of individual-based modeling as a clear alternative to classical approach; an example is presented illustrating the design of propagation models on two scenarios: general IoT networks and drone swarms. Finally, the conclusions are presented in Sect. 4.

2 Characterization of Malware Propagation

2.1 *Malware Propagation Using Classical and Global Epidemiological Models*

The design of mathematical models for malware propagation is based on the general framework stated by (classic) Mathematical Epidemiology. In this sense, the nature of the great majority of (deterministic and/or stochastic) models is global on the complete network K_n (see, for example, [7, 14], and references therein).

The paradigmatic example of global models (and the cornerstone of the modern Mathematical Epidemiology) is given by the Kermack-McKendrick model [10]. This is a SIR (Susceptible-Infectious-Recovered) compartmental model described by differential equations. If $x(t)$ is the number of susceptible devices at t, $y(t)$ the number of infectious devices, and $z(t)$ represents the number of recovered devices at t, then its simplest formulation is as follows:

$$x'(t) = -\beta x(t)y(t),$$

$$y'(t) = \beta x(t)y(t) - \gamma y(t), \tag{3}$$
$$z'(t) = \gamma y(t),$$

where β stands for the infectious rate, and γ represents the recovery rate (per unit of time).

Modern compartmental models including different compartments (Susceptible, Exposed, Infectious, Recovered, etc.) are based on Kermack and McKendrick proposal and from a general point of view they can be described as follows. Suppose that the population of devices can be classified into $n + m$ compartments such that the first n compartments, $X_1, \ldots, X_n$, correspond to infected devices (each of them representing a state within the evolution of the epidemic process: latents, infectious with symptoms, infectious without symptoms, etc.), and the last m compartments, $X_{n+1}, \ldots, X_{n+m}$, stand for uninfected compartments (susceptible, vaccinates, inmune, etc.). Let $x_i(t) \in [0, 1]$ be the density of devices in the i-th compartment, $1 \le i \le n + m$, at time t, then $x(t) = (x_1(t), \ldots, x_{n+m}(t))$ for each time t.

Let $\mathcal{F}_i(x(t))$, $1 \le i \le n + m$, be the rate at which secondary infections (the incidence) modify the i-th compartment, and $\mathcal{V}_i(x(t))$ be the rate at which disease progression, remove and recovery modify the i-th compartment such that $\mathcal{V}_i(x(t)) = \mathcal{V}_i^+(x(t)) - \mathcal{V}_i^-(x(t))$ where $\mathcal{V}_i^+$ represents the inflow to the i-th compartment, and $\mathcal{V}_i^-$ stands for the outflow from the i-th compartment. The dynamics of the epidemiological (mathematical) model is described by means of the following IVP:

$$x_i'(t) = \mathcal{F}_i(x(t)) + \mathcal{V}_i(x(t)), \quad 1 \le i \le n + m, \tag{4}$$
$$x(0) = \left(x_1^0, \ldots, x_{n+m}^0\right), x_i(t) \in [0, 1], 1 \le i \le n + m.$$

where the following assumptions are done:

(1) $\mathcal{F}_i(x(t)) > 0$ if $1 \le i \le n$, and $\mathcal{F}_i(x(t)) = 0$ if $n + 1 \le i \le n + m$.
(2) $\mathcal{V}_i^+(x(t)) \ge 0$, $\mathcal{V}_i^-(x(t)) \ge 0$ for $1 \le i \le n + m$. In fact, if $x_i(t) = 0$ then $\mathcal{V}_i^-(x(t)) = 0$.
(3) If $x_j(t) = 0$ for $1 \le j \le n$, then:

$$\mathcal{F}_i(x(t)) = 0, \quad \mathcal{V}_i^+(x(t)) = 0, \quad 1 \le i \le n. \tag{5}$$

In this case the qualitative theory of differential equations is used to analyze the behavior of the solutions of the system (4). Specifically results about the (local and global) stability of the steady states of the system (usually the disease free equilibrium point and the endemic equilibrium point) are obtained. This is characterized by the basic reproductive number: If $\mathcal{R}_0 < 1$ (*resp.*, $\mathcal{R}_0 > 1$) the disease free equilibrium point (*resp.*, the endemic equilibrium point) is asymptotically stable.

2.2 A Brief Critical Analysis

As is illustrated in the last section, mathematical models for malware propagation are based on the design of mathematical models for biological agents spreading. Consequently global (deterministic or stochastic) compartmental global models are considered. The main characteristics of this type of models are two:

(1) An homogeneous contact topology is considered, that is, the complete graph K_n (where n is the total number of devices) characterizes from the mathematical point of view the topology of communication connections between devices.
(2) The epidemiological coefficients involved in the model are the same for every pairwise communication contact between devices of the same types, and for every device of the same compartment.

As a consequence, each device is in contact with the rest of the devices at every step of time. This assumption could be realistic if the propagation phenomenon occurs in a very large networks such as Internet but really it is unrealistic if the propagation happens on other type of networks where the contact topology plays a fundamental role: IoT networks or drone swarms, for example.

On the other hand, and due to the nature of the mathematical tools used in the design of global models, heterogeneity and diversity cannot be considered in an efficient way. That is, it is not possible to take into account the local connections of the i-th device (its adjacent nodes, $\mathcal{N}_i$, using the graph theory terminology). At most, in some global models defined on complex networks, the degree distribution of such complex networks, $P(k)$, can be explicitly reflected in the mathematical expression of the incidence [9]. Furthermore, the particular specifications of devices (computational resources, role, energy consumption, etc.) are not taken into account either: note that the epidemiological coefficients involved in global models refer to characteristics of a general representative device of each compartment.

The global approach is correct and mathematically well founded. Consequently one can use the classical mathematical epidemiology (understood as the study of the models for biological agents propagation) as the theoretical framework for designing (and analyzing) mathematical models for malware spreading. However, for the sake of efficiency and realism, it is very important to point out that this approach exhibits some drawbacks than must be overcome. For example we can mention the following:

1. The nature and structure of models must be improved: the individual-based paradigm [1] should be used instead of the global paradigm.
2. The specifications of malicious code (which distinguish it from biological agents) must be considered in the determination of both epidemiological coefficients and structural relations (equations which involve variables and coefficients, and which describe the dynamics of the propagation process).
3. The characteristics of both, devices and network, must be also considered for designing mathematical models. They are usually very different from those showed by humans and the contact networks.

In this sense we should answer the following questions:

- What are the differences between malware propagation process and biological agents propagation process?
- What are the differences between malware and biological agents infection process (in the hosts)?
- What are the differences between device communication networks and human contact networks?

3 The Individual-Based Paradigm

In this section the individual-based paradigm is introduced and an illustrative example of an individual-based model (IB-model for short) for malware propagation is presented.

3.1 The Framework

In the individual-based paradigm the fundamental unit of modeling is the individual (that is, the devices). Consequently, as is mentioned in Sect. 2.2, in the design of mathematical models for malware propagation one has to take into account the following: the particular characteristics of the devices related with the propagation and infection processes, and the particular contact topology (that yields to considering complex networks topologies). In this sense, the scenario can be mathematically characterized by a directed and weighted complex network $\mathcal{G} = (\mathcal{V}, \mathcal{E})$, where the set of nodes $\mathcal{V} = \{v_1, \ldots, v_n\}$ represents the population of devices, and the edges stand for the communication connection between such devices. The dynamics of an IB-model is defined by means of the following:

- The time dependent variables that represent the states of the devices at each step of time: $s_i^t \in \mathcal{S}$.
- The state set $\mathcal{S}$ where the last mentioned variables take values. This set could be finite (if each device can be endowed at every step of time with a finite number of possible states), or infinite.
- The state of each device is updated at discrete steps of time accordingly to a local transition rule defined by continuous or discrete mathematical tools.
- The complex network, $\mathcal{G} = (\mathcal{V}, \mathcal{E})$, that mathematically describes the contact topology defined by the communication network.

Different local transition functions can be defined. Depending on the nature of this functions the model can be deterministic or stochastic, and the state set could be finite or infinite.

3.2 *An IB-Model Based on Differential Equations*

In this subsection an illustrative example of a deterministic individual based compartmental model whose dynamics is defined by differential equations. This is a SI (Susceptible-Infectious) compartmental model: population is divided into susceptible devices and infectious devices such that when the malware reaches a device, it remains on it indefinitely.

Description of the Model

Set $\mathcal{G} = (\mathcal{V}, \mathcal{E})$ the directed and weighted graph that defines the contact communication network. Let $v_i \in \mathcal{V}$ be the i-th node/device, then $k_i \leq n$ stands for its degree and $\mathcal{N}_i = \{v_{j_1}, \ldots, v_{j_{k_i}}\} \subset \mathcal{V}$ represents its neighborhood, that is the adjacent nodes from v_i: $v_j \in \mathcal{N}_i$ if and only if $e_{ij} = (v_i, v_j) \in \mathcal{E}$. Moreover each edge, for example $e_{ij} \in \mathcal{E}$, is endowed with a weight: $a_{ij} \in [0, 1]$, which stands for the strength of infection from the i-th device to the j-th device. Thus $A = (a_{ij})_{1\leq i,j\leq n}$ is the adjacency matrix associated to $\mathcal{G}$.

The state of the i-th device at step of time t is $s_i^t = (x_i(t), y_i(t)) \in [0, 1] \times [0, 1]$ where $x_i(t)$ represents the probability that the i-th device is susceptible at t, and $y_i(t) = 1 - x_i(t)$ stands for the probability that the i-th device is infectious at t. The dynamics is described by means of the following initial value problem [13]:

$$
\begin{aligned}
x_i'(t) &= -\sum_{v_j\in\mathcal{N}_i} a_{ji} x_i(t) y_j(t), \quad 1 \leq i \leq n, \\
y_i'(t) &= \sum_{v_j\in\mathcal{N}_i} a_{ji} x_i(t) y_j(t), \quad 1 \leq i \leq n, \qquad (6) \\
x_i(0) &= 1 - y_i^0,\ y_i(0) = y_i^0.
\end{aligned}
$$

Note that this system of $2n$ ordinary differential equations can be reduced to the following IVP with n ordinary differential equations:

$$
y_i'(t) = \sum_{v_j\in\mathcal{N}_i} a_{ji}\,(1 - y_i(t))\, y_j(t), \quad y_i(0) = y_i^0 \quad 1 \leq i \leq n, \qquad (7)
$$

If $y(t) = (y_1(t), y_2(t), \ldots, y_n(t))^\mathsf{T}$, system (7) that can be expressed in matrix form as follows:

$$
y'(t) = [Id_n - \operatorname{diag}(y(t))]\, A\, y(t), \quad y(0) = \left(y_1^0, y_2^0, \ldots, y_n^0\right). \qquad (8)
$$

3.3 *Illustrative Simulations*

In what follows two illustrative study cases are shown: the first one deals with malware propagation on a general IoT network whereas in the second one some simulations on a drone swarm are shown. In both scenarios the heterogeneity of devices are, obviously, considered and this is reflected in the strength of infection between the nodes, $0 \leq a_{ij} \leq 1$, $1 \leq i, j \leq n$, which depends on the characteristics of the sender device, the receiving device, the malware specimen and the communication channel. For the sake of simplicity it is considered that these coefficients do not depend on time, that is, they remain constants throughout time.

IoT Networks

In this case it is supposed that the device network is static: devices are stationary and they will not change the position after the initial deployment. Moreover, due to the inherent diversity of possible devices connected in this type of networks, the values of a_{ij} could be very different and vary greatly between the different pairs of connected nodes. For the sake of simplicity and visibility a grid network with 10×10 nodes/devices is considered and the coefficients representing the strength of infection are taken at random. Suppose that at $t = 0$ there is only one infectious device: the node with highest de degree centrality. In Fig. 1 the evolution of the infection probability of each device is shown: $y_i(t)$, $1 \leq i \leq n$.

Note that the infection probabilities (which are all zero at $t = 0$ with the exception of the probability of the unique infectious device at the onset of the outbreak—which, obviously, is constant an equal to 1 throughout the time—) rapidly grow until reaching the maximum value corresponding to the infectious

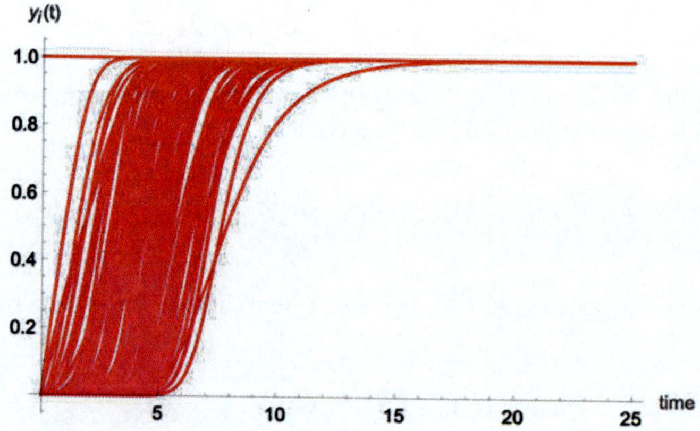

Fig. 1 Evolution of the infection probabilities of all devices

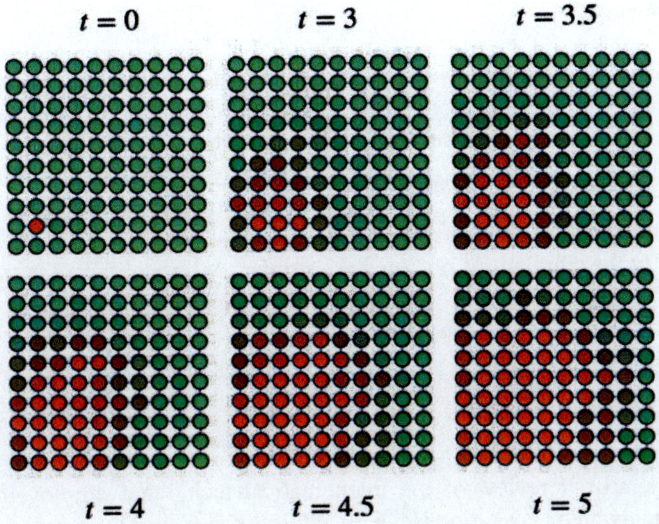

Fig. 2 Spatio-temporal evolution of malware propagation on the grid network at $t = 0, 3, 3.5, 4, 4.5, 5$

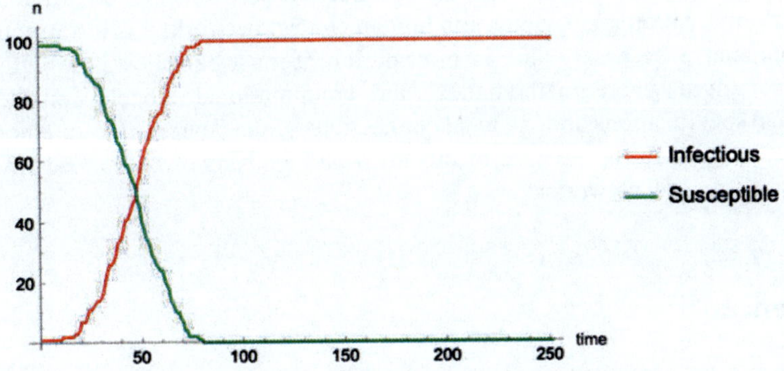

Fig. 3 Global evolution of the number of susceptible and infectious devices on the grid network

state of the device. Consequently an endemic steady state is obtained. This result, obtained empirically can be theoretically proved [13].

In Fig. 2 the spatio-temporal propagation of malware on the grid network is shown at some steps of time. Note that individual models not only allow us to determine the individual evolution of the state of each device but also the global evolution. In Fig. 3 the evolution of the estimated number of susceptible and infectious devices is introduced where the criterion that has been used is the following: the state of the i-th device at t is susceptible (*resp.* infectious) if $x_i(t) > y_i(t)$ (*resp.* $x_i(t) < y_i(t)$).

Swarm Networks

Finally, in this case it is supposed that the network topology is dynamic: due to the constant movement of drones the contact communication networks changes over time. This can be represented by means of the of complex networks with the same set of nodes: $\mathcal{G} = \{\mathcal{G}_t = (\mathcal{V}, \mathcal{E}_t), t \geq 0\}$. As the great majority of devices have the same characteristics and specifications, it could be supposed that all epidemiological coefficients are very similar: $a_{ij} \sim a_{rs}$ with $e_{ij}, e_{rs} \in \mathcal{E}$. Consequently the simulations obtained are similar to those obtained in the case of a general IoT network.

4 Conclusions

In this work a brief review of the mathematical techniques employed to design models for malware propagation is presented. It is shown that the methodology used is based on classical mathematical epidemiology devoted to the study of biological agents propagation. This approach is not completely efficient and realistic since the characteristics of both malware and communication networks are too different from those exhibit by biological agents and human contact networks. As a consequence a new approach is necessary and, in our opinion, it must be follows the individual-based paradigm where mathematical and computational models considers the individual specifications and the local connections of the devices. This methodology is particularly useful in the case of the study and analysis of "advanced" malware propagation on IoT networks.

References

1. Batista Guerra, F., Martin del Rey, A., Queiruga-Dios, A.: A new individual-based model to simulate malware propagation in wireless sensor networks. Mathematics **8**(3), 410 (2020). https://doi.org/10.3390/math8030410
2. Diekmann, O., Heesterbeek, J., Metz, J.: On the definition and the computation of the basic reproduction rate r_0 in models for infectious diseases in heterogeneous populations. J. Math. Biol. **28**, 365–382 (1990). https://doi.org/10.1007/BF00178324
3. Dietz, K.: Transmission and control of arboviruses. In: Ludwig, D., Cooke, K. (eds.) Epidemiology, pp. 104–121. SIAM, Philadelphia (1975)
4. Gleissner, W.: A mathematical theory for the spread of computer viruses. Comput. Secur. **8**, 35–41 (1989). https://doi.org/10.1016/0167-4048(89)90037-0
5. Goffman, W., Newill, V.: Generalization of epidemic theory: An application to the transmission of ideas. Nature **204**, 225–228 (1964). https://doi.org/10.1038/204225a0
6. Hamer, W.: Epidemic disease in England - the evidence of variability and of persistence. Lancet **167**, 733–738 (1906)
7. Hernandez Guillen, J.D., Martin del Rey, A.: A mathematical model for malware spread on WSNs with population dynamics. Physica A **545**, 123609 (2020). https://doi.org/10.1016/j.physa.2019.123609

8. Hethcote, H.: Mathematical models for the spread of infectious diseases. In: Ludwig, D., Cooke, K. (eds.) Epidemiology, pp. 122–131. SIAM, Philadelphia (1975)
9. Karyotis, V., Khouzani, M.: Malware Diffusion Models for Modern Complex Networks. Theory and Applications. Morgan Kaufmann (2016)
10. Kermack, W., McKendrick, A.: A contribution to the mathematical theory of epidemics. Proc. R. Soc. Lond. A Math. Phys. Eng. Sci. **115**, 700–721 (1927). https://doi.org/10.1098/rspa.1927.0118
11. MacDonald, G.: The analysis of equilibrium in malaria. Trop. Dis. Bull. **49**, 813–829 (1952)
12. Martin del Rey, A.: Mathematical modeling of the propagation of malware: a review. Secur. Commun. Netw. **8**(15), 2561–2579 (2015). https://doi.org/10.1002/sec.1186
13. Mei, W., Mohagheghi, S., Zampieri, S., Bullo, F.: On the dynamics of deterministic epidemic propagation over networks. Annu. Rev. Control **44**, 116–128 (2017). https://doi.org/10.1016/j.arcontrol.2017.09.002
14. Nwokoye, C.H., Madhusudanan, V.: Epidemic models of malicious-code propagation and control in wireless sensor networks: An indepth review. Wirel. Pers. Commun. **125**, 1827–1856 (2022). https://doi.org/10.1007/s11277-022-09636-8
15. Ross, R.: The Prevention of Malaria. John Murray, London (1911)
16. Ross, R., Hudson, H.: An application of the theory of probabilities to the study of a priori pathometry - part III. Proc. R. Soc. Lond. A **43**, 225–240 (1917). https://doi.org/10.1098/rspa.1917.0015

Forms of Assessment in view of Development of Mathematical Competencies

Daniela Richtarikova

Abstract During the past twenty years, the need to define outcomes of education as Competences has been recognized in many countries. The European Parliament and the Council of the European Union acknowledged the mathematical competence in 2006 to be a key competence for everyday life (Recommendation of the European Parliament and of the Council of 18 December 2006 on key competences for lifelong learning (2006/962/EC). https://eur-lex.europa.eu/legal-content/EN/TXT/PDF/?uri=CELEX:32006H0962&from=EN. Last accessed 10 July 2023) and it was incorporated into the national educational programs. The engineering tertiary educational bodies grouped around SEFI, Mathematics Special Interest Group adopted the concept of eight overlaying mathematical competencies introduced by Danish KOM project (Niss M, Mathematical competencies and the learning of mathematics: the Danish KOM project. In: A. Gagatsis, S. Papastravidis (eds.) 3rd Mediterranean conference on mathematics education. Hellenic Mathematical Society and Cyprus Mathematical Society, Athens, Greece, pp 115–124) in 2003 which were linked to the carefully specified core content, and formulated as the core content-related competencies defining learning outcomes in the third edition of "A Framework for Mathematics Curricula in Engineering Education" (Alpers, B. et al., A Framework for Mathematics Curricula in Engineering Education. SEFI, Brussels, (2013)).

The paper deals with appropriate forms of summative assessment, applicable in the higher engineering education. Special attention is paid to mutual communication in a class, a phenomenon that was negatively affected to a large extent by pandemic restrictions. We introduce discussion as a part of summative assessment, not only providing the space for evaluating the mathematical competencies, but more, as an aspect that essentially contributes to their cultivation. The results of a study in which 230 students participated is presented.

D. Richtarikova (✉)
Faculty of Mechanical Engineering, Slovak University of Technology in Bratislava, Bratislava, Slovakia
e-mail: daniela.richtarikova@stuba.sk

V. Gayoso Martínez et al. (eds.), *Mathematical Methods for Engineering Applications*, Springer Proceedings in Mathematics & Statistics 439,
https://doi.org/10.1007/978-3-031-49218-1_15

The strategy of deploying more assessment forms is discussed, taking into account building of comprehensive Mathematical Competency in gradual continuous and permanent manner.

1 Introduction

1.1 Teaching Mathematics in the Beginning of 2020s

Present time can be characterized by massive spread of information among population. People enjoy, but also face everyday usage of internet where they consume, as well as, post their own pieces of text, images, or videos. How to find and identify relevant information in the overwhelming amount available is not easy, and there is a reason to suggest that it will not get any easier in the future either.

Mathematics has always belonged to the basic teaching subjects and is taught across all compulsory levels of national educational systems. In Slovakia, at the turn of the millennium, a sequence of several changes fundamentally changed the general perception of its importance.

Mathematics was decided to no longer be the subject of obligatory leaving exam at secondary schools and it was no longer required in the entrance exams for admission procedure at universities. Consequently, it lost its privilege position among subjects when the amount of teaching hours at primary and secondary schools dramatically fell, and mathematics irretrievably lost its importance appreciation also in public. At tertiary engineering schools, the number of teaching hours was desperately restricted also, approximately to one third, due to the Bologne transformation process. On the contrary, demands on using mathematics in engineering have not decreased but principally changed. For the present, it is more about understanding the context and setting up suitable calculation methods, which of course require knowledge of the mechanism and conditions of use. To teach all necessary content is impossible. Taking into account the fact that the students coming to study the engineering field at university:

- are poorly prepared in mathematics due to the abovementioned changes in the system
- mostly belong to average and weak students (with the exception of information technology studies)
- perceive mathematics just to be a calculation technique

to teach mathematics in means to achieve a satisfactory level of *competencies related to the most essential but still controllable content* is a possible solution. It also appears to be the attainable solution that would enable students to overcome the pitfalls of new coming artificial intelligence, and to use widely available information systems, software technologies and products in reasonable way.

1.2 Educational Outcomes as Competencies

The development of pedagogical science is undoubtedly related to the rapid development of psychology in the twentieth century.

An attempt to isolate the outcomes of mathematical education in the matter of what learners should know, and what they should be able to do after completing the given course for primary and secondary schools, could be seen in middle European countries already after the World War II. Czechoslovak archived state education programs revealed that (see more in [8])

- "skills" and "content" were separated in learning plans and curricula already in 1948 [11]
- ten years later, educational goals had to be reached using real economic data in the neighborhoods, solving practical problems, acquiring logical thinking by solving judgment tasks. Pupils were supposed to learn to be precise, critical and persistent, thereby creating elements of a "scientific worldview" [5]
- in 1987, the outcomes were formulated as [10]:

  targeted knowledge on:

  - defined mathematical concepts and relationships between them, including inferences
  - mathematical terminology, phraseology and symbolism
  - methods of mathematical work
  - mathematizing a real situation
  - graphical-algebraic approaches to the solution
  - application of the mathematics curriculum in the given field

  targeted skills:

  - to know how to find, correctly evaluate, sort and use mathematical information
  - to work actively with mathematical terminology, phraseology and symbolism
  - to know how to use acquired knowledge in solving tasks and problems
  - to master algorithms for solution of mathematical problems
  - to solve tasks and problems rationally and accurately in hand as well as aided by calculators, tables, and computers
  - to know how to apply mathematics in natural sciences and in the particular field of study
  - to interpret correctly obtained mathematical results
  - to read mathematical text with understanding appropriately to learners' age and level of their mathematical knowledge/e.g. text in the textbook of mathematics

The European Parliament and the Council of the European Union (EC) acknowledged the mathematical competence in 2006 to be a key competence for everyday life [4] and not long after, it was incorporated into all EU national educational programs. In Slovakia, the learning outcomes have been formulated as educational standards in contents and performance in order to develop *the mathematical competence* in the way how it was formulated by EC. Heavy emphasis is now placed on the way of teaching, when "Teaching should be conducted in such a way as to develop students' logical and critical thinking, their ability to argue, and to enable each of them to gain knowledge through discovery. Sufficient attention and time must be devoted to the use of acquired knowledge in solving real tasks". "The use of ICT is also an important part of teaching ..." [3].

The engineering tertiary educational bodies grouped around The European Society for Engineering Education (SEFI) Mathematics Special Interest Group (MSIG) adopted the concept uniting the target knowledge and skills around eight overlaying mathematical competencies introduced by Danish KOM project [4] in 2003, which were linked to the carefully specified core content, and formulated as the core content-related competencies defining learning outcomes in the third edition of "A Framework for Mathematics Curricula in Engineering Education" [1].

Deviation from content, and transition to goals formulated as active skills in action is apparent all over the world. The importance of developing "mathematical competence" for forming of mathematical literacy is nowadays recognized across all continents, even though differing a little in partial competencies classification. (e.g. [2, 4, 6]). Though research papers and case studies on methods and processes of developing particular competencies are relatively abundant, the assessment possibilities with respect to particular competencies are not so frequent subject of research.

2 Notes on Some Common Forms of Summative Course Assessment

One of the most spread and popular assessment forms is a multiple choice test. In connection with a school information system, it represents a very effective and variable tool, which makes it possible to assemble selection of one or more correct answers and some number of distractors randomly from a set of possible answers. The tool is reliable allowing to pick up questions as well as closed answers individually in random manner. Evaluation is automatic and objective, which is its main advantage. In general, it does not need but enables external intervention. The questions are content oriented with dichotomous (true, false) evaluation, which can be considered to be its main weak point. The level of reached competency can be only presumed with respect to the character of questions. In case the answers differ in minimum, high demand for attention obscures conceptual understanding or basis of calculation.

A test with open answers has much better possibilities for assessment of competencies. Students create answers by themselves. An evaluator manually grades the performance elements representing quality of examined competence or content. Using competence determined learning outputs as scoring category allows to calculate scores (simple or weighted) of specific competencies across all tasks (see details in [7, 9]) and to give feedback to the teacher as well to the student about their achieved level, and to focus more on competencies they want to improve. The biggest disadvantage is examiner's high time consumption. Influence of personal subjectivity can be eliminated by strictly predefined scale of the quality grading.

A test with semi-open answers asks to add words, numbers or expressions on specific places. The evaluation can also be automated if the correct answers are precisely specified. Dealing with assessing competencies, it stands in between previously mentioned kinds of test.

At technical universities, oral exams usually serve to assess theoretical parts. The most conventional way of its execution includes answering randomly picked questions on particular topic, or answering supplementary questions about a work or problem done within a course or within a part of the exam. Although oral examination should be taken at secondary schools minimally twice a year, at tertiary engineering schools, it used to be replaced by written test (usually in the form of closed multiple choice test), and in some cases, theory questions are even omitted at all. The reason is prosaic, a large number of students in a course and large time requirement. On the other hand, the level of particular competency can be detected easier and more, articulation in one's own words stimulates further clarification, comprehension, and memory.

Projects or semester works belong to traditional forms of learning and assessment at universities. Since 1990s, they have become very popular also at primary and secondary schools. On contrary to previous forms, they are usually conducted as home, open books teamwork, executed in pairs or small groups, sometimes also individually. Regarding to the competence development, they are very valuable connecting effectively real life with school using interdisciplinarily information and skills of various kind. They require a high degree of student's ability to work autonomously in team, and they are always associated with the highest aims of taxonomies of learning objectives, like specific and nonspecific transfer of Nemierko's taxonomy, or analysis, synthesis and evaluation of Bloom's taxonomy; developing critical thinking skills and competences in mathematical, technical and social areas. In appropriate way, they teach to work with literature and how to cite avoiding plagiarism.

3 Discussion as a Part of the Exam

Exams are one of the experiences that usually become deeply etched on students' memories, often for the rest of their lives. They represent an indivisible and very important part of the pedagogical process. We took advantage of this feature, and in order to achieve an overall better understanding of the course sense and of the main principles of the taught methods we decided to make the examination a little bit special and to include, experimentally, discussion into the summative assessment of two compulsory-elective mathematical courses. The first one was the subject Applied mathematics in the first year of master's degree. After impressive results, we have applied the exam structure also in the second year of the bachelor degree in the subject of Numerical mathematics.

The exam consisted of three components:

The first one covered the compulsory written part of open problems. Besides the selection and application of suitable methods, correct computations and results, the problems usually required also verbal reasoning and explanation item of the method itself, its limitations, conditions of use, or process calculation, result, etc.

The second one was the project work dealing with real simplified problems that were given or subject to student's choice, but posed in other specialist courses, theses or areas of interest. It was compulsory for master's degree and optional for bachelor's degree.

The exam was completed by the oral part executed in the form of discussion. Students were entitled to participate it, if they had achieved at least 60% in the written part and had submitted the project with a satisfactory evaluation at a minimum.

The final mark was set by all three parts.

The discussion was conducted in five steps:

1. The discussion started with speaking about the topics of the course, which parts students consider to be interesting and why. This opening encouraged students to make quick revision of the content and focus their mind on the substance and issues which were not obvious for them, that surprised them or that caused them difficulties. All students took turns in the free talk which helped to create a pleasant ambience.
2. The conversation went into a deeper discussion on some of the above issues. Students informed about basic characteristics, principles, necessary and sufficient conditions of use, pitfalls, what made difficulties, how to solve them, and applications in engineering.
3. From the previous conversation, a problem used to emerge, or it could be posed by the examiner, which had not been addressed within a course but the students could come to solution by a combination of their knowledge and skill experiences.
4. At the end of the discussion, students exchanged their ideas about the most important conclusions of the course, the messages to other fields, and ways how to treat with learned methods in the engineering area.
5. The discussion finished with their self-assessment or peer-assessment in the discourse and overall exam assessment.

4 Discussion Organization, Comments, Observations and Results

Discussions were held in a class with blackboard and chairs arranged into a circle. The blackboard was a common place for notes and visual explanations. The usual number of students was up to 8. The maximal time of discussion was one hour.

The discussion required a high level of knowledge on the side of students and a very responsive approach on the side of the examiner, taking into account that it was based on direct immediate reaction. The students had to be attentive, listen to others and react in real time. They did not have the time for preparation, so they had to think aloud and present their style of thinking, to reveal what they really knew and how they were able to work with their pieces of knowledge and skills, especially when looking for solution of unknown problem in the step three. Controlling a framework of content, the examiner encouraged students to join the discussion with their opinion and their way of solution. To keep the discussion in flow, it was important to avoid direct negative evaluation, and to focus on facts and act positively. In a positive atmosphere and when the students were well prepared, the discussions were fluent and required only minimal intervention from the teacher. Forming discussion groups from students of approximately the same competence level allowed examiners to adequately adapt the methodology of conducting the discussion: to challenge individual students to reactions, to pose the problem and supplementary questions of appropriate difficulty, or if necessary, only let students intentionally formulate questions on the most important issues and answer them.

Naturally, before such unusual form of assessment, the students were trained to discuss in classes during the semester. Regularly, a short whole class discussions were inserted into the lessons, where students were to articulate their mathematical opinion, to give an agreeing or disagreeing statement and justify it. Students had also the opportunity to discuss in groups when developing projects outside the lessons or when they prepared themselves for midterm tests.

The feedback from students was exceptional. The students considered this way of examination to be less stressed, although they were exposed to direct comparison, immediate reactions and in the end they had to evaluate themselves or peer to peer. Despite the demands described above, no one wanted to be assessed by traditional oral form or by written theory test. The following items were considered to be the most appreciated benefits of discussion:

- gaining self-confidence, that even when faced with an unknown problem, students were able to analyze and synthetize the learned knowledge and skills and come to new conclusions and opinions
- observation, how "thinking in a team" is effective
- the summary of important facts and relations in mutual conversation helped many students to understand relations they were not able to comprehend alone
- consciousness increase of the relationship between mathematical methods and engineering
- the way the students were treated

5 Conclusion

Overall, 80 students of master's degree and 150 students of bachelor's degree took the exam containing the discussion exam. We have to conclude that each discussion was different depending on the level of knowledge and skills of the students who met in the exam group, what affected mainly the difficulty level of discussed problems. Nevertheless, comparing the behavior and reactions of same students in discussions within a course, as well as the results of their tests, we noted an increase mainly in critical thinking and in ability of abstraction. We also saw an increase in the ability to articulate a mathematical opinion, in the way the students give information to others, in the way that students were able to chain arguments and to defend their opinion, as well as in the ability to listen to others. Although to conduct the assessment using the discussion is not easy for examiners and it requires good planning, hard communication work with students during semester, and prompt reactions on the spot, its contribution to developing the mathematical competence is evident.

Exam structure consisting of several examination forms creates a balanced model that facilitates appraisal of all kinds of competencies. It includes activities preferred by various cognitive and learning styles, combines theoretical with hands-on items and gives complex information about the student's competencies in a field. Moreover, it is lively, for students attractive, and each time providing some action point.

References

1. Alpers, B., et al.: A Framework for Mathematics Curricula in Engineering Education. SEFI, Brussels (2013)
2. McGehee, N.: Competency-Based Education in Mathematics. https://michiganvirtual.org/blog/competency-based-education-in-mathematics/ (2021). Last accessed 10 July 2023
3. National educational program "Štátny vzdelávací program", (in Slovak), [National educational program]. https://www.statpedu.sk/sk/svp/statny-vzdelavaci-program/. Last accessed 10 July 2023
4. Niss, M.: Mathematical competencies and the learning of mathematics: the Danish KOM project. In: Gagatsis, A., Papastravidis, S. (eds.) 3rd Mediterranean Conference on Mathematics Education, pp. 115–124. Hellenic Mathematical Society and Cyprus Mathematical Society, Athens, Greece (2003)
5. O úzkom spojení školy so životom a o ďalšom rozvoji výchovy a vzdelania v ČSR. (in Slovak) [The resolution on close connection of school with life and the further development of education in Czechoslovakia], Bratislava 1959
6. Recommendation of the European Parliament and of the Council of 18 December 2006 on key competences for lifelong learning (2006/962/EC). https://eur-lex.europa.eu/legal-content/EN/TXT/PDF/?uri=CELEX:32006H0962&from=EN. Last accessed 10 July 2023
7. Richtarikova, D.: Mathematical competence assessment and work in groups. In: Proceedings of the 20th SEFI SIG in Mathematics, pp. 13–18. European Society for Engineering Education, Brussels (2021) ISBN 978-2-87352-022-9

8. Richtarikova, D.: On grounds for competence oriented teaching and assessment. In: Proceedings of the 19th SEFI Mathematics Working Group Seminar, pp. 59–69. European Society for Engineering Education, Brussels (2018) ISBN: 978-2-87352-017-5
9. Richtarikova, D.: Mathematical competency oriented assessment – rules math guides on complex numbers. In: ICMASE 2020 Proc. Salamanca University Press. https://doi.org/10.14201/0AQ0302. Last accessed 10 July 2023
10. Učebné osnovy pre študijné odbory stredných odborných učilíšť (in Slovak) [Curricula for secondary vocational schools], pp.199–233, SPN, (1987)
11. Zákon č. 95/1948 Zb. Zákon o základní úpravě jednotného školství (školský zákon). (in Czech) [Law on basic adjustment of unique education. School Law]. http://www.zakonypreludi.sk/zz/rocnik/1948. Last accessed 10 July 2023

Deep-Control of Memory via Stochastic Optimal Control and Deep Learning

Emel Savku

Abstract In this survey work, we introduce Stochastic Differential Delay Equations and their impacts on Stochastic Optimal Control problems. We observe *time delay* in the dynamics of a state process that may correspond to inertia or memory in a financial system. For such systems, we demonstrate two special approaches to handle delayed control problems by applying the Dynamic Programming Principle. Moreover, we clarify the technical challenges rising as a consequence of the conflict between the path-dependent, infinite-dimensional nature of the problem and the necessity of the Markov property. Furthermore, we present two different Deep Learning algorithms to solve targeted delayed control tasks and illustrate the results for a complete memory portfolio optimization problem.

1 Introduction

Stochastic Optimal Control studies on decision-making under uncertainties to maximize or minimize an objective criterion that describes the desired control task. The solution of the control problem designs the time path of controlled variables corresponding to the state processes. So far, the control theory has gotten significant attention from the researchers and been utilized in a wide range of problems in finance, insurance, physics, biology, and management systems (for comprehensive monographs with applications in finance, see [38, 42]).

It is well known that two mainly applied techniques of Stochastic Optimal Control arise as *Dynamic Programming Principle* (**DPP**) and *Stochastic Maximum Principle*. In this survey work, we mainly focus on DPP and its challenges in a *time-delayed* environment. The fundamental concern of DPP is to find the optimal control processes and the value function of targeted control problems, both of which have to satisfy some technical verification conditions. The value function appears as a

E. Savku (✉)
University of Oslo, Department of Mathematics, Oslo, Norway
https://www.mn.uio.no/math/english/people/aca/emelsa/index.html

V. Gayoso Martínez et al. (eds.), *Mathematical Methods for Engineering Applications*, Springer Proceedings in Mathematics & Statistics 439,
https://doi.org/10.1007/978-3-031-49218-1_16

solution of a Partial Differential Equation (**PDE**), called *Hamilton-Jacobi-Bellman Equation* (**HJBE**). The main technical issues of using this method properly is to keep *Markov property* and face the *curse of dimension*. Moreover, we need to justify the *smoothness* of the value function; otherwise, we may need to find a solution in the viscosity sense [7, 18].

Furthermore, one of our principal curiosities in this work is to introduce the impact of *memory*. This phenomenon appears in the dynamics of the state processes as a significant mathematical challenge called *time delay* and lead us to work on *Stochastic Differential Delay Equations* (**SDDEs**) [35]. At the very first step, the main difference between the models with and without delay components arises in the mathematical approach of obtaining a solution for such a system.

Let us explain this technical fact for a diffusion process as in Mohammed [35]:

We have a complete probability space $(\Omega, \mathbb{F}, (\mathscr{F}_t)_{t\geq 0}, \mathbb{P})$, which satisfies the usual assumptions, i.e., $(\mathscr{F}_t)_{t\geq 0}$ is a right-continuous filtration, and for each $t \geq 0$, $\mathscr{F}_t$ contains all $\mathbb{P}$-null sets in $\mathbb{F}$. Then, the state process is defined as:

$$dX(t) = \sigma X(t-\delta)dW(t), \qquad t \geq 0. \tag{1}$$

If $\delta = 0$, then very well-known closed-form solution of Eq. (1) can be obtained by applying Itô 's formula once, and we get:

$$X(t) = X(0)e^{\sigma W(t)-\sigma^2/2}, \qquad t \geq 0.$$

On the other hand, if we assume $\delta > 0$, we need an initial path $\phi(\cdot)$ to solve Eq. (1) such that

$$X(t) = \phi(t), \qquad -\delta \leq t \leq 0.$$

Then, if terminal time T is not finite, by recursive Itô integrations over steps of length δ, we observe that there is no closed form solution:

$$X(t) = \phi(0) + \sigma \int_0^t \phi(u-\delta)dW(u), \qquad 0 \leq t \leq \delta,$$

$$X(t) = X(r) + \sigma \int_r^t \left[\phi(0) + \sigma \int_0^{v-\delta} \phi(u-\delta)dW(u)\right] dW(v), \qquad \delta < t \leq 2\delta,$$

$$\dots = \dots, \qquad 2\delta < t \leq 3\delta,$$

$$\vdots$$

Here, the solution process $\{X(t) : t \geq -\delta\}$ is still an $\mathscr{F}_t$-martingale, but it is not Markovian anymore.

From another side, for such delayed models, in stochastic control theory, it becomes impossible to establish an immediate application of DPP since we lose the Markov property. Furthermore, the initial condition of the control task appears as an initial segment rather than a constant value. Hence, our control problem is defined as a *path-dependent* and *infinite-dimensional* control task. Consequently, such equations impose the researchers to reformulate targeted problems with new, advanced theoretical structures and to implement sophisticated computational techniques (for an extensive and well-developed theory of control problems in infinite dimension, see [11]).

In our work, we aim to introduce some special approaches to find the value function and the control processes in such a time-delayed environment. As we indicated, firstly, we have to derive HJBEs, i.e., the corresponding PDEs, which are in an infinite dimensional space for such a setting. The first approach focuses on finding a *smooth solution*, i.e., a value function with second-order continuous derivatives, but in general, value functions do not have this property. Therefore, a good approach appears as the concept of viscosity solutions [7], which does not require the smoothness property to describe the solution and can be studied in the case of infinite dimension. In the latter one, the HJBEs involve Fréchet derivatives; consequently, our purpose is to show that the value function is the unique viscosity solution of the corresponding PDE; see [5, 13–15, 21, 24, 25, 40] and references therein. One of the very recent methodologies to describe an infinite-dimensional control problem for which the goal is to apply the theory of viscosity solutions in a Hilbert space has been provided in [15] (a very nice summary of the efforts made so far for the viscosity solutions and a detailed literature review can be found in this work).

Another important approach followed by researchers is to study on the techniques, which reduce the dimension and solve the delayed control problem in a Markovian setting; see [9, 10, 13, 33, 34]. Many authors have focused on the solutions of path-dependent PDEs and viscosity solutions, and moreover, several different theoretical perspectives have been developed for arising nonlinear HJBEs [1, 8, 13, 29]. Here, we can explicitly conclude that the DPP approach for delayed control problems is *still* a very hot topic in stochastic control theory, even after many years since its first appearance in literature; see [8, 13, 21, 23, 25, 33, 40, 44].

Furthermore, the delayed models respond to the needs of *financial* and *actuarial* problems in a more realistic way. It is well known that investors tend to look at *historical performance* of risky assets before making their investment decisions. In financial markets, even small delays in information flow may affect the bid and ask prices, for which we see the rising importance of high-frequency trading technologies. Besides, it is clearly observed that for an accurately priced policy, the insurance companies or agencies have to take into account the data collected from past events related to the insured or to the relevant environment. These kinds of facts have been studied in a diversified list of control works such as [1, 5, 6, 9, 12, 14, 22, 24, 34, 39, 43, 45].

On the other hand, we see efficient implementations of Deep Learning in Stochastic Optimal Control theory more and more. Many practical decision-making problems can be nonlinear and *high dimensional*. Deep Learning utilizes *Artificial Neural Networks* to transform a set of very large inputs into a set of outputs. In addition, the word *deep* expresses the use of multiple layers in the network and corresponds to nonlinear systems. So far, several attempts have arisen to develop solution methods for HJBEs via neural networks; see [2, 3, 19, 20, 26, 27, 30, 37] and references therein. We see that some of these works are for deterministic models, but we still believe that they can be inspiring to develop techniques for stochastic models. Also, we should emphasize that these papers focus on diffusion models *without delay*. Just in [26], we see that Recurrent Neural Networks have been applied for a delayed diffusion process. When we focus on learning to store information over extended time intervals by recurrent back propagation, the efficiency of *Long-Short Term Memory* networks (**LSTM**) has been studied in several different fields, such as handwriting recognition, speech recognition, machine translation, speech activity detection, robot control, video games, and healthcare [28, 46]. The LSTM architecture can create an additional module in a neural network that can effectively learn which information can be utilized in a sequence and when that information is no longer needed.

Finally, we would like to emphasize that both Stochastic Optimal Control and Deep Learning provide solutions by applying different methods with their own limitations and strengths. The goal of our survey work is to attract the researchers' attention to this very new, demanding, and promising connection. We will present some special approaches that attempt to handle the memory impact with the help of very sophisticated mathematical techniques. This chapter is organized as follows: In Sect. 2, we introduce the theoretical development of the solution set-up of a stochastic control problem for a delayed diffusion process and show how to reduce the dimension of the problem. In Sect. 3, we provide the verification results and mathematical nature of time-delayed problems in an infinite-dimensional setting. In Sect. 4, we focus on two types of Deep Learning algorithms to handle stochastic control problems with memory. The last section is devoted to the conclusion and future work.

2 A General Formulation of Bellman's Principle of Optimality with Memory

In this section, we will follow the assumptions and technical construction of [33] to present a version of *Bellman's principle of optimality* for stochastic control problems with delay.

Let us fix $0 \leq T < \infty$ and $0 \leq \delta < \infty$. The integers $n, r \geq 1$ hold and $C[-\delta, 0]^n$ denotes the separable Banach space of continuous paths $\phi : [-\delta, 0] \to \mathbb{R}^n$ with the norm $\|\phi\| := \sup_{-\delta \leq s \leq 0} |\phi(s)|$, where $|\cdot|$ is the Euclidean norm in $\mathbb{R}^n$. Here, the segment of the path of a function $\phi_t : [-\delta, T] \to \mathbb{R}^n$ from $t - \delta$ to t, $0 \leq t \leq T$, is defined as follows:

$$\phi_t(s) = \phi(t+s), \quad -\delta \leq s \leq 0.$$

Let us introduce the SDDE:

$$\begin{aligned} dX(t) &= b(t, X_t)dt + \sigma(t, X_t)dW(t), \\ X(s, \omega) &= \phi(s, \omega), \qquad \text{for} \quad -\delta \le s \le 0, \end{aligned} \tag{2}$$

where $\phi : \Omega \to C[-\delta, 0]^n$ is a random variable, independent of Brownian motion W with distribution:

$$\mu(B) = \mathbb{P}(\phi \in B), \qquad B \in \mathscr{B}(C[-\delta, 0]^n).$$

Here, $\mathscr{B}(C[-\delta, 0]^n)$ represents the σ-algebra generated by $C[-\delta, 0]^n$. Moreover, we define $\mathscr{G}_t := \sigma(\phi,\ W(s);\ 0 \le s \le t)$ for $0 \le t \le T$, and $\mathscr{N}$ denotes the collection of $\mathbb{P}$-null sets in Ω. Then, $\{W(t), \mathscr{F}_t; 0 \le t \le T\}$ is an r-dimensional Brownian motion, and

$$\mathscr{F}_t := \sigma(\mathscr{G}_t \cup \mathscr{N}), \qquad 0 \le t \le T, \tag{3}$$

justifies the right continuity and completeness conditions.

The authors adapt a method provided by Yong and Zhou [47] in Chapter 4 to prove DPP. For this purpose, authors need to develop the weak uniqueness of solutions of SDDEs. Hence, firstly, let us present the following definitions to describe one of the main results of Larssen [33].

Definition 1 (Larssen [33]) A strong solution of Eq. (2) on a given probability space $(\Omega, \mathscr{F}, \mathbb{P})$ with a fixed Brownian motion W and initial path ϕ is a process $X = \{X(t); -\delta \le t \le T\}$ with continuous sample paths and the characteristics:

S1 $X(t)$ is $\mathscr{F}_t$-measurable for $0 \le t \le T$ and $\mathscr{F}_0$-measurable for $-\delta \le t \le 0$,
S2 $\mathbb{P}[X_0 = \phi] = 1$,
S3 for $1 \le i \le n$ and $1 \le j \le r$, we have

$$\mathbb{P}\left[\int_0^T \left\{|b_i(t, X_t)| + \sigma_{ij}^2(t, X_t)\right\} dt < \infty\right] = 1,$$

S4 Equation (2) can be represented in integral form:

$$\begin{aligned} X(t) &= X_0(0) + \int_0^t b(s, X_s)ds + \int_0^t \sigma(s, X_s)dW(s), \qquad 0 \le t \le T, \\ X(t) &= X_0(t), \qquad -\delta \le t \le 0 \end{aligned}$$

holds almost surely.

Let us give another definition that describes strong uniqueness of a solution of the SDDE (2):

Definition 2 (Larssen [33]) Let $b(t, \xi)$ and $\sigma(t, \xi)$ be given. Suppose that whenever W is an r-dimensional Brownian motion on some probability space $(\Omega, \mathscr{F}, \mathbb{P})$ and $\{\mathscr{F}_t\}$ is given by Eq. (3), ϕ is an initial path independent of W and X, $\tilde{X}$ are two strong solutions of Eqs. (2) relative to W with initial path ϕ, then

$$\mathbb{P}[X(t) = \tilde{X}(t) \text{ for all } 0 \leq t \leq T] = 1.$$

Now we can state that under these assumptions, strong uniqueness holds for (b, σ). For further development of the theory, let us provide the concept of weak solution of an SDDE (2):

Definition 3 (Larssen [33]) A weak solution of Eq. (2) is a triple (X, W), $(\Omega, \mathscr{F}, \mathbb{P})$, $\{\mathscr{F}_t\}$ where

W1 $(\Omega, \mathscr{F}, \mathbb{P})$ is a probability space and $\{\mathscr{F}_t\}$ is a filtration with $\mathscr{F}_t \subset \mathscr{F}, 0 \leq t \leq T$, satisfying the usual assumptions,

W2 $X = \{X(t); -\delta \leq t \leq T\}$ is a continuous process and $X(t)$ is $\mathscr{F}_t$-measurable for $0 \leq t \leq T$ and $\mathscr{F}_0$-measurable for $-\delta \leq t \leq 0$,

W3 for $1 \leq i \leq n$ and $1 \leq j \leq r$, we have

$$\mathbb{P}\left[\int_0^T \left\{|b_i(t, X_t)| + \sigma_{ij}^2(t, X_t)\right\} dt < \infty\right] = 1,$$

W4 the integral version of Eq. (2):

$$\begin{aligned} X(t) &= X_0(0) + \int_0^t b(s, X_s)ds + \int_0^t \sigma(s, X_s)dW(s), \qquad 0 \leq t \leq T, \\ X(t) &= X_0(t), \qquad -\delta \leq t \leq 0 \end{aligned}$$

holds almost surely.

Finally, let us give the following definition to present one of the fundamental theorems of [33].

Definition 4 (Larssen [33]) We say that weak uniqueness holds for Eq. (2) if, for any two weak solutions (X, W), $(\Omega, \mathscr{F}, \mathbb{P})$, $\{\mathscr{F}_t\}$ and $(\tilde{X}, \tilde{W})$, $(\tilde{\Omega}, \tilde{\mathscr{F}}, \tilde{\mathbb{P}})$, $\left\{\tilde{\mathscr{F}}_t\right\}$ with the same initial distribution

$$\mathbb{P}[X_0 \in B] = \tilde{\mathbb{P}}[\tilde{X}_0 \in B] \quad \text{for all } B \in \mathscr{B}(C[-\delta, 0]^n),$$

the two processes $Y := X|_{[0,T]}$ and $\tilde{Y} := \tilde{X}|_{[0,T]}$ also have the same law, i.e.,

$$\mathbb{P}[Y \in B] = \tilde{\mathbb{P}}[\tilde{Y} \in B] \quad \text{for all } B \in \mathscr{B}(C[0, T]^n).$$

Now we can state the following theorem:

Theorem 1 (Larssen [33]) *Strong uniqueness implies weak uniqueness.*

Let us introduce the strong formulation of the stochastic control problem in a delayed environment as in [33]:

$$\begin{aligned} dX(t) &= b(t, X_t, u(t))dt + \sigma(t, X_t, u(t))dW(t), \qquad t \in [0, T], \\ X(t) &= \phi(t), \qquad t \in [-\delta, 0], \end{aligned} \tag{4}$$

where u is the control process such that $u : [0, T] \times \Omega \to U, U \subset \mathbb{R}^m, m \geq 1$, and $\phi \in C[-\delta, 0]^n$ is a deterministic function.

The authors define the cost functional of the targeted control problem as follows:

$$J(u) = E\left[\int_0^T f(t, X_t, u(t))dt + h(X_T)\right]. \tag{5}$$

Here, we represent the class of $\{\mathscr{F}_t\}_{0 \leq t \leq T}$-adapted control processes $u : \Omega \times [0, T] \to U$ by $\mathscr{U}^s[0, T]$ such that the system (4) has a unique strong solution.

However, the main purpose is to minimize the cost functional (5) over the class $\mathscr{U}^s[0, T]$ with respect to the state processes (4), the authors in [33] follow a weak formulation to have more flexibility to be able to work in different probability spaces as follows:

$$\begin{aligned} dX(t) &= b(t, X_t, u(t))dt + \sigma(t, X_t, u(t))dW(t), \qquad t \in [s, T], \\ X(t) &= \phi(t - s), \qquad t \in [s - \delta, s], \end{aligned} \tag{6}$$

for any $(s, \phi) \in [0, T) \times C[-\delta, 0]^n$. Then, the control problem is to find 5-tuples $\tilde{u} = (\tilde{\Omega}, \tilde{\mathscr{F}}, \tilde{\mathbb{P}}, \tilde{W}, \tilde{u}) \in \mathscr{U}[s, T]$ such that

$$J(s, \phi; \tilde{u}) = \inf_{u \in U[s,T]} J(s, \phi; u)$$

for any given $(s, \phi) \in [0, T) \times C[-\delta, 0]^n$.

As we see, in stochastic control, we aim to minimize the expectation of the involved processes according to their distributions. If the solution of Eq. (6) has the same distribution in different probability spaces, then we can work in a more convenient probability space.

Remark 1 Please see Definition 4.1. of [33] for a technical description of admissible control processes.

We state the following assumptions as in [33], which help to prove the existence of unique strong solution of the weak formulation (6):

A1 The maps $b(t, \eta, u)$ and $\sigma(t, \eta, u)$ are Lipschitz on bounded sets in $C[-\delta, 0]^n$ uniformly in second variable.

A2 There is a constant $K > 0$ such that

$$|b(t, \eta, u)|_{\mathbb{R}^n} + |\sigma(t, \eta, u)|_{\mathbb{R}^{n\times r}} \leq K(1 + \|\eta\|_{C[-\delta,0]^n})$$

for all $0 \leq t \leq T$, $\eta \in C[-\delta, 0]^n$ and $u \in U$.

A3 The initial path process $\phi : \Omega \to C[-\delta, 0]^n$ is $\mathcal{F}_t^s$-measurable and belongs to the space $L^2(\Omega; C[-\delta, 0]^n)$, that is,

$$\|\phi\|^2_{L^2(\Omega;C[-\delta,0]^n)} := E[\|\phi(\omega)\|^2_{C[-\delta,0]^n}] < \infty.$$

A4 The maps $f : [0, T] \times C[-\delta, 0]^n \times U \to \mathbb{R}$ are uniformly continuous and there exists a constant $L > 0$ such that

$$|f(t, \eta, u) - f(t, \tilde{\eta}, u)| + |h(\eta) - h(\tilde{\eta})| \leq L \|\eta - \tilde{\eta}\|,$$

for all $t \in [0; T]$, $\eta, \tilde{\eta} \in C[-\delta, 0]^n$, $u \in U$, and

$$|f(t, 0, u)| + |h(0)| \leq L, \quad \text{for all } (t, u) \in [0, T] \times U.$$

Then, we can present the following theorem:

Theorem 2 (Larssen [33]) *Under the assumptions A1–A3, for any* $(s, \phi) \in [0, T) \times L^2(\Omega; C[-\delta, 0]^n)$ *and* $u \in \mathcal{U}[s, T]$*, Eq. (6) admits a unique strong solution* $X(\cdot) = X^{s,\phi,u}(\cdot) \in L^2(\Omega; C[-\delta, 0]^n)$.

Finally, we can present the main result of [33], which describes the *Dynamic Programming Principle*:

Theorem 3 (Larssen [33]) *Let A1–A4 hold. Then, for any* $(s, \phi) \in [0, T) \times C[-\delta, 0]^n$ *and* $\mathcal{F}_t^s$*-stopping time* $\hat{s} \in [s, T]$,

$$V(s, \phi) = \inf_{u \in U[s,T]} E\left[\int_s^{\hat{s}} f(t, X_t^{s,\phi,u}, u(t))dt + V(\hat{s}, X_{\hat{s}}^{s,\phi,u})\right].$$

The value function of such problems depends on the initial path in a complicated way. Since the initial path ϕ is in $C[-\delta, 0]^n$, the control problem is infinite-dimensional and hence, classical approaches to solve the HJBEs do not work.

Now, we will provide the HJBEs for a special setting, which demonstrates the past dependence in a particularly simple way. Here, we will explain this construction for a one-dimensional system. First, let us introduce the linear functional via [33]:

$$x(\xi) := \xi(0), \quad y(\xi) := \int_{-\delta}^0 e^{\lambda s}\xi(s)ds, \quad z(\xi) := \xi(-\delta),$$

where λ is a constant. For any $X \in C[-\delta, T]$ for $t \in [0, T]$, we represent:

$$X(t) = x(X_t), \tag{7}$$

$$Y(t) := y(X_t) = \int_{-\delta}^{0} e^{\lambda s} X(t+s)ds, \tag{8}$$

$$Z(t) := z(X_t) = X(t-\delta). \tag{9}$$

Hence, let us state the corresponding model of [33]:

$$\begin{aligned} dX(t) &= b(t, X(t), Y(t), Z(t), u(t))dt \\ &\quad + \sigma(t, X(t), Y(t), Z(t), u(t))dW(t), \quad t \in [s, T], \\ X(t) &= \phi(t-s), \quad t \in [s-\delta, s], \quad \phi \in C[-\delta, 0]. \end{aligned} \tag{10}$$

$G(t) = g(t, x(X_t), y(X_t))$ is defined where g is in $C^{1,2,1}(\mathbb{R}^3)$ and Itô's formula is developed for such a delayed system:

Lemma 1 (Elsanosi and Larssen [9])

$$dG(t) = \mathscr{L}_u g dt + \frac{\partial g}{\partial x}\sigma(t, x, y, z, u)dW(t),$$

where

$$\begin{aligned} \mathscr{L}_u g &= \mathscr{L}_u g(t, x, y, z) \\ &= \frac{\partial g}{\partial t} + b(t, x, y, z, u)\frac{\partial g}{\partial x} + \frac{1}{2}\sigma^2(t, x, , y, z, u)\frac{\partial^2 g}{\partial x^2} + \frac{\partial g}{\partial y}[x - e^{-\lambda\delta}z - \lambda y] \end{aligned}$$

which is evaluated at

$$x = x(X_t) = X(t), \quad y = y(X_t), \quad z = z(X_t).$$

In [9], the authors showed that for a system defined as in (10), the value function depends on ϕ only through the functional $x(\phi)$ and $y(\phi)$. Therefore, we assume that

$$V(s, \phi) = V(s, x(\phi), y(\phi)) = V(s, x, y). \tag{11}$$

Finally, the DPP can be defined as follows:

$$V(s, x, y) = \inf_{u \in U[s,T]} E\left[\int_s^{\hat{s}} f(t, X(t), Y(t), u(t))dt + V(\hat{s}, X(\hat{s}), Y(\hat{s}))\right]$$

for all $\mathscr{F}_t^s$-stopping times $\hat{s} \in [s, T)$ and $(x, y) \in \mathbb{R}^2$, where $\phi \in C[-\delta, 0]$ is such that $x = x(\phi) = X(s)$ and $x = y(\phi) = Y(s)$.

Finally, we can present the main theorem for HJBEs as follows:

Theorem 4 (Larssen [33]) *If we assume that Eq. (11) holds and that* $V(s,x,y) \in C^{1,2,1}([0,T] \times \mathbb{R}^2)$ *then* $V(s,x,y)$ *solves the following HJBE:*

$$\inf_{u \in U} \{\mathscr{L}_u V(s,x,y,z) + f(s,x,y,u)\} = 0 \quad \textit{for all} \quad z \in \mathbb{R},$$

with boundary condition

$$V(T,x,y) = h(x,y).$$

Remark 2 In [9], the authors showed that the system (10) can be converted to a system without delay since b and σ in Eq. (10) are linear in x, y and z with the relations (7), (8), and (9). Furthermore, under certain conditions, the control problem can be reduced to a finite-dimensional one. Then, the DPP can be applied as usual. We invite the reader to [34] for the technical conditions that ensure the HJBEs to be finite-dimensional when the system (10) is nonlinear.

3 A Stochastic Control Approach for Infinite Dimensional Hamilton-Jacobi-Bellman Equations

In this section, we follow the technical assumptions and mathematical formulations developed in Chang et al. [5]. The authors obtain an infinite-dimensional HJBE by utilizing the Bellman type DPP, Theorem 3, which was first demonstrated by Larssen [33]. Different from the previous section, here Chang, Pang and Pemy [5] prove that the value function is the unique viscosity solution of the introduced delayed control problem.

In [5], the authors study on a finite time horizon, $0 < T < \infty$, stochastic control problem for a delayed diffusion process. Let $\delta > 0$ be a fixed constant, and $\mathbb{J} = [-\delta, 0]$ denote the duration of the bounded memory of the delayed state process given below by Eq. (12). The inner product $(\cdot|\cdot)$ in $L^2(\mathbb{J}, \mathbb{R}^n)$, $n \in \mathbb{N}$ is defined as:

$$(\phi|\psi) = \int_{-\delta}^{0} \langle \phi(s), \psi(s) \rangle \, ds, \quad \text{and} \quad \|\phi\|_2 = (\phi|\phi)^{\frac{1}{2}}, \quad \forall \phi, \psi \in \mathbf{C} = C(\mathbb{J}, \mathbb{R}^n),$$

where $\langle \cdot, \cdot \rangle$ denotes the inner product in $\mathbb{R}^n$.

Let $(\Omega, \mathscr{F}, \mathbb{P}, \mathbb{F})$ be a complete filtered probability space, where $\mathbb{F} = \{\mathscr{F}(t),\ t \geq 0\}$ is the $\mathbb{P}$-augmentation of $\mathscr{F}^W(t) = \sigma\{W(s);\ 0 \leq s \leq t\}$, $t \geq 0$. Furthermore, $L_2(\Omega, \mathbf{C})$ represents the space of $\mathbf{C}$-valued random variables $\Xi : \Omega \to \mathbf{C}$ such that

$$\|\Xi\|_{L_2} = \left\{ \int_{\Omega} \|\Xi\|^2 \, dP(\omega) \right\}^{\frac{1}{2}} < \infty.$$

In addition, the authors show the space of $\mathscr{F}_t$-measurable, $\Xi \in L_2(\Omega, \mathbf{C})$ by $L_2(\Omega, \mathbf{C}; \mathscr{F}_t)$. In this section, the system of the following SDDE is considered as:

$$dX(s) = f(s, X_s, u(s))ds + g(s, X_s, u(s))dW(s), \quad s \in [t, T] \tag{12}$$
$$X_t = \phi_t, \qquad \forall \phi_t \in L_2(\Omega, \mathbf{C}; \mathscr{F}_t)$$

Remark 3 Please see Definition 2.3 of [5] for technical assumptions of an admissible control process.

$\mathscr{U}[t, T]$ denotes the collection of admissible controls over the interval $[t, T]$. It is assumed that $f : [0, T] \times \mathbf{C} \times U \to \mathbb{R}^n$ and $g : [0, T] \times \mathbf{C} \times U \to \mathbb{R}^{n \times m}$ are continuous and satisfy the following assumptions, see [5]:

Assumption 1

$$|f(t, \phi, u) - f(t, \varphi, u)| + |g(t, \phi, u) - g(t, \varphi, u)| \leq \Lambda \|\phi - \varphi\|, \tag{13}$$

for all (t, ϕ, u) *and* (t, φ, u) *are in* $[0, T] \times \mathbf{C} \times U$.

Assumption 2 *There exists a constant* K *such that*

$$|f(t, \phi, u)| + |g(t, \phi, u)| \leq K(1 + \|\phi\|), \tag{14}$$

for all $(t, \phi, u) \in [0, T] \times \mathbf{C} \times U$.

Remark 4 We know that under Assumptions (13) and (14), $\{X_s = X_s(\cdot; t, \phi, u(\cdot)), s \in [t, T]\}$ is a $\mathbf{C}$-valued Markov process (see Mohammed [35, 36]).

Let L and Ψ be two continuous functions on $[0, T] \times \mathbf{C} \times U$ and $[0, T] \times \mathbf{C}$ such that both have at most polynomial growth in $L_2(\mathbb{J}, \mathbb{R})$. Then, our problem is to find the optimal control process $u(\cdot) \in \mathscr{U}[t, T]$ and the value function $V : [0, T] \times \mathbf{C} \to \mathbb{R}$ for each initial datum $(t, \phi) \in [0, T] \times \mathbf{C}$ such that

$$\begin{aligned} V(t, \phi) &= \sup_{u(\cdot) \in \mathscr{U}[t,T]} J(t, \phi, u(\cdot)) \\ &= \sup_{u(\cdot) \in \mathscr{U}[t,T]} E\Bigg[\int_t^T e^{-\rho(s-t)} L(s, X_s(t, \phi, u(\cdot)), u(s))ds \\ &\qquad + e^{-\rho(T-t)} \Psi_T(X_T(t, \phi, u(\cdot)))\Bigg] \end{aligned} \tag{15}$$

The space of bounded functional is represented $\Phi : \mathbf{C} \to \mathbb{R}$ by C^* and the space of bounded bilinear functional $\tilde{\Phi} : \mathbf{C} \times \mathbf{C} \to \mathbb{R}$ by $C^\dagger$. Moreover, the authors define $\mathbf{B} = \{v 1_{\{0\}} : v \in \mathbb{R}^n\}$, where $1_{\{0\}} : [-\delta, 0] \to \mathbb{R}^n$ is as follows:

$$1_{\{0\}}(\theta) = \begin{cases} 0 \text{ for } & \theta \in [-\delta, 0), \\ 1 \text{ for } & \theta = 0. \end{cases}$$

Then, the direct sum is denoted by:

$$\mathbf{C} \oplus \mathbf{B} = \left\{ \phi + v 1_{\{0\}} | \phi \in \mathbf{C}, \ v \in \mathbb{R}^n \right\}$$

and is equipped with the norm:

$$\left\| \phi + v 1_{\{0\}} \right\| = \sup_{\theta \in [-\delta, 0]} |\phi(\theta)| + |v|, \quad \phi \in \mathbf{C}, \ v \in \mathbb{R}^n.$$

We know that for each sufficiently smooth function Φ, its first order Fréchet derivative $D\Phi(\phi) \in \mathbf{C}^*$ has a unique and continuous extension $\overline{D\Phi(\phi)} \in (\mathbf{C} \oplus \mathbf{B})^*$. Similarly, we can show its second order Fréchet derivative by $D^2\Phi(\phi) \in \mathbf{C}^\dagger$ and its unique continuous linear extension is $\overline{D^2\Phi(\phi)} \in (\mathbf{C} \oplus \mathbf{B})^\dagger$ (see [35], Lemma 3.1 and Lemma 3.2 for more details).

Furthermore, S operator is defined for all $\phi \in \mathbf{C}$ as follows:

$$S(\Phi)(\phi) = \lim_{h \to 0^+} \frac{1}{h} [\Phi(\tilde{\phi}_h) - \Phi(\phi)], \tag{16}$$

where we have $\Phi : \mathbf{C} \to \mathbb{R}$ and $\tilde{\phi} : [-\delta, T] \to \mathbb{R}^n$ is an extension of ϕ denoted by

$$\tilde{\phi}(t) = \begin{cases} \phi(t) \text{ if } & t \in [-\delta, 0), \\ \phi(0) \text{ if } & t \geq 0. \end{cases}$$

Moreover, $\tilde{\phi}_t \in \mathbf{C}$ is presented by $\tilde{\phi}_t(\theta) = \tilde{\phi}(t + \theta)$, $\theta \in [-\delta, 0]$. Let $\hat{\mathscr{D}}(S)$ denote the domain of the operator S and be the set of $\Phi : \mathbf{C} \to \mathbb{R}$ such that the limit (16) exists for each $\phi \in \mathbf{C}$. In addition, we represent the set of all functions $\Phi : [0, T] \times \mathbf{C} \to \mathbb{R}$ by $\mathscr{D}(S)$ such that $\Phi(t, \cdot) \in \hat{\mathscr{D}}(S)$, for all $t \in [0, T]$.

Furthermore, let us show the space of functions $\Phi : [0, T] \times \mathbf{C} \to \mathbb{R}$ as $C^{1,2}_{lip}([0, T] \times \mathbf{C})$ such that $\frac{\partial \Phi}{\partial t} : [0, T] \times \mathbf{C} \to \mathbb{R}$ and $D^2\Phi : [0, T] \times \mathbf{C} \to \mathbf{C}^\dagger$ exist and continuous. Then, we have:

$$\left\| D^2\Phi(t, \phi) - D^2\Phi(t, \varphi) \right\|^\dagger \leq K \left\| \phi - \varphi \right\|, \quad \forall t \in [0, T], \ \phi, \varphi \in \mathbf{C}.$$

Now, let us present the following results that will lead us to the main goal of the work [5]:

Theorem 5 (Chang et al. [5]) *Suppose $\Phi \in C^{1,2}_{lip}([0, T] \times \mathbf{C}) \cap \mathscr{D}(S)$. Let $u(\cdot) \in \mathscr{U}[t, T]$ and $\{X_s : s \in [t, T]\}$ be the* **C***-valued Markov solution process of Eq. (12) with the initial data $(t, \phi_t) \in [0, T] \times \mathbf{C}$. Then,*

$$\lim_{\epsilon \downarrow 0} \frac{E[\Phi(t+\epsilon, X_{t+\epsilon})] - \Phi(t, \phi_t)}{\epsilon}$$

$$= \frac{\partial}{\partial t}\Phi(t,\phi_t) + S(\Phi)(t,\phi_t) + \overline{D\Phi(t,\phi_t)}(f(t,\phi_t,u(t))1_{\{0\}})$$

$$+ \frac{1}{2}\sum_{j=1}^{m} \overline{D^2\Phi(t,\phi_t)}(g(t,\phi_t,u(t))e_j 1_{\{0\}}, g(t,\phi_t,u(t))e_j 1_{\{0\}}),$$

where e_j, $j = 1, 2, \ldots, m$ is the jth unit vector of the standard basis in $\mathbb{R}^m$.

The following theorem will help us to define the HJBE heuristically:

Theorem 6 (Chang et al. [5]) *Let Assumptions (13) and (14) hold. Then, for any $(t, \phi) \in [0, T] \times \mathbf{C}$ and $\mathbb{F}_t$-stopping time $\tau \in [t, T]$,*

$$V(t,\phi) = \sup_{u(\cdot)\in\mathscr{U}[t,T]} E\Bigg[\int_t^{\tau} e^{-\rho(s-t)} L(s, X_s(t,\phi,u(\cdot)), u(s))ds$$
$$+ e^{-\rho(\tau-t)} V(\tau, X_\tau(t,\phi,u(\cdot)))\Bigg].$$

Let $v \in U$ be. We define:

$$\mathscr{A}^v V(t,\phi) = S(V)(t,\phi) + \overline{DV(t,\phi)}(f(t,\phi,v)1_{\{0\}})$$
$$+ \frac{1}{2}\sum_{i=1}^{m} \overline{D^2V(t,\phi)}(g(t,\phi,v)e_i 1_{\{0\}}, g(t,\phi,v)e_i 1_{\{0\}}).$$

By using Theorem 6, the following result is obtained that specifies the value function of targeted control problem:

Theorem 7 (Chang et al. [5]) *Suppose V is the unique solution defined by (15) and satisfies $V \in C^{1,2}_{lip}([0,T]\times\mathbf{C}) \cap \mathscr{D}(S)$. Then, the value function V satisfies the following HJB equation:*

$$\rho V(t,\phi) - \frac{\partial V}{\partial t}(t,\phi) - \max_{v\in U}[\mathscr{A}^v V(t,\phi) + L(t,\phi,v)] = 0 \tag{17}$$

on $[0,T]\times\mathbf{C}$, and $V(T,\phi) = \Psi(\phi)$, $\forall\phi \in \mathbf{C}$.

Note that we do not know whether the value function ensures the necessary smoothness condition. Hence, at this step, we have to consider viscosity solutions rather than a classical solution of HJBEs, and in fact, we will see that the value function is the unique viscosity solution of the HJBE (17).

Definition 5 (Chang et al. [5]) Let $w \in C([0,T]\times\mathbf{C})$ be. We say that w is a viscosity subsolution of (17) if for every $\Gamma \in C^{1,2}_{lip}([0,T]\times\mathbf{C}) \cap \mathscr{D}(S)$, for $(t,\phi) \in$

$[0, T] \times \mathbf{C}$ satisfying $\Gamma \geq w$ on $[0, T] \times \mathbf{C}$ and $\Gamma(t, \phi) = w(t, \phi)$, we have

$$\rho\Gamma(t, \phi) - \frac{\partial \Gamma}{\partial t}(t, \phi) - \max_{v \in U}[\mathscr{A}^v \Gamma(t, \phi) + L(t, \phi, v)] \leq 0.$$

We say that w is a viscosity super solution of (17) if for every $\Gamma \in C^{1,2}_{lip}([0, T] \times \mathbf{C}) \cap \mathscr{D}(S)$, and for $(t, \phi) \in [0, T] \times \mathbf{C}$ satisfying $\Gamma \leq w$ on $[0, T] \times \mathbf{C}$ and $\Gamma(t, \phi) = w(t, \phi)$, we have

$$\rho\Gamma(t, \phi) - \frac{\partial \Gamma}{\partial t}(t, \phi) - \max_{v \in U}[\mathscr{A}^v \Gamma(t, \phi) + L(t, \phi, v)] \geq 0.$$

We say that w is a viscosity solution of (17) if it is both a viscosity supersolution and a viscosity subsolution of (17).

Hence, for the value function (15), we have the following properties:

Lemma 2 (Chang et al. [5]) *The value function V defined in (15) is continuous and there exists a constant $\Lambda > 0$ a positive integer k such that for every $(t, \phi) \in [0, T] \times \mathbf{C}$,*

$$|V(t, \phi)| \leq \Lambda(1 + \|\phi\|_2)^k,$$

and there exists a constant $K > 0$ such that

$$|V(t, \phi) - V(t, \varphi)| \leq K \|\phi - \varphi\|, \quad \forall (s, \phi), (s, \varphi) \in [0, T] \times \mathbf{C}.$$

Finally, we have the following theorem:

Theorem 8 (Chang et al. [5]) *The value function V is a viscosity solution of the HJB Equation*

$$\rho V(t, \phi) - \frac{\partial V}{\partial t}(t, \phi) - \max_{v \in U}[\mathscr{A}^v V(t, \phi) + L(t, \phi, v)] = 0$$

on $[0, T] \times \mathbf{C}$, and $V(T, \phi) = \Psi(\phi), \quad \forall \phi \in \mathbf{C}$.

Let us provide the following comparison principle which will immediately lead us to the uniqueness result of the described viscosity solution.

Theorem 9 (Chang et al. [5]) *Assume that $V_1(t, \phi)$ and $V_2(t, \phi)$ are both continuous with respect to the argument (t, ϕ) and are viscosity subsolution and supersolution of (17) respectively with at most a polynomial growth. In other terms, there exists a real number $\Lambda > 0$ and a positive integer $k > 0$ such that*

$$|V_i(t, \phi)| \leq \Lambda(1 + \|\phi\|_2)^k, \quad \textit{for} \quad (t, \phi) \in [0, T] \times \mathbf{C}, \ i = 1, 2.$$

Then,

$$V_1(t,\phi) \le V_2(t,\phi) \quad \textit{for all} \quad (t,\phi) \in [0,T] \times \mathbf{C}.$$

A computational approach for the related HJBE can be found in [4], where a finite difference scheme was implemented to approximate the viscosity solution. Furthermore, the convergence of the scheme was demonstrated using the Banach fixed point theorem. In this survey work, rather than such numerical methods, we aim to introduce two Deep Learning methods for computational experiments.

4 Deep Learning Approach for Stochastic Control Problems with Memory

We know that delayed stochastic control problems have remained intractable and attractive since the first day that they were started to be studied in literature. Different from the theoretical methods presented in Sects. 2 and 3, several authors attempted to implement numerical methods to solve delayed control systems; see [16, 17, 31, 32]. As a consequence of their path-dependent and infinite-dimensional nature, we need advanced numerical algorithms to handle such problems. Therefore, in this section, we will present the neural network architectures developed by Han and Hu [26] to address this challenge. Firstly, let us remember the state processes:

$$\begin{aligned} dX(t) &= b(t, X_t, \pi(t))dt + \sigma(t, X_t, \pi(t))dW(t), \quad t \in [0,T], \\ X(t) &= \phi(t), \qquad t \in [-\delta, 0], \end{aligned} \tag{18}$$

where $\delta \ge 0$ is a fixed delay, $W(\cdot)$ is an $\mathbb{R}^l$-dimensional standard Brownian motion, and the control process π will be chosen from an admissible set of processes $\mathscr{A}$ in $\mathbb{R}^m$, and l, m are positive integers. Here, we follow the technical assumptions of Sect. 2.

In [26], however the delay term δ is assumed to be deterministic and known as a prior, authors observed that Recurrent Neural Networks may capture the notion of unknown δ implicitly as a consequence of their architecture design. Here, in this work, the aim is to minimize the expected cost as in Sect. 2. Let us remind the cost functional:

$$J(u) = E\left[\int_0^T f(t, X_t, u(t))dt + g(X_T)\right]. \tag{19}$$

Now, we can present the numerical algorithm of [26]. The authors assume that $0 = t_0 < t_1 < \ldots < t_{N_T} = T$ be an equidistributed partition of size $N_T \in \mathbb{N}$ on $[0, T]$. Furthermore, let $\delta < \infty$ cover N_δ subintervals as follows:

$$-\delta = t_{-N_\delta} \le t_{-N_\delta+1} \le \ldots \le t_0 = 0 \quad \text{with} \quad t_{k+1} - t_k \equiv h, \ \forall k = -N_\delta, \ldots, -1.$$

Then, the discretized versions of Eqs. (18) and (19):

$$X(t_{k+1}) = X(t_k) + b(t_k, X_{t_k}, \pi_{t_k})h + \sigma(t_k, X_{t_k}, \pi_{t_k})\Delta W(t_k), \tag{20}$$

and

$$\inf_{\{\pi(t_k)\}_{k=0}^{N_T-1}} E\left[\sum_{k=0}^{N_T-1} f(t_k, X_{t_k}, \pi_{t_k})h + g(X_T)\right], \tag{21}$$

where $X_{t_k} = (X(t_{k-N_\delta}), \ldots, X(t_k))$ is the path with N_δ lags and $\Delta W(t_k) = W(t_{k+1}) - W(t_k)$ represents the increments of the Brownian motion. Here, with these discretizations, as $h \to 0$, the value function corresponding to (20) and (21) may converge to the value function of the original problem (18) and (19). Since X_{t_k} is near optimal to the original one, we can expect that the value function is near optimal associated with (18).

Now let us introduce two different neural network architectures to approximate $\pi(t_k)$ as in [26].

4.1 Feedforward Neural Networks and Long-Short Term Memory Algorithms

Firstly, let us describe a *feedforward neural network* that is defined as fully connected layers of F_{d_1,d_2} as in [26]:

$$F_{d_1,d_2} : \mathbb{R}^{d_1} \to \mathbb{R}^{d_2},$$
$$x \to \rho(Ax + b),$$

where $x \in \mathbb{R}^{d_1}$ is the input, $F_{d_1,d_2}(x) \in \mathbb{R}^{d_2}$ is the output, $A \in \mathbb{R}^{d_1 \times d_2}$ is the weight matrix, b is the bias vector, and $\rho(\cdot)$ is the activation function with common choices:

$$\rho_{ReLU}(x) = \max\{x, 0\}, \quad \rho_{Id}(x) = x, \quad \rho_s(x) = \frac{1}{1+e^{-x}}, \quad \rho_{\tanh}(x) = \tanh(x).$$

It is assumed that $\bar{\delta} = N_{\bar{\delta}} h$ for which $\bar{\delta} \neq \delta$ exists and $\bar{X}_{t_k} \equiv (X(t_{k-N_{\bar{\delta}}}), \ldots, X(t_k), t_k) \in \mathbb{R}^{n \times (N_{\bar{\delta}}+1)+1}$. Here, the idea is to approximate $\pi(t_k)$ by a feedforward neural network for which the input is the state history up to lag $\bar{\delta}$. Then,

$$\pi(t_k) \approx F_{d_l,m} \circ F_{d_{l-1},d_l} \ldots F_{d_1,d_2} \circ F_{n \times (N_{\bar{\delta}}+1)+1, d_1)}(\bar{X}_{t_k}), \tag{22}$$

where l is the number of hidden layers. This algorithm is expected to produce controls adapted to the canonical filtration $\mathcal{F}_t^X$. By using a fixed input dimension and adding a time variable to the input, it becomes possible to share the parameters of sub-neural networks and reduce the number of parameters by a certain factor.

Now, we can introduce the second deep learning algorithm which is called *Long-Short Term Memory* (**LSTM**). To approximate $\pi(t_k)$, let us define the following components of this special recurrent neural network as described in [26]:

$$\begin{aligned}
&\text{forget gate:} \quad f_k = \rho_s(W_f x_k + U_f h_{k-1} + b_f),\\
&\text{input gate:} \quad i_k = \rho_s(W_i x_k + U_i h_{k-1} + b_i),\\
&\text{output gate:} \quad o_k = \rho_s(W_o x_k + U_o h_{k-1} + b_o),\\
&\text{cell:} \quad c_k = f_k \odot c_{k-1} + i_k \odot \rho_{\tanh}(W_c x_k + U_c h_{k-1} + b_c),\\
&\text{output of the } k\text{-th unit:} \quad h_k = o_k \odot \rho_{\tanh}(c_k),
\end{aligned} \tag{23}$$

where $\odot$ represents the Hadamard product. The cell information is stored in c_k and the k-th input is denoted by x_k. Moreover, $h_k,\ f_k,\ i_k,\ o_k$ and c_k are all d_h-dimensional vectors.

Here, in [26], the initial information of $h_k,\ c_k$ are determined with respect to the discretized initial path ϕ and $(X(t_0), t_0), (X(t_1), t_1), \ldots$ are considered as the input sequence $x_0, x_1, \ldots$. Furthermore, the authors define an affine transformation of h_k as the proxy of $\pi(t_k)$ as follows:

$$\pi(t_k) \approx Wh_k + b.$$

It is indicated that since discretized state values, (20), of length $N_{\bar{\delta}} + 1$ are used in (22), we need prior knowledge of δ to use. Hence, firstly, a good estimate of $\bar{\delta}$ has to be obtained.

For the technical details of the implementation see Section 3.4 in [26] and for Python codes, we invite the reader to check the GitHub repository.[1]

Now, we will present a benchmark example of portfolio optimization with complete memory, which is a result from [40].

4.2 An Application to Finance

In [40], the authors consider a Merton's type portfolio problem with complete memory over a finite time horizon. The problem is formulated to provide optimal investment and consumption controls, such that the stochastic control problem aims to maximize total expected discounted utility. The state process is described with complete memory on a finite time horizon. The authors develop explicit analytical solutions for the associated HJB equations in a finite-dimensional space

[1] https://github.com/frankhan91/RNN-ControlwithDelay.

for exponential, logarithmic, and power utility functions under certain technical conditions. Furthermore, they prove the verification theorems such that for those utility functions, it is ensured that the solutions are equal to the value functions. Also, they derive the optimal controls.

Let us introduce the dynamics of the wealth process $X(\cdot)$ and the objective criterion as follows:

$$\begin{aligned} dX(t) &= [((\mu_1 - r)\pi(t) - c(t) + r)X(t) + \mu_2 Y(t)]dt \\ &\quad + \sigma\pi(t)X(t)dW(t), \quad t \in [s, T], \\ X(t) &= \phi(t), \qquad t \in (-\infty, s], \quad \phi \in L^2(\Omega, C((-\infty, 0], \mathbb{R}))), \\ &\max_{c(t)\in\mathbb{R}^+, \pi(t)\in\mathbb{R}} E_\phi\left[\int_0^T e^{-\beta t} U_1(c(t)X(t))dt + e^{-\beta T} U_2(X(T), Y(T))\right], \end{aligned} \tag{24}$$

where $\pi(\cdot)$ and $c(\cdot)$ are the control processes describing the investment and the consumption respectively. Let us define:

$$Y(t) := \int_{-\infty}^0 e^{\lambda s} X(t+s)ds.$$

In [40], the authors assume that the value function depends only on $X(0) \equiv \phi(0)$ and $Y(0) \equiv \int_{-\infty}^0 e^{\lambda s}\phi(t+s)ds$.

Theorem 10 (Pang and Hussain [40]) *Consider the portfolio optimization problem with infinite delay (24) with the complex history $X_0 = \phi \in L^2(\Omega, C((-\infty, 0], \mathbb{R})))$ under log utility:*

$$U_1(x) = \log(x), \qquad U_2(x, y) = \frac{1}{\beta}\log(x + \eta y),$$

and

$$\eta = \frac{1}{2}\left(\sqrt{(r+\lambda)^2 + 4\mu_2} - (r+\lambda)\right),$$

and require a state constraint $X(t) \geq 0$, $\forall t \in [0, T]$. Then, the optimal investment and consumption rates are

$$\begin{aligned} \pi^*(t) &= \frac{(\mu_1 - r)(X^*(t) + \eta Y^*(t))}{\sigma^2 X^*(t)}, \\ c^*(t) &= \frac{\beta(X^*(t) + \eta Y^*(t))}{X^*(t)} X^*(t), \end{aligned}$$

where $X^*(t)$ solves the SDDE (24) associated with π^* and c^*. The value function is of the form

$$V(t, x, y) = p(t) + \frac{1}{\beta}\log(x + \eta y)$$

with

$$p(t) = \frac{\Lambda_2}{\beta}(1 - e^{-\beta(T-t)}), \quad \Lambda_2 = \frac{1}{2}\frac{(\mu_1 - r)^2}{\beta\sigma^2} + \log(\beta) - 1 - \frac{1}{\beta}(r + \eta),$$

and the optimal cost $V_0 = V(0, X(0), Y(0))$.

The stochastic control problem (24) is one of the benchmark examples studied in [26], for which the authors compare the results obtained from feedforward neural networks, the LSTM algorithm introduced by (23), and the results provided by analytical solutions described in Theorem 10.

Firstly, the authors set some special values as $\lambda = 0.3$, $\beta = 0.2$, $\mu_1 = 0.1$, $\mu_2 = 0.2$, $r = 0.05$, and $\sigma = 0.4$. In the nature of the problem, as we see above $\delta = \infty$, and the initial path (segment) ϕ is assumed to be constant for $t \leq -4$. Furthermore, the state process evolves for $t \in [-4, 0]$.

The LSTM model has a dimension of hidden state $d_h = 60$, which gives 15,242 parameters in total. On the other side, the feedforward model has 2 hidden layers with a width of 9, which gives 17,474 parameters. Moreover, the state history up to lag $\bar{\delta} = 4 < \infty$ with $N_{\bar{\delta}} = 80$ is taken in the feedforward algorithm (one can find the other technical assumptions in Section 5.3 of [26]).

The authors observe that the LSTM model converges to a higher utility faster than the feedforward algorithm and moreover, with a more stable convergence. For the feedforward model, we have an initial input, and then the inputs retain the same sliding window length of the state history, and gradually lose the initial information. Hence, it is concluded that the feedforward model performs suboptimal; however, the LSTM is capable of naturally capturing infinite memory. All the conclusions and graphical illustrations can be found for three different carefully selected financial applications in Section 5 of [26].

5 Conclusion and Outlook

In this survey work, we intended to demonstrate a hot topic at the intersection of Stochastic Optimal Control and Deep Learning within the context of *memory*. In financial engineering, several researchers have paid significant attention to time-delayed models so far, and now we see rapidly emerging interest from the side of a newly developing discipline, Deep Learning.

As we presented in Sects. 2 and 3, it is highly complicated to approach delayed control problems by DPP. Since we lose the Markov property, the application of DPP does not become immediately applicable. Still in our work, we introduced some effective ways to handle such control problems. However, it is out of scope of this survey article, but we would like to mention that another way of solving delayed systems can be achieved by Anticipated Backward Stochastic Differential Equations. This type of Backward Stochastic Differential Equations (**BSDEs**) was first developed by Peng and Yang [41], and then many authors applied them to Stochastic Optimal Control by applying the Stochastic Maximum Principle (**SMP**). The methods based on BSDEs promise to significantly improve the current simulation capabilities when we study high-dimensional linear and semi-linear parabolic problems. Here, the advantage rises since we do not need to impose any Markovian property when we benefit SMP; see [39, 43, 44].

On the other side, the inevitable importance of LSTM is clearly seen in a wide range of areas such as time series applications, natural language processing, sentiment analysis, image and video captioning, computer vision, text recognition, and robotics (see the survey [28] and references therein). Moreover, to the best of our knowledge, implementing such algorithms is quite new in stochastic optimal control.

Finally, we would like to highlight that by this work, we hope to make researchers familiar with Deep Learning that is another powerful aspect of handling computational challenges in Stochastic Optimal Control.

Funding This project is supported by SCROLLER:A Stochastic ContROL approach to Machine Learning with applications to Environmental Risk models, Project 299897 from the Norwegian Research Council.

Conflict of Interest The authors declare no conflict of interest.

References

1. Bayraktar, E., Keller, C.: Path-dependent Hamilton–Jacobi equations in infinite dimensions. J. Funct. Anal. **275**, 2096–2161 (2018)
2. Blechschmidt, J., Ernst, O.: Three ways to solve partial differential equations with neural networks—A review. GAMM-Mitteilungen **44**, e202100006 (2021)
3. Castro, J.: The Kolmogorov infinite dimensional equation in a Hilbert space via deep learning methods. Preprint. ArXiv:2206.06451 (2022)
4. Chang, M., Pang, T., Pemy, M.: Finite difference approximations for stochastic control systems with delay. Stoch. Anal. Appl. **26**, 451–470 (2008)
5. Chang, M., Pang, T., Pemy, M.: Optimal control of stochastic functional differential equations with a bounded memory. Stochastics Int. J. Probab. Stoch. Process. **80**, 69–96 (2008)
6. Chen, C., Sun, D., Chang, C.: Numerical solution of time-delayed optimal control problems by iterative dynamic programming. Optim. Control Appl. Methods **21**, 91–105 (2000)
7. Crandall, M., Lions, P.: Viscosity solutions of Hamilton-Jacobi equations. Trans. Am. Math. Soc. **277**, 1–42 (1983)
8. Ekren, I., Touzi, N., Zhang, J.: Viscosity solutions of fully nonlinear parabolic path dependent PDEs: Part I. Ann. Probab. Off. J. Inst. Math. Stat. **44**, 1212–1253 (2016)
9. Elsanosi, I., Larssen, B.: Optimal consumption under partial observations for a stochastic system with delay. Preprint Series. Pure Mathematics. Nb. No/URN: NBN: No-8076 (2001)
10. Elsanosi, I., Øksendal, B., Sulem, A.: Some solvable stochastic control problems with delay. Stochastics Int. J. Probab. Stoch. Process. **71**, 69–89 (2000)

11. Fabbri, G., Gozzi, F., Swiech, A.: Stochastic Optimal Control in Infinite Dimension. Probability and Stochastic Modelling. Springer (2017)
12. Federico, S.: A stochastic control problem with delay arising in a pension fund model. Finance Stoch. **15**, 421–459 (2011)
13. Federico, S., Tankov, P.: Finite-dimensional representations for controlled diffusions with delay. Appl. Math. Optim. **71**, 165–194 (2015)
14. Federico, S., Goldys, B., Gozzi, F.: HJB equations for the optimal control of differential equations with delays and state constraints, I: regularity of viscosity solutions. SIAM J. Control Optim. **48**, 4910–4937 (2010)
15. Feo, F., Federico, S., Świech, A.: Optimal control of stochastic delay differential equations and applications to path-dependent financial and economic models. Preprint. ArXiv:2302.08809 (2023)
16. Fischer, M., Nappo, G.: Time discretisation and rate of convergence for the optimal control of continuous-time stochastic systems with delay. Appl. Math. Optim. **57**, 177–206 (2008)
17. Fischer, M., Reiss, M.: Discretisation of stochastic control problems for continuous time dynamics with delay. J. Comput. Appl. Math. **205**, 969–981 (2007)
18. Fleming, W., Soner, H.: Controlled Markov Processes and Viscosity Solutions. Springer Science & Business Media (2006)
19. Fouque, J., Zhang, Z.: Deep learning methods for mean field control problems with delay. Front. Appl. Math. Stat. **6**, 11 (2020)
20. Germain, M., Pham, H., Warin, X.: Neural networks-based algorithms for stochastic control and PDEs in finance. Preprint. ArXiv:2101.08068 (2021)
21. Gozzi, F., Masiero, F.: Stochastic optimal control with delay in the control I: Solving the HJB equation through partial smoothing. SIAM J. Control Optim. **55**, 2981–3012 (2017)
22. Gozzi, F., Masiero, F.: Stochastic optimal control with delay in the control II: Verification theorem and optimal feedbacks. SIAM J. Control Optim. **55**, 3013–3038 (2017)
23. Gozzi, F., Masiero, F.: Errata: Stochastic optimal control with delay in the control I: Solving the HJB equation through partial smoothing, and stochastic optimal control with delay in the control II: verification theorem and optimal feedbacks. SIAM J. Control Optim. **59**, 3096–3101 (2021)
24. Gozzi, F., Roma, S., Marinelli, C.: 13 Stochastic Optimal Control of Delay Equations Arising in Advertising Models. Stochastic Partial Differential Equations and Applications-VII, pp. 133 (2005)
25. Gozzi, F., Marinelli, C., Savin, S.: On controlled linear diffusions with delay in a model of optimal advertising under uncertainty with memory effects. J. Optim. Theory Appl. **142**, 291–321 (2009)
26. Han, J., Hu, R.: Recurrent neural networks for stochastic control problems with delay. Math. Control Signals Syst. **33**, 775–795 (2021)
27. Han, J., Jentzen, A., et al.: Deep learning-based numerical methods for high-dimensional parabolic partial differential equations and backward stochastic differential equations. Commun. Math. Stat. **5**, 349–380 (2017)
28. Hochreiter, S., Schmidhuber, J.: Long short-term memory. Neural Comput. **9**, 1735–1780 (1997)
29. Ishii, H., Lions, P.: Viscosity solutions of fully nonlinear second-order elliptic partial differential equations. J. Differential Equations **83**, 26–78 (1990)
30. Kim, J., Yang, I.: Hamilton-Jacobi-Bellman equations for q-learning in continuous time. In: Learning For Dynamics and Control, pp. 739–748 (2020)
31. Kushner, H.: Numerical approximations for stochastic systems with delays in the state and control. Stochastics Int. J. Probab. Stoch. Process. **78**, 343–376 (2006)
32. Kushner, H.: Numerical Methods for Controlled Stochastic Delay Systems. Springer (2008)
33. Larssen, B.: Dynamic programming in stochastic control of systems with delay. Stochastics Int. J. Probab. Stoch. Process. **74**, 651–673 (2002)

34. Larssen, B., Risebro, N.: When are HJB-equations in Stochastic Control of Delay Systems Finite Dimensional? Taylor & Francis (2003)
35. Mohammed, S.: Stochastic Functional Differential Equations. Pitman Advanced Publishing Program (1984)
36. Mohammed, S.: Stochastic differential systems with memory: theory, examples and applications. In: Stochastic Analysis and Related Topics VI: Proceedings of the Sixth Oslo—Silivri Workshop Geilo 1996, pp. 1–77 (1998)
37. Nakamura-Zimmerer, T., Gong, Q., Kang, W.: Adaptive deep learning for high-dimensional Hamilton–Jacobi–Bellman equations. SIAM J. Sci. Comput. **43**, A1221–A1247 (2021)
38. Øksendal, B., Sulem, A.: Applied Stochastic Control of Jump Diffusions. Springer (2007)
39. Øksendal, B., Sulem, A., Zhang, T.: Optimal control of stochastic delay equations and time-advanced backward stochastic differential equations. Adv. Appl. Probab. **43**, 572–596 (2011)
40. Pang, T., Hussain, A.: A stochastic portfolio optimization model with complete memory. Stoch. Anal. Appl. **35**, 742–766 (2017)
41. Peng, S., Yang, Z.: Anticipated backward stochastic differential equations. Ann. Probab. **37**, 877–902 (2009)
42. Pham, H.: Continuous-time Stochastic Control and Optimization with Financial Applications. Springer Science & Business Media (2009)
43. Savku, E.: Memory and Anticipation: Two main theorems for Markov regime-switching stochastic processes. Preprint. ArXiv:2302.13890 (2023)
44. Savku, E., Weber, G.: A stochastic maximum principle for a Markov regime-switching jump-diffusion model with delay and an application to finance. J. Optim. Theory Appl. **179**, 696–721 (2018)
45. Savku, E., Weber, G.: A regime-switching model with applications to finance: Markovian and Non-Markovian cases. In: Dynamic Economic Problems with Regime Switches, pp. 287–309 (2021)
46. Van Houdt, G., Mosquera, C., Nápoles, G.: A review on the long short-term memory model. Artif. Intell. Rev. **53**, 5929–5955 (2020)
47. Yong, J., Zhou, X.: Stochastic Controls: Hamiltonian Systems and HJB Equations. Springer Science & Business Media (1999)

Exponentiated Weibull Mixture Cure Model to Handle Right-Censored Data Set

Mohamed A. S. Ishag, Anthony Wanjoya, Aggrey Adem, and Ahmed Z. Afify

Abstract Survival analysis is an important statistical tool for analyzing time-to-event data, such as the time to failure of a product or the time to death in medical research. Often, these data sets are subject to right-censoring, where some individuals are still alive or have not experienced the event of interest at the end of the study period. However, the cure rate models are frequently used to model this type of data. The models provide estimates of the fraction of patients cured of disease as well as the distribution of survival times for uncured patients. In this paper, we propose an exponentiated Weibull mixture cure model, to handle right-censored data sets in survival analysis. The model assumes that the population is a mixture of two sub-populations: a cured population that will never experience the event of interest, and a susceptible population that will experience the event of interest at some point in time. The proposed model applied to the real-life data set and compared its performance with other commonly used survival models. The results show that the exponentiated Weibull mixture cure model provides a better fit to the data and more accurate predictions of survival than other models.

M. A. S. Ishag (✉)
Pan African University, Institute for Basic Sciences, Technology, and Innovation, Nairobi, Kenya
e-mail: ishag.mohamed@students.jkuat.ac.ke

A. Wanjoya
Jomo Kenyatta University of Agriculture and Technology, Nairobi, Kenya

A. Adem
Technical University of Mombasa, Mombasa, Kenya

A. Z. Afify
Benha University, Benha, Egypt
e-mail: ahmed.afify@fcom.bu.edu.eg

V. Gayoso Martínez et al. (eds.), *Mathematical Methods for Engineering Applications*, Springer Proceedings in Mathematics & Statistics 439,
https://doi.org/10.1007/978-3-031-49218-1_17

1 Introduction

Survival analysis is a statistical technique used to analyze time-to-event data. It is widely applied in medical research, engineering, and social sciences, among other fields, to study the occurrence of events such as death, disease, or failure of a mechanical component. Often, these data sets are subject to right-censoring, where some individuals are still alive or have not experienced the event of interest at the end of the study period. However, a small percentage of subjects will never experience the event of interest. These subjects are typically thought to have endless survival times and are stated to be cured. In recent times, there has been an increase in interest in applying mixture models to handle right-censored data sets in survival analysis. A mixture model assumes that the population is a mixture of two sub-populations, each with its own probability distribution. Mixture models have been used to model complex survival data, including those with non-proportional hazards, time-varying covariates, and competing risks. To account for this factor, classic survival models have developed into what are now often known as cure models. Some patients will never relapse from a specific condition and are therefore cured of the disease [2]. When the distribution of survival times for vulnerable patients is known, researchers frequently favor cure fraction models over parametric models. Cure fraction models, which are an extended variant of standard survival models, play an essential role in the study of survival data with long-term survivors. In many instances, analyzing the overall survival rate relative to survival in the population as a whole is of interest [8]. Models of cure proportion are classified into two categories; mixture and unmixture cure models. According to [4] the mixture cure model posits that an individuals is divided into two categories: cured and uncured. A number of researchers have extensively investigated a model of mixture cure, for example [21] and [17]. According to [5], the mixture cure model, does not validate the proportional hazard property for all individuals. Omer et al. [16] proposed the cure models using exponentiated exponential Weibull distribution for cancer patients with melanoma. López-Cheda et al. [10] presented a package of R for estimating the functions described in the mixture cure model using a nonparametric approach that takes right-censored survival data into account. As a result, the main goal of this model is to estimate the fraction of individuals as well as the distribution of survival times for those who survive. Furthermore, some authors have developed the maximum likelihood estimation approach for evaluating mixture cure models like [7, 18] and [20].

In this paper, we propose an exponentiated Weibull mixture cure model to handle right-censored data sets. The exponentiated Weibull distribution is a flexible distribution that can model a wide range of hazard functions, including monotonic, bathtub-shaped, and unimodal hazard functions. The purpose of this study is to provide a maximum likelihood estimation approach for investigating the parameter of the proposed model where cured subjects are present. We apply the proposed model to a real-world data set and compare its performance with other commonly used survival models. The findings show that the exponentiated Weibull mixture cure model provides a better fit and more accurate predictions than other models.

2 The Cure Models

The cure models are a class of survival models that allow for the possibility of a proportion of individuals who have a probability of never experiencing the event of interest, even if they are followed up indefinitely. These individuals are considered "cured" and are not at risk of the event of interest. The remaining individuals in the population are considered "uncured" and are at risk of the event of interest. Cure models are useful in situations where there may be a sub-population of individuals who are less susceptible to the event of interest, or where the event may not be observable for some individuals due to limitations in the study design or follow-up time. In survival analysis, it is frequently assumed that every participant in the study is susceptible to the occurrence of interest. However, this assumption could be inappropriate in some cases where a group of individuals is susceptible to the occurrence of such an event. To assess the capacity of a certain "treatment" to "cure," survival regression models must include a cure proportion [11]. The current statistical methods in order to deal with such data are extensive and typically known as cure rate models [9]. The cure rate, which is the proportion of individuals who are cured, is an important parameter in mixture cure models. The cure rate can be estimated using maximum likelihood estimation, and a high cure rate may indicate the presence of a sub-population with a lower risk of the event of interest.

2.1 The Mixture Cure Model

The mixture cure model (MCM) is a popular approach for modeling data with long-term survivors. The advantage of the (MCM) model is that it allows factors to effect cured individuals differently than those who are vulnerable. The population split into the two categories, in the MCM model; uncured (susceptible) individuals who will experience the event of interest and cured (unsusceptible) individuals who won't. The cured individuals are those who do not develop the event of interest. Let T denote individuals chance of survival and ω be an indicator of the cure, where $\omega = 0$ implies the subject is cured, and if $\omega = 1$ means the subject is uncured. Furthermore, assume that $S^*(t)$ is survival function, $f^*(t)$ is density function and $F^*(t)$ is cumulative density function for susceptible subjects. As a result, the entire population cumulative density function is

$$F_{MCM}(t) = (1 - \omega)F^*(t), \tag{1}$$

The survival function is

$$S_{MCM}(t) = 1 - (1-\omega)F^*(t) = \omega + (1-\omega)S^*(t), \tag{2}$$

And probability density function is

$$f_{MCM}(t) = (1-\omega)f^*(t) \tag{3}$$

2.2 Maximum Likelihood Estimation of MCM Model

To estimate the model parameters in mixture cure model, the maximum likelihood approach is used. Assume that t_i is the right-censored survival time for subject i, additionally $t_i = min(T_i, C_i)$, where T_i is the ith individual's failure time and C_i is the ith individual's right-censored, $i = 1, 2, ...,$n. And δ_i represents the censoring indicator variable, which has values zero and one for censored and uncensored observations, respectively, where is $i = 1, 2, \ldots,$ n. The likelihood function of the MCM model is

$$l_{MCM}(\vartheta) = \prod_{i=1}^{n}[f^*(t)]^{\delta_i}[S^*(t)]^{1-\delta_i}, = \prod_{i=1}^{n}[(1-\omega)f^*(t)]^{\delta_i}[\omega + (1-\omega)S^*(t)]^{1-\delta_i} \tag{4}$$

And the log-likelihood function is

$$l_{MCM}(\vartheta) = ln(1-\omega)\sum_{i=1}^{n}\delta_i + \sum_{i=1}^{n}\delta_i ln(f^*(t)) + \sum_{i=1}^{n}(1-\delta_i)ln(\omega + (1-\omega)S^*(t)) \tag{5}$$

The estimates of parameters can be obtained directly by maximizing the total log-likelihood function.

3 Exponentiated Weibull Baseline Distribution

The Exponentiated Weibull distribution (EW) is a flexible distribution for modeling survival data, which can capture a variety of shapes of the hazard function. The EW distribution was first introduced by Mudholkar and Srivastava [12], as an extension of the Weibull distribution. The EW distribution has three parameters: a shape parameter, a scale parameter, and a location parameter. The distribution, is a flexible, and robust model for fitting survival data that can be employed effectively depending on the data structure [15]. Assume that the EW is a distribution of vulnerable subjects' survival times. The probability density function (pdf) of the Exponentiated Weibull distribution is given by:

$$f(t) = \rho k \upsilon (kt)^{\rho-1}(1 - exp\{-(kt)^{\rho}\})^{\upsilon-1}exp\{-(kt)^{\rho}\}, \tag{6}$$

where $\rho, \upsilon > 0$ represents two shape parameters and $k > 0$ is the scale parameters.

The hazard rate function (HRF) of the EW distribution is given by

$$h(t) = \frac{\rho k \upsilon (kt)^{\rho-1}(1 - exp\{-(kt)^{\rho}\})^{\upsilon-1} exp\{-(kt)^{\rho}\}}{1 - (1 - exp\{-(kt)^{\rho}\})^{\upsilon}}, \tag{7}$$

The cumulative density function (CDF) of the EW distribution is given by

$$F(t) = (1 - exp\{-(kt)^{\rho}\})^{\upsilon}, \tag{8}$$

The survival function (SF) of the EW model is given by

$$S(t) = 1 - (1 - exp\{-(kt)^{\rho}\})^{\upsilon}, \tag{9}$$

And the cumulative hazard function (CHF) of the EW distribution is given by

$$H(t) = -log\{1 - (1 - exp\{-(kt)^{\rho}\})^{\upsilon}\} \tag{10}$$

4 Proposed Mixture Cure Model

The exponentiated Weibull mixture cure model (EWMCM) is proposed with the goal of taking into consideration the fraction of cured subjects in the study. which can accommodate different hazard function shapes, including bathtub-shaped. The survival function of the EWMCM is

$$S_{EWMCM}(t) = \omega + (1 - \omega).1 - (1 - exp\{-(kt)^{\rho}\})^{\upsilon}, \tag{11}$$

The probability density function of the EWMCM is

$$f_{EWMCM}(t) = (1 - \omega).\rho k \upsilon (kt)^{\rho-1}(1 - exp\{-(kt)^{\rho}\})^{\upsilon-1} exp\{-(kt)^{\rho}\}, \tag{12}$$

And cumulative density function of the EWMCM is

$$F_{EWMCM}(t) = (1 - \omega).(1 - exp\{-(kt)^{\rho}\})^{\upsilon}. \tag{13}$$

4.1 Parameters Estimation of the EW-MCM Model

In this section, the EW-MCM parameters are estimated using maximum likelihood. Let $\vartheta = (\rho, k, \upsilon, \omega)$, then the log-likelihood function for ϑ is

$$l_{MCM}(\vartheta) = ln(1-\omega)\sum_{i=1}^{n}\delta_i + \sum_{i=1}^{n}\delta_i ln(f^*(t)) + \sum_{i=1}^{n}(1-\delta_i)ln(\omega + (1-\omega)S^*(t)) \tag{14}$$

$$\begin{aligned} l(\rho, k, \upsilon, \omega) = {} & ln(1-\omega)\sum_{i=1}^{n}\delta_i \\ & + \sum_{i=1}^{n}\delta_i ln\left[\rho k\upsilon(kt)^{\rho-1}\left(1-e^{\{-(kt)^{\rho}\}}\right)^{\upsilon-1} e^{\{-(kt)^{\rho}\}}\right] \\ & + \sum_{i=1}^{n}(1-\delta_i)ln\left[\omega + (1-\omega)\left(1-\left(1-e^{\{-(kt)^{\rho}\}}\right)^{\upsilon}\right)\right] \end{aligned} \tag{15}$$

The following scoring equations are obtained from partial derivatives of the log-likelihood function in Eq. (15) with regard to

$$\begin{aligned} \frac{\partial L}{\partial \rho} = \sum_{i=1}^{n} \Bigg[& \delta_i \left(\frac{\begin{array}{c} k\upsilon(kt)^{\rho-1}\left(1-e^{-(kt)^{\rho}}\right)^{\upsilon-1} e^{-(kt)^{\rho}} \\ +\rho k\upsilon(kt)^{\rho-1}\left(1-e^{-(kt)^{\rho}}\right)^{\upsilon-1} e^{-(kt)^{\rho}} \cdot \ln(kt)\cdot(kt)^{\rho-1} \end{array}}{\rho k\upsilon(kt)^{\rho-1}\left(1-e^{-(kt)^{\rho}}\right)^{\upsilon-1} e^{-(kt)^{\rho}}} \right) \\ & +(1-\delta_i)\left(\frac{(1-\omega)\upsilon\left(1-e^{-(kt)^{\rho}}\right)^{\upsilon-1} e^{-(kt)^{\rho}} \cdot \ln(kt)\cdot(kt)^{\rho-1}}{\omega + (1-\omega)\left(1-\left(1-e^{-(kt)^{\rho}}\right)^{\upsilon}\right)} \right) \Bigg] \end{aligned} \tag{16}$$

$$\frac{\partial L}{\partial k} = \sum_{i=1}^{n} \left[\delta_i \left(\frac{\begin{array}{c} \rho\upsilon(kt)^{\rho-1}\left(1-e^{-(kt)^{\rho}}\right)^{\upsilon-1} e^{-(kt)^{\rho}} \\ +\rho k\upsilon(kt)^{\rho-1}\left(1-e^{-(kt)^{\rho}}\right)^{\upsilon-1} e^{-(kt)^{\rho}} \cdot \ln(kt)\cdot(kt)^{\rho-1} \end{array}}{\rho k\upsilon(kt)^{\rho-1}\left(1-e^{-(kt)^{\rho}}\right)^{\upsilon-1} e^{-(kt)^{\rho}}} \right) \right] \tag{17}$$

$$\begin{aligned} \frac{\partial L}{\partial \upsilon} = \sum_{i=1}^{n} \Bigg[& \delta_i \left(\frac{k\rho(kt)^{\rho-1}\left(1-e^{-(kt)^{\rho}}\right)^{\upsilon-1} e^{-(kt)^{\rho}} \cdot \ln\left(1-e^{-(kt)^{\rho}}\right)}{\rho k\upsilon(kt)^{\rho-1}\left(1-e^{-(kt)^{\rho}}\right)^{\upsilon-1} e^{-(kt)^{\rho}}} \right) \\ & -(1-\delta_i)\left(\frac{(1-\omega)\left(1-e^{-(kt)^{\rho}}\right)^{\upsilon-1} e^{-(kt)^{\rho}} \cdot \ln\left(1-e^{-(kt)^{\rho}}\right)}{\omega + (1-\omega)\left(1-\left(1-e^{-(kt)^{\rho}}\right)^{\upsilon}\right)} \right) \Bigg] \end{aligned} \tag{18}$$

$$\frac{\partial L}{\partial \omega} = \sum_{i=1}^{n} \left[\delta_i \left(\frac{1}{1-\omega} \right) - (1-\delta_i)\left(\frac{1-\left(1-e^{-(kt)^{\rho}}\right)^{\upsilon}}{\omega + (1-\omega)\left(1-\left(1-e^{-(kt)^{\rho}}\right)^{\upsilon}\right)} \right) \right] \tag{19}$$

To maximize the log-likelihood function, the Newton-Raphson method is used to compute the numerical solution since it is a nonlinear system.

5 The Model Selection

In this section, the proposed distribution is fully applied to a real-world data set obtained from the literature to demonstrate the new model's capability and compare to the other four parametric survival distributions; Exponential-Weibull (ExW), Generalized Log-logistic (GLL), Log-logistic (LL), and the Weibull distribution (W). Certain analytical criteria are considered in order to determine which distribution best fits the applied data. The comparison of MCM based on various distributions was evaluated using the Akaike Information Criteria (AIC) proposed by Akaike [1], and the Bayesian Information Criteria (BIC) introduced by Schwarz [19]. The (AIC) and the (BIC) are given respectively:

$$AIC = -2\log L(\hat{\vartheta}) + 2k \tag{20}$$

and

$$BIC = -2\log L(\hat{\vartheta}) + k\ln n \tag{21}$$

where Lϑ is the likelihood function, k represents the number of four parameters in the model and n is the number of observations. The best model is the one with has the lowest AIC, and BIC.

6 Application to Clinical Trails Data

To demonstrate the applicability of the suggested mixture cure model, the rebuilt IPASS clinical trial data which is provided by Argyropoulos and Unruh [3] was examined. The data set is available in AHSurv [14]. For more information about the data see [6, 13]. The data set consisting of 1217 patients in this data set and the censoring rate for this clinical study is approximately 20%. The descriptive statistics for the data set are shown in Table 1.

Figure 1 illustrates plots for the survival time of the IPASS clinical trial dataset, histogram and box plot, indicating the data's growing hazard rate shape, which shows that the HRF is appropriate to analyze this data.

Table 1 Descriptive statistics of the IPASS clinical trails data set

Mean	Median	Variance	Sd	Minimum	Maximum
5.931	5.504	16.568	4.070	0.051	21.000

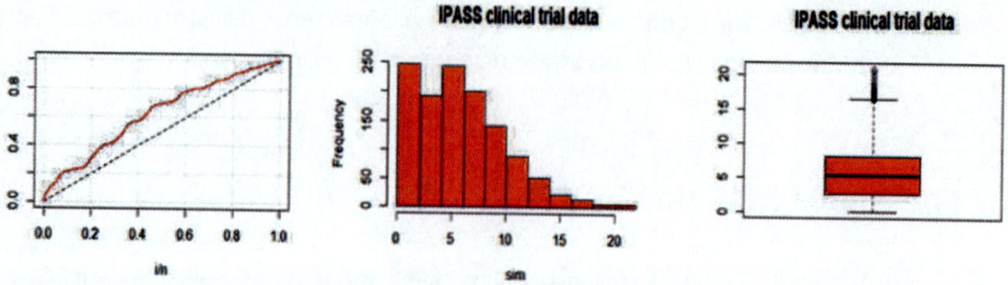

Fig. 1 The TTT plot, histogram, and the box plot of the IPASS clinical trial data

Table 2 Estimated parameters and goodness-of-fit statistics for different survival models

Model	$\hat{\rho}$	$\hat{\kappa}$	$\hat{\upsilon}$	$\hat{\omega}$	AIC	BIC
EW	0.964	0.190	1.685	0.017	5706.794	5723.211
	0.206	0.050	0.562	0.026		
W	7.452	1.370	–	0.014	5708.135	5727.447
	0.243	0.041	–	0.013		
LL	1.566	6.591	–	0.145	5712.893	5728.205
	0.058	0.324	–	0.025		
ExW	0.018	0.056	1.393	0.016	5714.956	5735.372
	0.014	0.014	0.058	0.012		
GLL	0.140	1.422	0.044	0.013	5708.614	5729.031
	0.007	0.072	0.035	0.030		

Table 2 provide the MLE, standard errors, and information criterion values for AIC and BIC obtained by MCM by applying the IPASS clinical trial data and probability distributions of interest. It is clear that the EW-MCM model provides the smallest values of selection criteria with AIC = 5706.794 and BIC = 5723.211, indicating that the MCM based on this distribution is preferred model to the ExW-MCM, GLL-MCM, W-MCM, and LL-MCM models. The EW-MCM model matches the data set more precisely with the lowest AIC and BIC values when compared to any other MCM model.This implies that the EW-MCM model provided the best fit for the data set.

Furthermore, the estimated parameters of the EW-MCM model are significant at the 1% significance level, comparison to other parametric models as shown in Table 3.

7 Discussion

The current study intends to determine the distribution of survival times in susceptible IPASS clinical trial data and investigate the impact of covariates on cure rates and survival times. To achieve this purpose, the cure fraction of MCM model based

Table 3 Shows the z-value, CI and p-value of the MCM for each model at level of significance 1%

Model	Parameter	Mean	z-value	CI 99%	p-value
EW	$\hat{\rho}$	0.964	4.681	(0.560–1.367)	<0.000
	$\hat{\kappa}$	0.190	3.764	(0.091–0.288)	<0.000
	$\hat{\upsilon}$	1.685	2.996	(0.583–2.788)	<0.003
	$\hat{\omega}$	0.017	0.666	(−0.067–0.033)	0.050
W	$\hat{\rho}$	7.452	30.609	(6.975–7.929)	<0.000
	$\hat{\kappa}$	1.370	33.072	(1.289–1.452)	<0.000
	$\hat{\omega}$	0.014	1.015	(−0.013–0.040)	0.310
LL	$\hat{\rho}$	1.566	27.213	(1.453–1.679)	<0.000
	$\hat{\kappa}$	6.591	20.334	(5.956–7.227)	<0.000
	$\hat{\omega}$	0.145	5.909	(−0.194–0.097)	<0.000
ExW	$\hat{\rho}$	0.018	1.223	(−0.011–0.046)	0.222
	$\hat{\kappa}$	0.056	3.964	(0.029–0.084)	<0.000
	$\hat{\upsilon}$	1.393	24.048	(1.280–1.507)	<0.000
	$\hat{\omega}$	0.016	1.299	(−0.008–0.040)	0.194
GLL	$\hat{\rho}$	0.140	19.199	(0.126–0.155)	<0.000
	$\hat{\kappa}$	1.422	19.829	(1.282–1.563)	<0.000
	$\hat{\upsilon}$	0.044	1.255	(−0.025–0.113)	0.210
	$\hat{\omega}$	0.013	0.434	(−0.072–0.046)	0.664

on the EW baseline distribution was developed for the analysis of the data. The cure fraction models can be used to estimate the chance of becoming cured. Furthermore, when there are no cure individuals in the population under the study, they can be reduced to classical survival models. The identification of the cure rate, the distribution of susceptible individuals' survival time, and the duration of follow-up time are important considerations in cure models. In Fig. 1, histogram and boxplot are demonstrates the existence of a cure fraction. As a result, the performance of the MCM model was compared with different susceptibility distributions for the IPASS data. Was noted that the MCM model based on the EW baseline distribution is the best fit for the provided data set with the lowest AIC and BIC values. It was further noted that the parameters of the EW-MCM model are significant at the 1% significance level, compared to other parametric distributions.

8 Conclusion

In conclusion, the exponentiated-Weibull mixture cure model is a flexible and powerful tool for analyzing survival data with a cured population. In the analysis of the IPASS clinical trial data, it was observed that the EW baseline distribution is the best distribution for the survival time of susceptible individuals with respect to the ExW, GLL, W, and LL distributions.

The study has identified an area on which additional research might be useful which is an analysis of the IPASS data. It could possible to employ alternative extended distributions using Bayesian inference techniques, because the experience of cured individuals can be incorporated into a prior distribution of the parameter of the cure rate, can result in more accurate inferences. Future research can focus on evaluating the model's performance using simulation. And to develop a Bayesian technique for the proposed models presented in this paper.

Overall, the exponentiated Weibull mixture cure model provides a new and effective approach to handle right-censored data sets in survival analysis. The model's flexibility and improved accuracy make it a valuable addition to the field of survival analysis.

References

1. Akaike, H.: A new look at the statistical model identification. IEEE Trans. Autom. Control **19**(6), 716–723 (1974). https://doi.org/10.1109/TAC.1974.1100705
2. Amico, M., Van Keilegom, I.: Cure models in survival analysis. Annu. Rev. Stat. Appl. **5**, 311–342 (2018). https://doi.org/10.1146/annurev-statistics-031017-100101
3. Argyropoulos, C., Unruh, M.L.: Analysis of time to event outcomes in randomized controlled trials by generalized additive models. PLoS One **10**(4), e0123784 (2015)
4. Boag, J.W.: Maximum likelihood estimates of the proportion of patients cured by cancer therapy. J. R. Stat. Soc. Ser. B (Methodol.) **11**(1), 15–53 (1949)
5. Chen, M.-H., Ibrahim, J.G., Sinha, D.: A new Bayesian model for survival data with a surviving fraction. J. Am. Stat. Assoc. **94**(447), 909–919 (1999). https://doi.org/10.1080/01621459.1999.10474196
6. Demarqui, F.N., Mayrink, V.D., Ghosh, S.K.: An unified semiparametric approach to model lifetime data with crossing survival curves. Preprint. arXiv:1910.04475 (2019)
7. Farewell, V.T.: The use of mixture models for the analysis of survival data with long-term survivors. Biometrics, 1041–1046 (1982). https://doi.org/10.2307/2529885
8. Jensen, R.K., et al.: Fitting parametric cure models in R using the packages cuRe and rstpm2. Comput. Methods Progr. Biomed. **226**, 107125 (2022). https://doi.org/10.1016/j.cmpb.2022.107125
9. Lambert, P.C.: Modeling of the cure fraction in survival studies. Stata J. **7**(3), 351–375 (2007). https://doi.org/10.1177/1536867X0700700304
10. López-Cheda, A., M Amalia Jácome, and Ignacio López-de-Ullibarri: npcure: An R package for nonparametric inference in mixture cure models. R J **13**(1), 21 (2021)
11. Mastor, A.B., et al: Extended exponential-Weibull mixture cure model for the analysis of cancer clinical trials. In: International Conference on Mathematics and its Applications in Science and Engineering, pp. 239–247. Springer (2022). https://doi.org/10.1007/978-3-031-21700-5_22
12. Mudholkar, G.S., Srivastava, D.K.: Exponentiated Weibull family for analyzing bathtub failure-rate data. IEEE Trans. Reliab. **42**(2), 299–302 (1993). https://doi.org/10.1109/24.229504
13. Muse, A.H., et al.: Flexible parametric accelerated hazard model: simulation and application to censored lifetime data with crossing survival curves. Math. Comput. Appl. **27**(6), 104 (2022). https://doi.org/10.3390/mca27060104
14. Muse, A.H., et al.: AHSurv: An R package for flexible parametric accelerated hazards (AH) regression models (2022). https://doi.org/10.3390/mca27060104

15. Nadarajah, S., Cordeiro, G.M., Ortega, E.M.M.: The exponentiated Weibull distribution: a survey. In: Statistical Papers, vol. 54, pp. 839–877 (2013). https://doi.org/10.1007/s00362-012-0466-x
16. Omer, M.E.A.M.E., et al.: Cure models with exponentiated Weibull exponential distribution for the analysis of melanoma patients. Mathematics **8**(11), 1926 (2020). https://doi.org/10.3390/math8111926
17. Peng, Y., Dear, K.B.G.: A nonparametric mixture model for cure rate estimation. Biometrics **56**(1), 237–243 (2000). https://doi.org/10.1111/j.0006-341X.2000.00237.x
18. Peng, Y., Dear, K.B.G., Denham, J.W.: A generalized F mixture model for cure rate estimation. Stat. Med. **17**(8), 813–830 (1998). https://doi.org/10.1002/(SICI)1097-0258(19980430)17:8
19. Schwarz, G.: Estimating the dimension of a model. Ann. Stat. **6**, 461–464 (1978). Find this article online
20. Sy, J.P., Taylor, J.M.G.: Estimation in a Cox proportional hazards cure model. Biometrics **56**(1), 227–236 (2000). https://doi.org/10.1111/j.0006-341X.2000.00227.x
21. Taylor, J.M.G.: Semi-parametric estimation in failure time mixture models. Biometrics, 899–907 (1995). https://doi.org/10.2307/2532991

On Strong Fuzzy Partial Metric Spaces

Elif Güner and Halis Aygün

Abstract In 2019, Gregori et al. (Int. J. Gen. Syst. 48(3):260–279, 2019), initiated the structure of fuzzy partial metric space (FPMS) based on the residuum operators. Since it is not easy to show different examples when one works with residuum operators, in recent papers, we face to lack of examples of FPMs. Motivated by this fact, in this work, we aim to generate different FPMs from partial metrics (PMs) and vice versa. For this aim, we first give the concept of strong FPMSs by investigating some properties of them and providing some examples by considering different t-norms. In continuation, we give some techniques to construct strong FPMs from existing PMs by means of the pseudo-inverse (p-inverse) of the additive generator (a-generator) of continuous Archimedean t-norms. We also show that FPMs can induce PMs via a-generators. Moreover, we study whether the topologies induced by the presented techniques coincide or not.

1 Introduction and Preliminaries

In 1994, Matthews [14] presented the structure of PMSs as a generalization of metric spaces (MSs) by relaxing the condition of zero self-distance as stated in the next definition:

Definition 1 Let $\mathcal{A}$ be a discourse of universe and $\mathcal{D} : \mathcal{A} \times \mathcal{A} \to [0, \infty)$ be a map fulfilling the next axioms for each $a, b, c \in \mathcal{A}$:

[PM1] $\mathcal{D}(a, a) = \mathcal{D}(a, b) = \mathcal{D}(b, b) \Leftrightarrow a = b$;
[PM2] $0 \leq \mathcal{D}(a, a) \leq \mathcal{D}(a, b)$;
[PM3] $\mathcal{D}(a, b) = \mathcal{D}(b, a)$;
[PM4] $\mathcal{D}(a, c) \leq \mathcal{D}(a, b) + \mathcal{D}(b, c) - \mathcal{D}(b, b)$.

E. Güner (✉) · H. Aygün
Department of Mathematics, Kocaeli University, Kocaeli, Turkey
e-mail: elif.guner@kocaeli.edu.tr; halis@kocaeli.edu.tr

V. Gayoso Martínez et al. (eds.), *Mathematical Methods for Engineering Applications*, Springer Proceedings in Mathematics & Statistics 439,
https://doi.org/10.1007/978-3-031-49218-1_18

In this case, $\mathcal{D}$ is called a PM on $\mathcal{A}$ and the pair $(\mathcal{A}, \mathcal{D})$ denotes the PMS.

We observe that a MS $(\mathcal{A}, d)$ is a PMS that holds, in addition to the axioms [PM1]–[PM2], the next property:

$$d(a, a) = 0 \text{ for every } a \in \mathcal{A}.$$

We know from the paper [14], each PM $\mathcal{D}$ on $\mathcal{A}$ generate a T_0 topology $\tau(\mathcal{D})$ on $\mathcal{A}$ that has as a base the collection of open balls $\{D_{\mathcal{D}}(a, \epsilon) : a \in \mathcal{A}, \epsilon > 0\}$, where $D_{\mathcal{D}}(a, \epsilon) = \{b \in \mathcal{A} : \mathcal{D}(a, b) < \mathcal{D}(a, a) + \epsilon\}$. Furthermore, in such a topology $\tau(\mathcal{D})$, a sequence $(a_n)_{n\in\mathbb{N}}$ is convergent to a point $a \in \mathcal{A}$ iff $\mathcal{D}(a, a) = \lim_{n\to\infty} \mathcal{D}(a, a_n)$. Moreover, a sequence $(a_n)_{n\in\mathbb{N}}$ in a PMS $(\mathcal{A}, \mathcal{D})$ is said to be a Cauchy sequence if $\lim_{n,m\to\infty} \mathcal{D}(a_n, a_m)$ exists. If $\lim_{n,m\to\infty} \mathcal{D}(a_n, a_m) = 0$, then $(a_n)_{n\in\mathbb{N}}$ is called a 0-Cauchy sequence. So, as usual, a PMS $(\mathcal{A}, \mathcal{D})$ is said to be complete if all Cauchy sequence $(a_n)_{n\in\mathbb{N}}$ in $\mathcal{A}$ converges to a point $a \in \mathcal{A}$, that is, $\mathcal{D}(a, a) = \lim_{n\to\infty} \mathcal{D}(a, a_n) = \lim_{n,m\to\infty} \mathcal{D}(a_n, a_m)$. In addition, we say that $(\mathcal{A}, \mathcal{D})$ is 0-complete if every 0-Cauchy sequence in $\mathcal{A}$ converges to a point $a \in \mathcal{A}$ such that $\mathcal{D}(a, a) = 0$. Notice that every complete PMS is 0-complete. Some more studies related to PMSs can be found in [2, 11].

On the other hand, after the fuzzy set theory was initiated by Zadeh [18] in 1965, the structure of fuzzy metric space (FMS) which is another extension of the MS was given by Kramosil and Michalek [13]. Before giving the definition of FMSs, we remind the concept of the continuous triangular norm (cont. t-norm) that is crucial in the concepts of FMSs and FPMSs with which we deal.

Definition 2 A binary operation $\circledast$ on $\mathcal{I}$ ($denote$ $[0, 1] = \mathcal{I}$) fulfilling the next axioms for each $a, b, c \in \mathcal{I}$, is called a triangular norm (shortly, t-norm):

(TN1) $a \circledast b = b \circledast a$; (Commutativity)
(TN2) $(a \circledast b) \circledast c = a \circledast (b \circledast c)$; (Associativity)
(TN3) $a \circledast b \leq a \circledast c$ whenever $b \leq c$; (Monotonicity)
(TN4) $a = a \circledast 1$. (Boundary Condition)

In addition to the above conditions, provided that $\circledast$ is continuous (according to the standard topology) as a function described on $\mathcal{I} \times \mathcal{I}$, $\circledast$ is said to be a cont. t-norm. The followings are examples of well-known cont. t-norms:

(i) $a \wedge b := min(a, b)$ (Minimum t-norm)
(ii) $a \circledast_P b := a.b$ (Product t-norm)
(iii) $a \circledast_{\mathfrak{L}} b := max(a + b - 1, 0)$ (Luckasievicz t-norm)
(iv) $a \circledast_D b := \begin{cases} min(a, b), & a = 1 \text{ or } b = 1 \\ 0, & \text{otherwise} \end{cases}$ (Drastic product t-norm)

A t-norm $\circledast$ is called Archimedean provided that for each $a, b \in \mathcal{I}$ there is a natural number $n \in \mathbb{N}$ such that $b > a^{(n)}$ where $a^{(n)} = a \circledast a \circledast ... \circledast a(n - times)$. Here, we also recall the definition of the a-generator of a cont. t-norm which we will use in the construction techniques.

A strictly decreasing function $h_{\circledast} : \mathcal{I} \to [0, \infty]$ fulfilling the conditions "right-cont. at 0" and "$h_{\circledast}(1) = 0$" and $a \circledast b = h_{\circledast}^{(-1)}(h_{\circledast}(a) + h_{\circledast}(b))$ for each $a, b \in \mathcal{I}$ is called an a-generator of $*$ where $h_{\circledast}^{(-1)}$ is the p-inverse of the function $h_{\circledast}$ and is defined as

$$h_{\circledast}^{(-1)}(a) = \begin{cases} h_{\circledast}^{-1}(a), & 0 \leq a < h_{\circledast}(0) \\ 0, & h_{\circledast}(0) \leq a \leq \infty \end{cases}.$$

Now, we continue with the notion of FMS due to Kramosil and Michalek given in the next definition in the version formulated by Grabiec in [5].

Definition 3 Let $\mathcal{A}$ be a discourse of universe, $\circledast$ a cont. t-norm and $\mathcal{W} : \mathcal{A} \times \mathcal{A} \times [0, \infty) \to \mathcal{I}$ a map fulfilling the next axioms for each $a, b, c \in \mathcal{A}$ and $s, \mathfrak{v} > 0$:

(FM1) $\mathcal{W}(a, b, 0) = 0$;
(FM2) $a = b$ iff $\mathcal{W}(a, b, \mathfrak{v}) = 1$ for each $\mathfrak{v} > 0$;
(FM3) $\mathcal{W}(a, b, \mathfrak{v}) = \mathcal{W}(b, a, \mathfrak{v})$;
(FM4) $\mathcal{W}(a, b, \mathfrak{v}) \circledast \mathcal{W}(b, c, s) \leq \mathcal{W}(a, c, \mathfrak{v} + s)$;
(FM5) $\mathcal{W}_{a,b} : (0, \infty) \to \mathcal{I}$ is a left-cont. function, where $\mathcal{W}_{a,b}(\mathfrak{v}) = \mathcal{W}(a, b, \mathfrak{v})$ for every $\mathfrak{v} > 0$.

In this case, $(\mathcal{W}, \circledast)$ (or simply $\mathcal{W}$) is called a fuzzy metric (FM) on $\mathcal{A}$ and the triple $(\mathcal{A}, \mathcal{W}, \circledast)$ denotes the FMS.

With a small modification of the preceding description, to obtain a Hausdorff topology from FMSs, was given by George and Veeramani in [4] as follows:

Definition 4 Let $\mathcal{A}$ be a discourse of universe, $\circledast$ a cont. t-norm and $\mathcal{W} : \mathcal{A} \times \mathcal{A} \times [0, \infty) \to \mathcal{I}$ a map fulfilling the axioms **(FM3)**, **(FM4)** and the followings for each $a, b, c \in \mathcal{A}$ and $s, \mathfrak{v} > 0$:

(FGV1) $0 < \mathcal{W}(a, b, \mathfrak{v})$;
(FGV2) $a = b$ iff $\mathcal{W}(a, b, \mathfrak{v}) = 1$;
(FGV5) $\mathcal{W}_{a,b} : (0, \infty) \to \mathcal{I}$ is a cont. function.

In this case, $(\mathcal{W}, \circledast)$ (or simply $\mathcal{W}$) is called a GV-fuzzy metric (GV-FM) on $\mathcal{A}$ and the triple $(\mathcal{A}, \mathcal{W}, \circledast)$ denotes the GV-FMS.

Here, we note that a GV-FM $\mathcal{W}$ can be considered as a FM by taking $\mathcal{W}(a, b, 0) = 0$ for each $a, b \in \mathcal{A}$. Hence, GV-FMSs can be taken as a particular case of FMSs. It is studied in the paper [4] that each GV-FM $\mathcal{W}$ on $\mathcal{A}$ generate a Hausdorff topology $\tau_{\mathcal{W}}$ on $\mathcal{A}$ that has as a base the collection of open balls $\{D_{\mathcal{W}}(a, r, \mathfrak{v}) : a \in \mathcal{A}, r \in (0, 1), \mathfrak{v} > 0)\}$, here, $D(a, r, \mathfrak{v}) = \{b \in \mathcal{A} : \mathcal{W}(a, b, \mathfrak{v}) > 1-r\}$. In a GV-FMS, convergent sequences according to this topology are characterized as: a sequence $(w_n)_{n \in \mathbb{N}}$ in a GV-FMS $(\mathcal{A}, \mathcal{W}, \circledast)$ converges to $a \in \mathcal{W}$ if $\lim_{n\to\infty} \mathcal{W}(a, a_n, \mathfrak{v}) = 1$ for each $\mathfrak{v} > 0$. This characterization is also valid for FMSs. Besides, some applications of the notion of FMSs were given in [3, 6, 7, 15–17].

In 2019, Gregori et al. [8] established the structure of FPM by considering the axioms of (GV-)FMS and PMS together. In this construction, they gave the definition of FPM based on residuum operators. The mentioned residuum operators are related to the notion of cont. t-norm which is stated in the next proposition.

Proposition 1 *If $\circledast$ is a cont. t-norm, then $\circledast$-residuum operator of $\circledast$, denoted by $\rightarrow_{\circledast}$, is determined uniquely by the next equation*

$$a \rightarrow_{\circledast} b = \begin{cases} \sup\{c \in \mathcal{I} : a \circledast c = b\}, & a > b; \\ 1, & a \leq b \end{cases} \tag{1}$$

Provided that $\circledast$ is a cont. Archimedean t-norm and $h_{\circledast} : \mathcal{I} \rightarrow [0, \infty]$ an a-generator of $\circledast$, then the $\circledast$-residuum operator $\rightarrow_{\circledast}$ can be found easily with the following equation:

$$a \rightarrow_{\circledast} b = h_{*}^{(-1)}(max(h_{*}(b) - h_{*}(a), 0)).$$

For the detailed studies related to the residuum operators, the paper [12] can be referred to. Moreover, the next results provided in [12] on the residuum operator will be used later.

Proposition 2 *If $\circledast$ is a cont. t-norm, then $a \circledast (a \rightarrow_{\circledast} b) = a \wedge b$ for each $a, b \in \mathcal{I}$.*

We can obtain the next corollary directly from the previous propositions:

Corollary 1 *If $\circledast$ is a cont. t-norm, then $b \leq a \rightarrow_{\circledast} b$ for each $a, b \in \mathcal{I}$.*

The structures of FPM and GV-FPM [8, 9] are recalled in the next definitions:

Definition 5 Let $\mathcal{A}$ be a discourse of universe, $\circledast$ a cont. t-norm and $\mathcal{E} : \mathcal{A} \times \mathcal{A} \times [0, \infty) \rightarrow \mathcal{I}$ a map fulfilling the next axioms for each $a, b, c \in \mathcal{A}$ and $s, \mathfrak{v} > 0$:

(FPM0) $\mathcal{E}(a, b, 0) = 0$;
(FPM1) $\mathcal{E}(a, b, \mathfrak{v}) \leq \mathcal{E}(a, a, \mathfrak{v})$;
(FPM2) $a = b$ iff $\mathcal{E}(a, b, \mathfrak{v}) = \mathcal{E}(a, a, \mathfrak{v}) = \mathcal{E}(b, b, \mathfrak{v})$ for each $\mathfrak{v} > 0$;
(FPM3) $\mathcal{E}(a, b, \mathfrak{v}) = \mathcal{E}(b, a, \mathfrak{v})$;
(FPM4) $\mathcal{E}(a, a, \mathfrak{v}) \rightarrow_{\circledast} \mathcal{E}(a, c, \mathfrak{v}+s) \geq (\mathcal{E}(a, a, \mathfrak{v}) \rightarrow_{\circledast} \mathcal{E}(a, b, \mathfrak{v})) \circledast (\mathcal{E}(b, b, s) \rightarrow_{\circledast} \mathcal{E}(b, c, s))$;
(FPM5) $\mathcal{E}_{a,b} : (0, \infty) \rightarrow \mathcal{I}$ is a left-cont. function, where $\mathcal{E}_{a,b}(\mathfrak{v}) = \mathcal{E}(a, b, \mathfrak{v})$ for each $\mathfrak{v} > 0$.

In this case, $(\mathcal{E}, \circledast)$ (or simply $\mathcal{E}$) is called a FPM on $\mathcal{A}$ and the triple $(\mathcal{A}, \mathcal{E}, \circledast)$ denotes the FPMS.

Definition 6 Let $\mathcal{A}$ be a discourse of universe, $\circledast$ a cont. t-norm and $\mathcal{E} : \mathcal{A} \times \mathcal{A} \times [0, \infty) \rightarrow \mathcal{I}$ a map fulfilling the axioms **(FPM3)**, **(FPM4)** and the followings for each $a, b, c \in \mathcal{A}$ and $s, \mathfrak{v} > 0$:

(FPG1) $0 < \mathcal{E}(a, b, \mathfrak{v}) \leq \mathcal{E}(a, a, \mathfrak{v})$;
(FPG2) $a = b$ iff $\mathcal{E}(a, b, \mathfrak{v}) = \mathcal{E}(a, a, \mathfrak{v}) = \mathcal{E}(b, b, \mathfrak{v})$;

(FPG5) $\mathcal{E}_{a,b} : (0, \infty) \to \mathcal{I}$ is a cont. function.

In this case, $(\mathcal{E}, \circledast)$ (or simply $\mathcal{E}$) is called a GV-FPM on $\mathcal{A}$ and the triple $(\mathcal{A}, \mathcal{E}, \circledast)$ denotes the GV-FPMS.

In the paper [8], it was proved that a FPM $(\mathcal{A}, \mathcal{D}, \circledast)$ can induce a T_0 topology $\tau_{\mathcal{E}}$ on $\mathcal{A}$ that has as a base the collection $\{D_{\mathcal{E}}(a, r, \mathfrak{v}) : a \in \mathcal{A}, r \in (0, 1), \mathfrak{v} > 0\}$ where, for each $a \in \mathcal{A}$, $r \in (0, 1)$ and $\mathfrak{v} > 0$, $D_{\mathcal{E}}(a, r, \mathfrak{v}) = \{b \in \mathcal{A} : \mathcal{E}'(a, b, \mathfrak{v}) > 1 - r\}$ and $\mathcal{E}'(a, b, \mathfrak{v}) = \sup\{\mathcal{E}(a, a, s) \to_{\circledast} \mathcal{E}(a, b, s) : s \in (0, \mathfrak{v})\}$.

Unlike FMSs, in FPMSs, the map $\mathcal{E}_{a,b} : (0, \infty) \to \mathcal{I}$ may not be non-decreasing as shown in [8]. To obtain this property, in [1], a necessary condition was obtained as follows:

Proposition 3 *A (GV-)FPMS $(\mathcal{A}, \mathcal{E}, \circledast)$ fulfills the property (FPM4*) iff the function $\mathcal{E}_{a,b} : (0, \infty) \to \mathcal{I}$ is non-decreasing, for each $a, b \in \mathcal{A}$:*

(FPM4)* $\mathcal{E}(a, c, \mathfrak{v} + s) \geq \mathcal{E}(a, b, \mathfrak{v}) \circledast (\mathcal{E}(b, b, s) \to_{\circledast} \mathcal{E}(b, c, s)).$

From now on, in this paper, we first introduce the notion of strong FPMS. Then we investigate some properties of strong FPMSs and provide different examples by considering distinct t-norms. In continuation of this work, we give some techniques to construct strong FPMs from existing PMs by means of p-inverse of the a-generator of cont. Archimedean t-norms. Besides, we take into consideration of the minimum t-norm which is cont. but not Archimedean to obtain a similar construction with the same technique. We also show that FPMs can induce PMs via a-generators. Moreover, we show that the topology of the PM obtained coincides with the topology of the FPM from which constructs and vice versa.

2 Strong Fuzzy Partial Metric Spaces

In this section, we give the definitions of strong and stationary FPMSs by providing some examples.

Definition 7 Let $(\mathcal{A}, \mathcal{E}, \circledast)$ be a (GV-)FPM. The (GV-)FPM $\mathcal{E}$ is said to be

(i) strong if $\mathcal{E}$ fulfills the next axiom for each $a, b, c \in \mathcal{A}$ and each $\mathfrak{v} > 0$:
(FPM6) $\mathcal{E}(a, c, \mathfrak{v}) \geq \mathcal{E}(a, b, \mathfrak{v}) \circledast (\mathcal{E}(b, b, \mathfrak{v}) \to_{\circledast} \mathcal{E}(b, c, \mathfrak{v}))$.
(ii) stationary if the function $\mathcal{E}_{a,b} : (0, \infty) \to \mathcal{I}$ given by $\mathcal{E}_{a,b}(\mathfrak{v}) = \mathcal{E}(a, b, \mathfrak{v})$ is constant for each $a, b \in \mathcal{A}$. If $(\mathcal{A}, \mathcal{E}, \circledast)$ is a stationary (GV-)FPMS, then we will write $\mathcal{E}(a, b)$ instead of $\mathcal{E}(a, b, \mathfrak{v})$.

Proposition 4 *If $(\mathcal{A}, \mathcal{E}, \circledast)$ is a stationary (GV-)FPMS, then $(\mathcal{A}, \mathcal{E}, \circledast)$ is a strong (GV-)FPMS.*

In the next lemma, unlike the FPMSs, we observe that the condition (FPM4) and the condition (FPM4*) are coincident for a strong FPMS (that is for a fixed parameter).

Lemma 1 *Let $(\mathcal{A}, \mathcal{E}, \circledast)$ be a (GV-)FPMS. Then $(\mathcal{A}, \mathcal{E}, \circledast)$ fulfills the condition (FPM6) iff $(\mathcal{A}, \mathcal{E}, \circledast)$ fulfills the next condition:*

(FPM6)* $\mathcal{E}(a, a, \mathfrak{v}) \rightarrow_{\circledast} \mathcal{E}(a, c, \mathfrak{v}) \geq (\mathcal{E}(a, a, \mathfrak{v}) \rightarrow_{\circledast} \mathcal{E}(a, b, \mathfrak{v}))\circledast(\mathcal{E}(b, b, \mathfrak{v}) \rightarrow_{\circledast} \mathcal{E}(b, c, \mathfrak{v}))$.

Proof Let $(\mathcal{A}, \mathcal{E}, \circledast)$ satisfy the condition (FPM6). Then we have for each $a, b, c \in \mathcal{A}$ and $\mathfrak{v} > 0$,

$$\mathcal{E}(a, a, \mathfrak{v}) \rightarrow_{\circledast} \mathcal{E}(a, c, \mathfrak{v}) \geq \mathcal{E}(a, a, \mathfrak{v}) \rightarrow_{\circledast} [\mathcal{E}(a, b, \mathfrak{v})\circledast(\mathcal{E}(b, b, \mathfrak{v}) \rightarrow_{\circledast} \mathcal{E}(b, c, \mathfrak{v}))]$$

since "$\rightarrow_{\circledast}$" is non-decreasing according to its second variable. Then, we obtain

$$\begin{aligned}
&\mathcal{E}(a, a, \mathfrak{v}) \rightarrow_{\circledast} \mathcal{E}(a, c, \mathfrak{v}) \geq \mathcal{E}(a, a, \mathfrak{v}) \\
&\rightarrow_{\circledast} [\mathcal{E}(a, b, \mathfrak{v})\circledast(\mathcal{E}(b, b, \mathfrak{v}) \rightarrow_{\circledast} \mathcal{E}(b, c, \mathfrak{v}))] \\
&\geq (\mathcal{E}(a, a, \mathfrak{v}) \rightarrow_{\circledast} \mathcal{E}(a, b, \mathfrak{v}))\circledast(\mathcal{E}(a, b, \mathfrak{v}) \\
&\rightarrow_{\circledast} [\mathcal{E}(a, b, \mathfrak{v})\circledast(\mathcal{E}(b, b, \mathfrak{v}) \rightarrow_{\circledast} \mathcal{E}(b, c, \mathfrak{v}))]) \\
&\geq (\mathcal{E}(a, a, \mathfrak{v}) \rightarrow_{\circledast} \mathcal{E}(a, b, \mathfrak{v}))\circledast((\mathcal{E}(b, b, \mathfrak{v}) \\
&\rightarrow_{\circledast} \mathcal{E}(b, c, \mathfrak{v})).
\end{aligned}$$

Conversely, suppose that $(\mathcal{A}, \mathcal{E}, \circledast)$ fulfills the condition (FPM6*). Then we have for each $a, b, c \in \mathcal{A}$ and $\mathfrak{v} > 0$,

$$\begin{aligned}
&\mathcal{E}(a, a, \mathfrak{v})\circledast(\mathcal{E}(a, a, \mathfrak{v}) \rightarrow_{\circledast} \mathcal{E}(a, c, \mathfrak{v})) \\
\geq\ &\mathcal{E}(a, a, \mathfrak{v})\circledast[(\mathcal{E}(a, a, \mathfrak{v}) \rightarrow_{\circledast} \mathcal{E}(a, b, \mathfrak{v}))\circledast(\mathcal{E}(b, b, \mathfrak{v}) \rightarrow_{\circledast} \mathcal{E}(b, c, \mathfrak{v}))].
\end{aligned}$$

since $\circledast$ is non-decreasing according to its variables. From Proposition 2, we obtain

$$\begin{aligned}
\mathcal{E}(a, c, \mathfrak{v}) &= \mathcal{E}(a, a, \mathfrak{v})\circledast(\mathcal{E}(a, a, \mathfrak{v}) \rightarrow_{\circledast} \mathcal{E}(a, c, \mathfrak{v})) \\
&\geq \mathcal{E}(a, a, \mathfrak{v})\circledast[(\mathcal{E}(a, a, \mathfrak{v}) \rightarrow_{\circledast} \mathcal{E}(a, b, \mathfrak{v}))\circledast(\mathcal{E}(b, b, \mathfrak{v}) \rightarrow_{\circledast} \mathcal{E}(b, c, \mathfrak{v}))] \\
&= [\mathcal{E}(a, a, \mathfrak{v})\circledast(\mathcal{E}(a, a, \mathfrak{v}) \rightarrow_{\circledast} \mathcal{E}(a, b, \mathfrak{v}))]\circledast(\mathcal{E}(b, b, \mathfrak{v}) \rightarrow_{\circledast} \mathcal{E}(b, c, \mathfrak{v})) \\
&= \mathcal{E}(a, b, \mathfrak{v})\circledast(\mathcal{E}(b, b, \mathfrak{v}) \rightarrow_{\circledast} \mathcal{E}(b, c, \mathfrak{v}))
\end{aligned}$$

which implies that $(\mathcal{A}, \mathcal{E}, \circledast)$ satisfies the condition (FPM6).

Now, we notice that the axiom (FPM6) cannot replace the axiom (FPGV4) in the definition of (GV-)FPM since in this case $\mathcal{E}$ may not be a (GV-)FPM on $\mathcal{A}$ as shown in the next example:

Example 1 Let us consider the map $\mathcal{E}_d : \mathcal{A} \times \mathcal{A} \times (0, \infty) \rightarrow \mathcal{I}$ defined by $\mathcal{E}_d(a, b, \mathfrak{v}) = \frac{\mathfrak{v}}{\mathfrak{v}+max\{a,b\}}$ for each $a, b \in \mathcal{A}$ and $\mathfrak{v} > 0$. In the paper [8], it is shown that $(\mathcal{E}_d, \circledast_{\mathfrak{L}})$ does not satisfy the condition (FPM4). Hence, $(\mathcal{A}, \mathcal{E}_d, \circledast_{\mathfrak{L}})$

is not a FPMS. But the map $\mathcal{E}_d$ fulfills the condition (FPM6) for the t-norm $\circledast_{\mathfrak{L}}$. Indeed, firstly, we recall that, from [8], the next equality is hold for each $a, b \in \mathcal{A}$ and $\mathfrak{v} > 0$,

$$\begin{aligned}\mathcal{E}_d(a, a, \mathfrak{v}) \rightarrow_{\circledast_{\mathfrak{L}}} \mathcal{E}_d(a, b, \mathfrak{v}) &= \frac{\mathfrak{v}}{\mathfrak{v} + a} \rightarrow_{\circledast_{\mathfrak{L}}} \frac{\mathfrak{v}}{\mathfrak{v} + max\{a, b\}} \\ &= 1 + \frac{\mathfrak{v}}{\mathfrak{v} + max\{a, b\}} - \frac{\mathfrak{v}}{\mathfrak{v} + a}.\end{aligned}$$

Next,

$$\begin{aligned}&(\mathcal{E}_d(a, a, \mathfrak{v}) \rightarrow_{\circledast_{\mathfrak{L}}} \mathcal{E}_d(a, b, \mathfrak{v}))\circledast_{\mathfrak{L}}(\mathcal{E}_d(b, b, \mathfrak{v}) \rightarrow_{\circledast_{\mathfrak{L}}} \mathcal{E}_d(b, c, \mathfrak{v}))\\ &= (1 + \frac{\mathfrak{v}}{\mathfrak{v} + max\{a, b\}} - \frac{\mathfrak{v}}{\mathfrak{v} + a})\circledast_{\mathfrak{L}}(1 + \frac{\mathfrak{v}}{\mathfrak{v} + max\{b, c\}} - \frac{\mathfrak{v}}{\mathfrak{v} + b})\\ &= max\{0, 1 + \frac{\mathfrak{v}}{\mathfrak{v} + max\{a, b\}} + \frac{\mathfrak{v}}{\mathfrak{v} + max\{b, c\}} - \frac{\mathfrak{v}}{\mathfrak{v} + a} - \frac{\mathfrak{v}}{\mathfrak{v} + b}\}.\end{aligned}$$

If the right side is equal to 0, then it is clear that $(\mathcal{E}_d(a, a, \mathfrak{v}) \rightarrow_{\circledast_{\mathfrak{L}}} \mathcal{E}_d(a, b, \mathfrak{v}))\circledast_{\mathfrak{L}}(\mathcal{E}_d(b, b, \mathfrak{v}) \rightarrow_{\circledast_{\mathfrak{L}}} \mathcal{E}_d(b, c, \mathfrak{v})) \leq \mathcal{E}_d(a, a, \mathfrak{v}) \rightarrow_{\circledast_{\mathfrak{L}}} \mathcal{E}_d(a, c, \mathfrak{v})$. In the other case, we need to show that

$$\begin{aligned}&(\mathcal{E}_d(a, a, \mathfrak{v}) \rightarrow_{\circledast_{\mathfrak{L}}} \mathcal{E}_d(a, b, \mathfrak{v}))\circledast_{\mathfrak{L}}(\mathcal{E}_d(b, b, \mathfrak{v}) \rightarrow_{\circledast_{\mathfrak{L}}} \mathcal{E}_d(b, c, \mathfrak{v}))\\ &= 1 + \frac{\mathfrak{v}}{\mathfrak{v} + max\{a, b\}} + \frac{\mathfrak{v}}{\mathfrak{v} + max\{b, c\}} - \frac{\mathfrak{v}}{\mathfrak{v} + a} - \frac{\mathfrak{v}}{\mathfrak{v} + b}\\ &\leq 1 + \frac{\mathfrak{v}}{\mathfrak{v} + max\{a, c\}} - \frac{\mathfrak{v}}{\mathfrak{v} + a} = \mathcal{E}_d(a, a, \mathfrak{v}) \rightarrow_{\circledast_{\mathfrak{L}}} \mathcal{E}_d(a, c, \mathfrak{v}).\end{aligned}$$

In fact, for each $a, b, c \in \mathcal{A}$ and $\mathfrak{v} > 0$, we have

$$\begin{aligned}\frac{\mathfrak{v}}{\mathfrak{v} + max\{a, c\}} + \frac{\mathfrak{v}}{\mathfrak{v} + b} &= \frac{2t^2 + \mathfrak{v}(max\{a, c\} + b)}{\mathfrak{v}^2 + (\mathfrak{v} + b)(max\{a, c\}) + ty}\\ &\geq \frac{2t^2 + \mathfrak{v}(max\{a, b\} + max\{b, c\})}{\mathfrak{v}^2 + (\mathfrak{v} + b)(max\{a, b\} + max\{b, c\} - b) + ty}\\ &\geq \frac{2t^2 + \mathfrak{v}(max\{a, b\} + max\{b, c\})}{\mathfrak{v}^2 + \mathfrak{v}(max\{a, b\} + max\{b, c\}) + max\{a, b\}max\{b, c\}}\\ &= \frac{\mathfrak{v}(\mathfrak{v} + max\{a, b\}) + \mathfrak{v}(\mathfrak{v} + max\{b, c\})}{(\mathfrak{v} + max\{a, b\})(\mathfrak{v} + max\{b, c\})}\\ &= \frac{\mathfrak{v}}{\mathfrak{v} + max\{a, b\}} + \frac{\mathfrak{v}}{\mathfrak{v} + max\{b, c\}}\end{aligned}$$

which follows that $(\mathcal{E}_d(a, a, \mathfrak{v}) \rightarrow_{\circledast_{\mathfrak{L}}} \mathcal{E}_d(a, b, \mathfrak{v}))\circledast_{\mathfrak{L}}(\mathcal{E}_d(b, b, \mathfrak{v}) \rightarrow_{\circledast_{\mathfrak{L}}} \mathcal{E}_d(b, c, \mathfrak{v})) \leq \mathcal{E}_d(a, a, \mathfrak{v}) \rightarrow_{\circledast_{\mathfrak{L}}} \mathcal{E}_d(a, c, \mathfrak{v})$. As a result, we have that $(\mathcal{E}_d, \circledast_{\mathfrak{L}})$ satisfies the condition (FPM6).

Example 2 Let $(\mathcal{A}, \mathcal{D})$ be a PMS.

(1) The FPMS $(\mathcal{A}, \mathcal{E}_e, *_p)$ where $\mathcal{E}_e(a, b, \mathfrak{v}) = e^{-\mathcal{D}(a,b)/\mathfrak{v}}$ for each $a, b \in \mathcal{A}$ and $\mathfrak{v} > 0$, is strong.
(2) The FPMS $(\mathcal{A}, \mathcal{E}_d, *_p)$ where $\mathcal{E}_d(a, b, \mathfrak{v}) = \frac{\mathfrak{v}}{\mathfrak{v}+\mathcal{D}(a,b)}$ for each $a, b \in \mathcal{A}$ and $\mathfrak{v} > 0$, is strong.

2.1 Generating Fuzzy Partial Metrics from Partial Metrics

In this section, we give a technique to construct (GV-)FPMSs from the given PMSs by means of the p-inverse of the a-generator of cont. (Archimedean) t-norm.

Proposition 5 *Let* $\{(\mathcal{E}_{\mathfrak{v}}, \circledast) : \mathfrak{v} > 0\}$ *be a collection of stationary FPM on a non-empty set* $\mathcal{A}$*. If* $\{(\mathcal{E}_{\mathfrak{v}}, \circledast) : \mathfrak{v} > 0\}$ *is an increasing collection and the map* $\mathcal{E}_{a,b} : (0, \infty) \rightarrow \mathcal{I}$ *defined by* $\mathcal{E}_{a,b}(\mathfrak{v}) = \mathcal{E}(a, b, \mathfrak{v})$, $\forall a, b \in \mathcal{A}$ *and* $\mathfrak{v} > 0$*, is a left-cont. function, then* $(\mathcal{A}, \mathcal{E}, \circledast)$ *is a strong FPMS where the map* $\mathcal{E} : \mathcal{A} \times \mathcal{A} \times (0, \infty) \rightarrow \mathcal{I}$ *defined as* $\mathcal{E}(a, b, \mathfrak{v}) = \mathcal{E}_{\mathfrak{v}}(a, b)$ $\forall a, b \in \mathcal{A}$ *and* $\mathfrak{v} > 0$.

Theorem 1 *Suppose that* $(\mathcal{A}, \mathcal{D})$ *is a PMS,* $\psi : (0, \infty) \rightarrow (0, \infty)$ *a left-cont. increasing function,* $\circledast$ *a cont. Archimedean t-norm and* $h_{\circledast}$ *an a-generator of t-norm* $\circledast$*. Then* $(\mathcal{A}, \mathcal{E}_{h_{\circledast},\mathcal{D}}, \circledast)$ *is a strong FPMS where the map* $\mathcal{E}_{h_{\circledast},\mathcal{D}} : \mathcal{A}\times\mathcal{A}\times(0, \infty) \rightarrow \mathcal{I}$ *is defined for each* $a, b \in \mathcal{A}$ *and* $\mathfrak{v} > 0$,

$$\mathcal{E}_{h_{\circledast},\mathcal{D}}(a, b, \mathfrak{v}) = h_{\circledast}^{(-1)}\left(\frac{\mathcal{D}(a, b)}{\psi(\mathfrak{v})}\right)$$

and $\mathcal{E}_{h_{\circledast},\mathcal{D}}(a, b, 0) = 0$*. Moreover, the topology induced by* $\mathfrak{D}$ *and the topology induced by* $\mathcal{E}_{h_{\circledast},\mathcal{D}}$ *coincides. i.e.,* $\tau_{\mathfrak{D}} = \tau_{\mathcal{E}_{h_{\circledast},\mathcal{D}}}$.

Proof To prove that $(\mathcal{A}, \mathcal{E}_{h_{\circledast},\mathcal{D}}, \circledast)$ is a strong FPMS, we need to construct a collection of stationary FPMs $\{((\mathcal{E}_{h_{\circledast},\mathcal{D}})_t, \circledast) : \mathfrak{v} > 0\}$ that fulfills the assumptions of Proposition 5.

Fix the parameter $\mathfrak{v} > 0$. We know that, from the paper [10], if $h_{\circledast}$ is an a-generator of $\circledast$, then the function $g : \mathcal{I} \rightarrow \mathcal{I}$ defined by $g(a) = \psi(\mathfrak{v})h_{\circledast}(a), \forall a \in \mathcal{I}$, is also an a-generator of $\circledast$. Furthermore, we know that $g^{(-1)}(a) = h_{\circledast}^{(-1)}\left(\frac{a}{\psi(\mathfrak{v})}\right)$. Then, the map $E^{\circledast}_{\mathcal{D},\mathfrak{v}} : \mathcal{A} \times \mathcal{A} \rightarrow \mathcal{I}$ given by $E^{\circledast}_{\mathcal{D},\mathfrak{v}}(a, b) = g^{(-1)}(\mathcal{D}(a, b)) = h_{\circledast}^{(-1)}\left(\frac{\mathcal{D}(a,b)}{\psi(\mathfrak{v})}\right)$ fullfils the next properties for each $a, b \in \mathcal{A}$ and $\mathfrak{v} > 0$:

(i) $E^{\circledast}_{\mathcal{D},\mathfrak{v}}(a, a) = h_{\circledast}^{(-1)}\left(\frac{\mathcal{D}(a,a)}{\psi(\mathfrak{v})}\right) \geq h_{\circledast}^{(-1)}\left(\frac{\mathcal{D}(a,b)}{\psi(\mathfrak{v})}\right) = E^{\circledast}_{\mathcal{D},\mathfrak{v}}(a, b)$.

(ii) Let $E^{\circledast}_{\mathcal{D},\mathfrak{v}}(a,a) = E^{\circledast}_{\mathcal{D},\mathfrak{v}}(a,b) = E^{\circledast}_{\mathcal{D},\mathfrak{v}}(b,b)$. Then we have

$$h_{\circledast}^{(-1)}\left(\frac{\mathcal{D}(a,a)}{\psi(\mathfrak{v})}\right) = h_{\circledast}^{(-1)}\left(\frac{\mathcal{D}(a,b)}{\psi(\mathfrak{v})}\right) = h_{\circledast}^{(-1)}\left(\frac{\mathcal{D}(b,b)}{\psi(\mathfrak{v})}\right)$$

which follows that $\mathcal{D}(a,a) = \mathcal{D}(a,b) = \mathcal{D}(b,b)$ and so $a = b$.

(iii) $E^{\circledast}_{\mathcal{D},\mathfrak{v}}(a,b) = h_{\circledast}^{(-1)}\left(\frac{\mathcal{D}(a,b)}{\psi(\mathfrak{v})}\right) = h_{\circledast}^{(-1)}\left(\frac{\mathcal{D}(b,a)}{\psi(\mathfrak{v})}\right) = E^{\circledast}_{\mathcal{D},\mathfrak{v}}(b,a)$.

(iv) For all $a, b, c \in \mathcal{A}$ and $\mathfrak{v}, s > 0$, we obtain

$$\begin{aligned}
&(E^{\circledast}_{\mathcal{D},\mathfrak{v}}(a,a) \rightarrow_{\circledast} E^{\circledast}_{\mathcal{D},\mathfrak{v}}(a,b)) \circledast (E^{\circledast}_{\mathcal{D},s}(b,b) \rightarrow_{\circledast} E^{\circledast}_{\mathcal{D},s}(b,c)) \\
&= h_{\circledast}^{-1}(min\{h_{\circledast}(0), h_{\circledast}(E^{\circledast}_{\mathcal{D},\mathfrak{v}}(a,a) \rightarrow_{\circledast} E^{\circledast}_{\mathcal{D},\mathfrak{v}}(a,b)) \\
&\quad + h_{\circledast}(E^{\circledast}_{\mathcal{D},s}(b,b) \rightarrow_{\circledast} E^{\circledast}_{\mathcal{D},s}(b,c))\}) \\
&= h_{\circledast}^{-1}(min\{h_{\circledast}(0), h_{\circledast}(E^{\circledast}_{\mathcal{D},\mathfrak{v}}(a,b)) - h_{\circledast}(E^{\circledast}_{\mathcal{D},\mathfrak{v}}(a,a)) \\
&\quad + h_{\circledast}(E^{\circledast}_{\mathcal{D},s}(b,c)) - h_{\circledast}(E^{\circledast}_{\mathcal{D},s}(b,c))\}) \\
&= h_{\circledast}^{-1}\left(min\{h_{\circledast}(0), \left(\frac{\mathcal{D}(a,b)}{\psi(\mathfrak{v})}\right) - \left(\frac{\mathcal{D}(a,a)}{\psi(\mathfrak{v})}\right) + \left(\frac{\mathcal{D}(b,c)}{\psi(s)}\right) - \left(\frac{\mathcal{D}(b,b)}{\psi(s)}\right)\}\right) \\
&\leq h_{\circledast}^{-1}\left(min\{h_{\circledast}(0), \frac{\mathcal{D}(a,c)}{\psi(\mathfrak{v}+s)} - \frac{\mathcal{D}(a,a)}{\psi(\mathfrak{v}+s)}\}\right) \\
&= h_{\circledast}^{-1}(min\{h_{\circledast}(0), h_{\circledast}(E^{\circledast}_{\mathcal{D},\mathfrak{v}+s}(a,c)) - h_{\circledast}(E^{\circledast}_{\mathcal{D},\mathfrak{v}+s}(a,a))\}) \\
&= h_{\circledast}^{-1}(min\{h_{\circledast}(0), h_{\circledast}(E^{\circledast}_{\mathcal{D},\mathfrak{v}+s}(a,a) \rightarrow_{\circledast} E^{\circledast}_{\mathcal{D},\mathfrak{v}+s}(a,c))\}) \\
&= E^{\circledast}_{\mathcal{D},\mathfrak{v}+s}(a,a) \rightarrow_{\circledast} E^{\circledast}_{\mathcal{D},\mathfrak{v}+s}(a,c).
\end{aligned}$$

Hence, we have that the map $(\mathcal{E}_{h_{\circledast},\mathcal{D}})_t : \mathcal{A} \times \mathcal{A} \times \rightarrow \mathcal{I}$ defined by $(\mathcal{E}_{h_{\circledast},\mathcal{D}})_t(a,b) = E^{\circledast}_{\mathcal{D},\mathfrak{v}}(a,b)$ is a stationary FPM. Now, consider the collection of the stationary FPMs $\{((\mathcal{E}_{h_{\circledast},\mathcal{D}})_t, \circledast) : \mathfrak{v} > 0\}$. We need to show that this collection is increasing. Take $0 < s < \mathfrak{v}$ and $a, b \in \mathcal{A}$. Then, we get

$$(\mathcal{E}_{h_{\circledast},\mathcal{D}})_t(a,b) = h_{\circledast}^{(-1)}\left(\frac{\mathcal{D}(a,b)}{\psi(\mathfrak{v})}\right) \geq h_{\circledast}^{(-1)}\left(\frac{\mathcal{D}(a,b)}{\psi(s)}\right) = (\mathcal{E}_{h_{\circledast},\mathcal{D}})_s(a,b)$$

since ψ is increasing and $h_{\circledast}$ is decreasing. Finally, the function $(\mathcal{E}_{h_{\circledast},\mathcal{D}})_{a,b} : (0,\infty) \rightarrow \mathcal{I}$ given by $(\mathcal{E}_{h_{\circledast},\mathcal{D}})_{a,b}(\mathfrak{v}) = \mathcal{E}_{h_{\circledast},\mathcal{D}}(a,b,\mathfrak{v})$ is left-cont. since ψ is left-cont. and $h_{\circledast}$ is cont. As a result, the collection $\{((\mathcal{E}_{h_{\circledast},\mathcal{D}})_t, \circledast) : \mathfrak{v} > 0\}$ fulfills all assumptions of Proposition 5 and thus $\mathcal{E}_{h_{\circledast},\mathcal{D}}$ is a strong FPM.

Example 3 Let $(\mathcal{A}, \mathcal{D})$ be a PMS.

(1) Take the product t-norm $\circledast_P$ which has an a-generator $h_{\circledast_P}(a) = -log(a)$, (suppose that $log(0) = -\infty$). Then the p-inverse of $h_{\circledast_P}$ is $h_{\circledast_P}^{(-1)}(a) = e^{-a}$, $\forall a \in [0, \infty]$. Hence we have, from Theorem 1, for each $a, b \in \mathcal{A}$ and $\mathfrak{v} > 0$,

$$\mathcal{E}_{h_{\circledast_P}, \mathcal{D}}(a, b, \mathfrak{v}) = e^{-\mathcal{D}(a,b)/\mathfrak{v}}.$$

(2) Take the Hamacher t-norm $\circledast_H$ which has an a-generator $h_{\circledast_H}(a) = \frac{1-a}{a}$, (assume that $h_{\circledast_H}(0) = \infty$). Then, the p-inverse of $h_{\circledast_H}$ is $h_{\circledast_H}^{(-1)}(a) = \frac{1}{1+a}$, $\forall a \in [0, \infty]$. Hence we have, from Theorem 1, for each $a, b \in \mathcal{A}$ and $\mathfrak{v} > 0$,

$$\mathcal{E}_{h_{\circledast_H}, \mathcal{D}}(a, b, \mathfrak{v}) = \frac{\mathfrak{v}}{\mathfrak{v} + \mathcal{D}(a, b)}.$$

(3) Take the Luckasievicz t-norm $\circledast_{\mathfrak{L}}$ which has an a-generator $h_{\circledast_H \mathfrak{L}}(a) = 1 - a$. Then, the p-inverse of $h_{\circledast_{\mathfrak{L}}}$ is $h_{\circledast_{\mathfrak{L}}}^{(-1)}(a) = max\{0, 1 - a\}$, $\forall a \in [0, \infty]$. Thus we have, from Theorem 1, for each $a, b \in \mathcal{A}$ and $\mathfrak{v} > 0$,

$$\mathcal{E}_{h_{\circledast_{\mathfrak{L}}}, \mathcal{D}}(a, b, \mathfrak{v}) = max\{0, 1 - \frac{\mathcal{D}(a, b)}{\mathfrak{v}}\}.$$

We can give the Proposition 5 for GV-FPMS as follows:

Proposition 6 *Let $\{(\mathcal{E}_t, \circledast) : \mathfrak{v} > 0\}$ be a collection of stationary GV-FPMs on a non-empty set $\mathcal{A}$. If $\{(\mathcal{E}_t, \circledast) : \mathfrak{v} > 0\}$ is an increasing collection and the map $\mathcal{E}_{a,b} : (0, \infty) \to \mathcal{I}$ defined by $\mathcal{E}_{a,b}(\mathfrak{v}) = \mathcal{E}(a, b, \mathfrak{v})$, $\forall a, b \in \mathcal{A}$ and $\mathfrak{v} > 0$, is a cont. function, then $(\mathcal{A}, \mathcal{E}, \circledast)$ is a strong GV-FPMS where the map $\mathcal{E} : \mathcal{A} \times \mathcal{A} \times (0, \infty) \to \mathcal{I}$ defined by $\mathcal{E}(a, b, \mathfrak{v}) = \mathcal{E}_t(a, b)$ $\forall a, b \in \mathcal{A}$ and $\mathfrak{v} > 0$.*

Next, we give the GV-FPM version of Theorem 1.

Theorem 2 *Suppose that $(\mathcal{A}, \mathcal{D})$ is a PMS, $\psi : (0, \infty) \to (0, \infty)$ a cont. increasing function, $\circledast$ a cont. Archimedean t-norm and $h_{\circledast}$ an a-generator of t-norm $\circledast$. Then $(\mathcal{A}, \mathcal{E}_{h_{\circledast}, \mathcal{D}}, \circledast)$ is a strong GV-FPMS where the map $\mathcal{E}_{h_{\circledast}, \mathcal{D}} : \mathcal{A} \times \mathcal{A} \times (0, \infty) \to \mathcal{I}$ is defined for each $a, b \in \mathcal{A}$ and $\mathfrak{v} > 0$,*

$$\mathcal{E}_{h_{\circledast}, \mathcal{D}}(a, b, \mathfrak{v}) = h_{\circledast}^{(-1)}\left(\frac{\mathcal{D}(a, b)}{\psi(\mathfrak{v})}\right).$$

Moreover, the topology induced by $\mathfrak{D}$ and the topology induced by $\mathcal{E}_{h_{\circledast}, \mathcal{D}}$ coincides. i.e., $\tau_{\mathfrak{D}} = \tau_{\mathcal{E}_{h_{\circledast}, \mathcal{D}}}$.

Now, we give a technique to construct a FPM from a given PM by considering the $\wedge$ t-norm which is not Archimedean.

Theorem 3 *Suppose that $(\mathcal{A}, \mathcal{D})$ is a PMS, $\psi : (0, \infty) \to (0, \infty)$ a left-cont. increasing superadditive function (i.e., $\psi(s + \mathfrak{v}) \geq \psi(s) + \psi(\mathfrak{v}), \forall s, \mathfrak{v} > 0$) and $g : [0, \infty] \to \mathcal{I}$ be a left-cont. decreasing function. Then $(\mathcal{A}, \mathcal{E}_{g,\mathcal{D}}, \circledast)$ is a FPMS where the map $\mathcal{E}_{g,\mathcal{D}} : \mathcal{A} \times \mathcal{A} \times (0, \infty) \to \mathcal{I}$ is defined for each $a, b \in \mathcal{A}$ and $\mathfrak{v} > 0$,*

$$\mathcal{E}_{g,\mathcal{D}}(a, b, \mathfrak{v}) = g\left(\frac{\mathcal{D}(a, b)}{\psi(\mathfrak{v})}\right)$$

and $\mathcal{E}_{g,\mathcal{D}}(a, b, 0) = 0$.

A similar version of Theorem 4 for GV-FPMS is given as follows:

Theorem 4 *Suppose that $(\mathcal{A}, \mathcal{D})$ is a PMS, $\psi : (0, \infty) \to (0, \infty)$ a cont. increasing superadditive function (i.e., $\psi(s + \mathfrak{v}) \geq \psi(s) + \psi(\mathfrak{v}), \forall s, \mathfrak{v} > 0$) and $g : [0, \infty] \to \mathcal{I}$ be a cont. decreasing function. Then $(\mathcal{A}, \mathcal{E}_{g,\mathcal{D}}, \circledast)$ is a GV-FPMS where the map $\mathcal{E}_{g,\mathcal{D}} : \mathcal{A} \times \mathcal{A} \times (0, \infty) \to \mathcal{I}$ is defined for each $a, b \in \mathcal{A}$ and $\mathfrak{v} > 0$,*

$$\mathcal{E}_{g,\mathcal{D}}(a, b, \mathfrak{v}) = g\left(\frac{\mathcal{D}(a, b)}{\psi(\mathfrak{v})}\right).$$

2.2 Generating Partial Metrics from Fuzzy Partial Metrics

Next, we show a construction for PMSs from given FPMSs by using a-generators of cont. Archimedean t-norms.

Theorem 5 *Suppose that $(\mathcal{A}, \mathcal{E}, \circledast)$ is a stationary FPMS where $\circledast$ is a cont. Archimedean t-norm and $h_\circledast$ is an a-generator of $\circledast$. Then $(\mathcal{A}, \mathcal{D}_{h_\circledast,\mathcal{E}})$ is a PMS where the map $\mathcal{D}_{h_\circledast,\mathcal{E}} : \mathcal{A} \times \mathcal{A} \to [0, \infty]$ is defined for each $a, b \in \mathcal{A}$,*

$$\mathcal{D}_{h_\circledast,\mathcal{E}}(a, b) = h_\circledast\Big(\mathcal{E}(a, b)\Big).$$

Moreover, the topology induced by $\mathcal{D}_{h_\circledast,\mathcal{E}}$ and the topology induced by $\mathcal{E}$ coincides. i.e., $\tau_{\mathcal{D}_{h_\circledast,\mathcal{E}}} = \tau_\mathcal{E}$.

Proof Let $(\mathcal{A}, \mathcal{E}, \circledast)$ be a stationary FPMS with the cont. Archimedean t-norm $\circledast$.

(PM1) Let $\mathcal{D}_{h_\circledast,\mathcal{E}}(a, b) = \mathcal{D}_{h_\circledast,\mathcal{E}}(a, a) = \mathcal{D}_{h_\circledast,\mathcal{E}}(b, b)$. Then we have $h_\circledast\Big(\mathcal{E}(a, b)\Big) = h_\circledast\Big(\mathcal{E}(a, a)\Big) = h_\circledast\Big(\mathcal{E}(b, b)\Big)$ which means that $\mathcal{E}(a, b) = \mathcal{E}(a, a) = \mathcal{E}(b, b)$. This follows that $a = b$ since $(\mathcal{A}, \mathcal{E}, \circledast)$ is a stationary FPMS.

(PM2) Let $a, b \in \mathcal{A}$. Then we obtain that $0 \leq \mathcal{D}_{h_\circledast,\mathcal{E}}(a, a) \leq \mathcal{D}_{h_\circledast,\mathcal{E}}(a, b)$ since $0 = h_\circledast(1) \leq h_\circledast(\mathcal{E}(a, a)) \leq h_\circledast(\mathcal{E}(a, b))$.

(PM3) It is obvious that $\mathcal{D}_{h_{\circledast},\mathcal{E}}(a,b) = h_(\mathcal{E}(a,b)) = h_{\circledast}(\mathcal{E}(b,a)) = \mathcal{D}_{h_{\circledast},\mathcal{E}}(b,a)$ for each $a, b \in \mathcal{A}$.

(PM4) The followings are valid for each $a, b, c \in \mathcal{A}$,

$$\begin{aligned}
\mathcal{D}_{h_{\circledast},\mathcal{E}}(a,b) &= h_{\circledast}(\mathcal{E}(a,c)) \leq h_{\circledast}(\mathcal{E}(a,a) \rightarrow_{\circledast} \mathcal{E}(a,c)) \\
&\leq h_{\circledast}((\mathcal{E}(a,a) \rightarrow_{\circledast} \mathcal{E}(a,b))\circledast(\mathcal{E}(b,b) \rightarrow_{\circledast} \mathcal{E}(b,c))) \\
&= h_{\circledast}(h_{\circledast}^{(-1)}(h_{\circledast}(\mathcal{E}(a,a) \rightarrow_{\circledast} \mathcal{E}(a,c)) \\
&\quad + h_{\circledast}(\mathcal{E}(b,b) \rightarrow_{\circledast} \mathcal{E}(b,c)))) \\
&\leq h_{\circledast}(\mathcal{E}(a,a) \rightarrow_{\circledast} \mathcal{E}(a,c)) + h_{\circledast}(\mathcal{E}(b,b) \rightarrow_{\circledast} \mathcal{E}(b,c)) \\
&= h_{\circledast}(\mathcal{E}(a,c)) - h_{\circledast}(\mathcal{E}(a,a)) \\
&\quad + h_{\circledast}(\mathcal{E}(b,c)) - h_{\circledast}(\mathcal{E}(b,b)) \\
&\leq h_{\circledast}(\mathcal{E}(a,c)) + h_{\circledast}(\mathcal{E}(b,c)) - h_{\circledast}(\mathcal{E}(b,b)) \\
&= \mathcal{D}_{h_{\circledast},\mathcal{E}}(a,c) + \mathcal{D}_{h_{\circledast},\mathcal{E}}(c,b) - \mathcal{D}_{h_{\circledast},\mathcal{E}}(b,b)
\end{aligned}$$

which implies that $\mathcal{D}_{h_{\circledast},\mathcal{E}}$ is a PM on $\mathcal{A}$.

Example 4 We know that the map $\mathcal{E} : [0,\infty) \times [0,\infty) \times [0,\infty) \rightarrow \mathcal{I}$ defined as $\mathcal{E}(a,b) = \frac{1}{1+max\{a,b\}}$ and $\mathcal{E}(a,b,0) = 0$ is a stationary FPM with the Hamacher t-norm $\circledast_H$. Also, an a-generator of $\circledast_H$ is given by $h_{\circledast_H}(a) = \frac{1-a}{a}$, (assume that $h_{\circledast_H}(0) = \infty$). Then, we have, from Theorem 5, for each $a, b \in [0,\infty)$ and $\mathfrak{v} > 0$, $\mathcal{D}_{h_{\circledast_H},\mathcal{E}}(a,b) = \frac{1-\frac{1}{1+max\{a,b\}}}{\frac{1}{1+max\{a,b\}}} = max\{a,b\}$ which is a well-known PM on $[0,\infty)$.

The non-stationary FPM version of the above theorem is given as follows:

Theorem 6 *Suppose that* $(\mathcal{A}, \mathcal{E}, \circledast)$ *is a FPMS where* $\circledast$ *is a cont. Archimedean t-norm and* $h_{\circledast}$ *is an a-generator of* $\circledast$. *Then* $(\mathcal{A}, \mathcal{D}_{h_{\circledast},\mathcal{E}})$ *is a PMS where the map* $\mathcal{D}_{h_{\circledast},\mathcal{E}} : \mathcal{A} \times \mathcal{A} \rightarrow [0,\infty]$ *is defined for each* $a, b \in \mathcal{A}$,

$$\mathcal{D}_{h_{\circledast},\mathcal{E}}(a,b) = h_{\circledast}\Big(\bigwedge_{\mathfrak{v}>0} \mathcal{E}(a,b,\mathfrak{v})\Big).$$

Moreover, the topology induced by $\mathcal{D}_{h_{\circledast},\mathcal{E}}$ *and the topology induced by* $\mathcal{E}$ *coincides. i.e.,* $\tau_{\mathcal{D}_{h_{\circledast},\mathcal{E}}} = \tau_{\mathcal{E}}$.

Proof Let $(\mathcal{A}, \mathcal{E}, \circledast)$ be a FPMS where $\circledast$ is a cont. Archimedean t-norm and $h_{\circledast}$ be an a-generator of $\circledast$. Then $\bigwedge_{\mathfrak{v}>0} \mathcal{E}(a,b,\mathfrak{v})$ fulfills the conditions (FPM1)-(FPM4) and thus, with a similar consideration of proof of Theorem 5, the map $\mathcal{D}_{h_{\circledast},\mathcal{E}} : \mathcal{A} \times \mathcal{A} \rightarrow [0,\infty]$ is defined for each $a, b \in \mathcal{A}$, $\mathcal{D}_{h_{\circledast},\mathcal{E}}(a,b) = h_{\circledast}\Big(\bigwedge_{\mathfrak{v}>0} \mathcal{E}(a,b,\mathfrak{v})\Big)$ is a PM on $\mathcal{A}$.

3 Conclusion

In this study, we introduce the notion of strong FPMSs which is a subclass of FPMSs. We observe that the condition FPM4* is coincident with the condition FPM4 in strong FPMSs. We also establish some constructions between PMSs and FPMSs by means of a-generators of t-norms that allow us to find different examples of the mentioned structures. For future work, one can study fixed point results in strong FPMSs and also can research the topological properties of strong FPMSs.

References

1. Aygün, H., Güner, E., Miñana, J.J., Valero, O.: Fuzzy partial metric spaces and fixed point theorems. Mathematics **10**(17), 3092 (2022). https://doi.org/10.3390/math10173092
2. Bugajewski, D., Mackowiak, P., Wang, R.D.: On compactness and fixed point theorems in partial metric spaces. Fixed Point Theory **23**(1), 163–178 (2022). https://doi.org/10.48550/arXiv.2005.06327
3. Camarena, J.G., Greogri, V., Morillas, S., Sapena, A.: Fast detection and removal of impulsive noise using peer groups and fuzzy metrics. J. Vis. Commun. Image Represent. **19**(1), 20–29 (2008). https://doi.org/10.1016/j.jvcir.2007.04.003
4. George, A., Veeramani P.: On some results in fuzzy metric spaces. Fuzzy Sets Syst. **64**(3), 395–399 (1994). https://doi.org/10.1016/0165-0114(94)90162-7
5. Grabiec, M.: Fixed points in fuzzy metric spaces. Fuzzy Sets Syst. **27**, 385–389 (1989). https://doi.org/10.1016/0165-0114(88)90064-4
6. Greçova, S., Morillas, S.: Perceptual similarity between color images using fuzzy metrics. J. Vis. Commun. Image Represent. **34**, 230–235 (2016). https://doi.org/10.1016/j.jvcir.2015.04.003
7. Gregori, V., Morillas, S., Sapena, A.: Examples of fuzzy metrics and applications. Fuzzy Sets Syst. **170**, 95–111 (2011). https://doi.org/10.1016/j.fss.2010.10.019
8. Gregori, V., Miñana, J.J., Miravet D.: Fuzzy partial metric spaces. Int. J. Gen. Syst. **48**(3), 260–279 (2019). https://doi.org/10.1080/03081079.2018.1552687
9. Gregori, V., Miñana, J.J., Miravet, D.: A duality relationship between fuzzy partial metrics and fuzzy quasi-metrics. Mathematics **8**(9), 1–16 (2020). https://doi.org/10.3390/math8091575
10. Grigorenko, O., Miñana, J.J., Valero, O.: Two new methods to construct fuzzy metrics from metrics. Fuzzy Sets Syst. **467**, 108483 (2023). https://doi.org/10.1016/j.fss.2023.02.004
11. Haghia, R.H., Rezapourb, S., Shahzad, N.: Be careful on partial metric fixed point results. Topol. Appl. **160**(3), 450–454 (2013). https://doi.org/10.1016/j.topol.2012.11.004
12. Klement, E.P., Mesiar, R., Pap, E.: Triangular Norms. Springer, Netherlands (2000)
13. Kramosil, I., Michálek, J.: Fuzzy metrics and statistical metric spaces. Kybernetika **11**(5), 336–344 (1975)
14. Matthews, S.G.: Partial metric topology. Ann. N. Y. Acad. Sci. **728**, 183–197 (1994). https://doi.org/10.1111/j.1749-6632.1994.tb44144.x
15. Miñana, J.J., Ortiz, A., Ortiz, E., Valero, O.: On the standard fuzzy metric: generalizations and application to model estimation. In: Proceedings of the Conference of the Spanish Association for Artificial Intelligence (CAEPIA), pp. 311–316 (2021)

16. Ortiz, A., Ortiz, E., Miñana, J.J., Valero, O.: On the use of fuzzy metrics for robust model estimation: A ransac-based approach. In: Advances in Computational Intelligence. Lecture Notes in Computer Science, vol. 12861, pp. 165–177 (2021)
17. Ralevic, N.M., Delic, M., Nedovic, L.: Aggregation of fuzzy metrics and its application in image segmentation. Iran. J. Fuzzy Syst. **19**(3), 19–37 (2022). https://doi.org/10.22111/IJFS.2022.6941
18. Zadeh, L.A.: Fuzzy sets. Inf. Control **8**(3), 338–353 (1965). https://doi.org/10.1016/S0019-9958(65)90241-X

An Application of Linear Diophantine Fuzzy Sets to the Edge Detection Techniques

Başak Aldemir, Elif Güner, and Halis Aygün

Abstract The utilization of fuzzy set theory within the domain of image processing provides a lot of advantages such as encompassing the management of uncertainty, adaptability to variations, noise tolerance and adaptive classification compared to the other techniques. These advantages contribute to heightened precision and adaptability in the realm of image processing, enabling more precise and versatile handling of visual data. The process of edge detection performs a pivotal role in the segmentation of foreground objects from the image background. So, it facilitates subsequent analysis and comprehension of the image's underlying structural properties through complex computational procedures. This complex process can be handled with the notion of fuzzy sets and their generalizations. The concept of linear Diophantine fuzzy sets is a generalization of fuzzy sets where the use of reference parameters corresponds to membership and non-membership grades. The aim of this study is to give an application of linear Diophantine fuzzy sets to edge detection of images. For this aim, we conduct a comprehensive evaluation to ascertain the similarity values using the linear Diophantine fuzzy similarity measure by leveraging the normalized membership values of the gray level associated with fundamental edge detection techniques.

1 Introduction

Fuzzy set (FS) theory [18] offers a flexible framework for effectively managing ambiguity and uncertainty, providing a versatile approach to represent and reason with indeterminate information. Therefore, a lot of generalizations for fuzzy sets,

B. Aldemir (✉)
Department of Mathematics, Afyon Kocatepe University, Afyonkarahisar, Turkey
e-mail: baldemir@aku.edu.tr

E. Güner · H. Aygün
Department of Mathematics, Kocaeli University, Kocaeli, Turkey
e-mail: elif.guner@kocaeli.edu.tr; halis@kocaeli.edu.tr

V. Gayoso Martínez et al. (eds.), *Mathematical Methods for Engineering Applications*, Springer Proceedings in Mathematics & Statistics 439,
https://doi.org/10.1007/978-3-031-49218-1_19

such as intuitionistic, Pythagorean, picture, spherical and linear Diophantine fuzzy sets, as well as more complex generalizations of these sets, have been extensively investigated. These generalized fuzzy sets have been the subject of numerous studies in the literature, and their applications have been extensively explored in various domains, as mentioned in [3, 5, 7, 9, 12, 15, 17, 19].

One such generalization is intuitionistic fuzzy sets, in which every element is defined by a degree of membership and a degree of non-membership, with the condition that the total of these levels does not surpass one. Intuitionistic fuzzy sets (IFS), introduced by Atanassov in 1986 [2], provide a more nuanced representation of uncertainty and ambiguity and there are a lot of applications such as decision-making, image processing, expert systems, pattern recognition and multi-criteria decision analysis [4, 6, 11, 16].

Building upon IFS, the structure of a linear Diophantine fuzzy set (LDFS) [14] incorporates reference parameters, offering a more flexible and efficient approach. LDFS fills gaps in the current frameworks and expands the range of membership and non-membership degrees through the use of reference parameters. Numerous application areas where linear Diophantine fuzzy sets find practical use have been described in the current literature. Aldring et al. [1], Gül and Aydoğdu [8], Kamacı [10], and Mohammad et al. [13] are some of them. These application areas cover a variety of fields and include a wide variety of problem-solving contexts.

Employing fuzzy set theory in image processing yields numerous benefits and advantages, including its ability to effectively address uncertainty, adapt to variations, tolerate noise, and enable adaptive classification, distinguishing it from other techniques. These advantages significantly enhance the precision and adaptability of image processing, enabling more accurate and flexible handling of visual data. Image processing encompasses multiple stages, with edge detection being a crucial initial step. Accurate edge detection is pivotal for subsequent stages to be executed correctly. This intricate process can be effectively managed and executed by leveraging the principles of fuzzy sets and their generalizations.

2 Preliminaries

This section includes two subsections that recall the notions of linear Diophantine fuzzy sets and some basic edge detection techniques.

2.1 Linear Diophantine Fuzzy Sets

We first recall the concepts of fuzzy sets, intuitionistic fuzzy sets, and linear Diophantine fuzzy sets. Then, we give some distance and similarity measures for linear Diophantine fuzzy sets. Furthermore, we give the definition of the Minkowski distance measure which is a generalization of the mentioned distance measures.

Definition 1 Let U be a non-empty set.

(i) [18] A fuzzy set (FS) F on U is given by

$$F = \{(u, \mu_F(u)) : u \in U\}$$

where $\mu_F : U \to [0, 1]$ is the function that represent the membership of u to the F.

(ii) [2] An intuitionistic fuzzy set (IFS) A on U is given by

$$A = \{(u, \mu_A(u), \nu_A(u)) : u \in U\}$$

where $\mu_A, \nu_A : U \to [0, 1]$ are the functions that represent the membership and non-membership function of u to the A, respectively, satisfying $0 \leq \mu_A(u) + \nu_A(u) \leq 1$ for all $u \in U$.

Remark 1 Each FS can be taken as an IFS by considering the non-membership function $\nu_A(u) = 1 - \mu_A(u)$. So, the collection of FSs is the subset of the collection of IFSs.

As a more general form, Riaz et al. [14] described the notion of linear Diophantine fuzzy sets in the following manner:

Definition 2 ([14]) Let U be a non-empty set. A linear Diophantine fuzzy set (LDFS) L on U is given by

$$L = \{(u, \langle \mu_L(u), \nu_L(u) \rangle, \langle \alpha_L, \beta_L \rangle) : u \in U\}$$

where $\mu_L, \nu_L : U \to [0, 1]$ are the functions that represent the membership and non-membership function of u to the L, respectively, and $\alpha_L, \beta_L \in [0, 1]$ denotes the reference parameters value satisfying $0 \leq \alpha_L \mu_L(u) + \beta_L \nu_L(u) \leq 1$, for all $u \in U$, with $0 \leq \alpha_L + \beta_L \leq 1$. This reference parameters help us to define or classify a particular system.

The hesitation value π_L is calculated by $\pi_L(u) = 1 - (\alpha_L \mu_L(u) + \beta_L \nu_L(u))$ for all $u \in U$. We use the pair $(\langle \mu, \nu \rangle, \langle \alpha, \beta \rangle)$ to denote the linear Diophantine fuzzy number LDFN if the conditions $0 \leq \alpha\mu + \beta\nu \leq 1$ and $0 \leq \alpha + \beta \leq 1$ are satisfied. The collection of all LDFSs on U will be represented by $\mathbf{L}(U)$.

Remark 2 Each IFS can be taken as an LDFS. That is, if $A = \{(u, \mu_A(u), \nu_A(u)) : u \in U\}$ is an IFS and the parameter values $\alpha_A, \beta_A \in [0, 1]$ satisfies $0 \leq \alpha_A + \beta_A \leq 1$, then we have that $0 \leq \alpha_A \mu_A(u) + \beta_A \nu_A(u) \leq \mu_A(u) + \nu_A(u) \leq 1$. Hence the set $A = \{(u, \langle \mu_A(u), \nu_A(u) \rangle, \langle \alpha_A, \beta_A \rangle) : u \in U\}$ satisfying $\alpha_A, \beta_A \in [0, 1]$ and $0 \leq \alpha_A + \beta_A \leq 1, \forall\, u \in U$, is an LDFS.

We obtain the following results from Remarks 1 and 2:

Corollary 1 *If* $F = \{(u, \mu_F(u)) : u \in U\}$ *is a FS, then we have that the set* $A = \{(u, \langle \mu_A(u), 1 - \mu_A(u) \rangle, \langle \alpha_A, \beta_A \rangle) : u \in U\}$ *satisfying* $0 \le \alpha_A + \beta_A \le 1$ *is an LDFS.*

Definition 3 ([14]) A LDFS on $\mathbf{L}(U)$ of the form $L_U = \{(u, \langle 1, 0 \rangle, \langle 1, 0 \rangle) : u \in U\}$ is called absolute LDFS and $L_0 = \{(u, \langle 0, 1 \rangle, \langle 0, 1 \rangle) : u \in U\}$ is called null(empty) LDFS.

Definition 4 ([14]) Let $L \in \mathbf{L}(U)$. Then the complement of L, represented by L^c, is given by $L^c = \{(u, \langle \nu_L(u), \mu_L(u) \rangle, \langle \beta_L(u), \alpha_L(u) \rangle) : u \in U\}$.

Definition 5 ([14]) Let $L, L_1, L_2 \in \mathbf{L}(U)$ and $k > 0$. Then,

(1) $L_1 \bigoplus L_2 = \{\big(u, \langle \mu_{L_1}(u) + \mu_{L_2}(u) - \mu_{L_1}(u)\mu_{L_2}(u), \nu_{L_1}(u)\nu_{L_2}(u) \rangle, \langle \alpha_{L_1}(u) + \alpha_{L_2}(u) - \alpha_{L_1}(u)\alpha_{L_2}(u), \beta_{L_1}(u)\beta_{L_2}(u) \rangle\big) : u \in U\}$,
(2) $L_1 \bigotimes L_2 = \{\big(u, \langle \mu_{L_1}(u)\mu_{L_2}(u), \nu_{L_1}(u) + \nu_{L_2}(u) - \nu_{L_1}(u)\nu_{L_2}(u) \rangle, \langle \alpha_{L_1}(u) \alpha_{L_2}(u), [4]\beta_{L_1}(u) + \beta_{L_2}(u) - \beta_{L_1}(u)\beta_{L_2}(u) \rangle\big) : u \in U\}$,
(3) $k.L = \{\big(u, \langle 1 - (1 - \mu_L(u))^k, \nu_L(u)^k \rangle, \langle 1 - (1 - \alpha_L(u))^k, \beta_L(u)^k \rangle\big) : u \in U\}$,
(4) $L^k = \{\big(u, \langle \mu_L(u)^k, 1 - (1 - \nu_L(u))^k \rangle, \langle \alpha_L(u)^k, 1 - (1 - \beta_L(u))^k \rangle\big) : u \in U\}$.

Proposition 1 ([14]) *For LDFSs* $L, L_1, L_2 \in \mathbf{L}(U)$ *and* $k > 0$, $L_1 \bigoplus L_2$, $L_1 \bigotimes L_2$, kL *and* L^k *are also LDFSs.*

Definition 6 ([8]) Let U be a non-empty set. Then a mapping $D : \mathbf{L}(U) \times \mathbf{L}(U) \to [0, 1]$ is said to be a distance measure on LDFSs if the following conditions hold for all $L_1, L_2 \in \mathbf{L}(U)$:

(i) $0 \le D(L_1, L_2) \le 1$,
(ii) $D(L_1, L_2) = D(L_2, L_1)$,
(iii) $D(L_1, L_2) = 0$ if $L_1 = L_2$,
(iv) If $L_1 \sqsubseteq L_2 \sqsubseteq L_3$ then $D(L_1, L_3) \ge D(L_1, L_2)$ and $D(L_1, L_3) \ge D(L_2, L_3)$.

Theorem 1 ([8]) *Let* $U = \{u_1, u_2, .., u_n\}$, $L_1, L_2 \in \mathbf{L}(U)$ *and the mappings* $D_H, D_E : \mathbf{L}(U) \times \mathbf{L}(U) \to [0, 1]$ *defined as follows:*

$$D_H(L_1, L_2) = \frac{1}{4n} \sum_{i=1}^{n} [|\mu_{L_1}(u_i) - \mu_{L_2}(u_i)| + |\nu_{L_1}(u_i) - \nu_{L_2}(u_i)| + |\alpha_{L_1}(u_i) - \alpha_{L_2}(u_i)| + |\beta_{L_1}(u_i) - \beta_{L_2}(u_i)|] \quad (1)$$

$$D_E(L_1, L_2) = \Big(\frac{1}{4n} \sum_{i=1}^{n} [(\mu_{L_1}(u_i) - \mu_{L_2}(u_i))^2 + (\nu_{L_1}(u_i) - \nu_{L_2}(u_i))^2 + (\alpha_{L_1}(u_i) - \alpha_{L_2}(u_i))^2 + (\beta_{L_1}(u_i) - \beta_{L_2}(u_i))^2]\Big)^{\frac{1}{2}}. \quad (2)$$

Then the mappings D_H *and* D_E *are distance measures on LDFSs. These distance measures are called the normalized Hamming distance and the normalized Euclidean distance, respectively.*

We can also define the Minkowski distance between LDFSs as follows:

Definition 7 Let $U = \{u_1, u_2, .., u_n\}$, $L_1, L_2 \in \mathbf{L}(U)$ and the mapping $D_M : \mathbf{L}(U) \times \mathbf{L}(U) \rightarrow [0, 1]$ defined as follows:

$$D_M(L_1, L_2) = \Big(\frac{1}{4n}\sum_{i=1}^{n}[|\mu_{L_1}(u_i) - \mu_{L_2}(u_i)|^a + |\nu_{L_1}(u_i) - \nu_{L_2}(u_i)|^a$$

$$+ |\alpha_{L_1}(u_i) - \alpha_{L_2}(u_i)|^a + |\beta_{L_1}(u_i) - \beta_{L_2}(u_i)|^a]\Big)^{\frac{1}{a}} \quad (3)$$

Then the mapping D_M is a distance measure on LDFSs. This distance measure is called the normalized Minkowski distance.

Note that, in the definition of the normalized Minkowski distance, if $a = 1$ and $a = 2$, then we have the definitions of the normalized Hamming distance and the normalized Euclidean distance, respectively.

Definition 8 ([8]) Let U be a non-empty set. Then a mapping $S : \mathbf{L}(U) \times \mathbf{L}(U) \rightarrow [0, 1]$ is said to be a similarity measure on LDFSs if the following conditions hold for all $L_1, L_2 \in \mathbf{L}(U)$:

(i) $0 \leq S(L_1, L_2) \leq 1$,
(ii) $S(L_1, L_2) = S(L_2, L_1)$,
(iii) $S(L_1, L_2) = 1$ if $L_1 = L_2$,
(iv) If $L_1 \sqsubseteq L_2 \sqsubseteq L_3$ then $S(L_1, L_3) \leq S(L_1, L_2)$ and $S(L_1, L_3) \leq S(L_2, L_3)$.

Theorem 2 *Let $f : [0, 1] \rightarrow [0, 1]$ be a decreasing function and D be a distance measure on LDFSs. Suppose that the mapping $S_D^f : \mathbf{L}(U) \times \mathbf{L}(U) \rightarrow [0, 1]$ is defined by*

$$S_D^f(L_1, L_2) = \frac{f(D(L_1, L_2)) - f(1)}{f(0) - f(1)}.$$

Then the mapping S_D^f is a similarity measure on LDFSs. This similarity measure is said to be an f-similarity measure based on the distance measure on LDFSs.

Example 1

(1) Let $f_1 : [0, 1] \rightarrow [0, 1]$ be defined by $f_1(u) = 1 - u$ and D be a distance measure on LDFSs. Then, we obtain

$$S_D^{f_1}(L_1, L_2) = \frac{f_1(D(L_1, L_2)) - f_1(1)}{f_1(0) - f_1(1)} = \frac{1 - D(L_1, L_2) - 0}{1 - 0} = 1 - D(L_1, L_2).$$

(2) Let $f_2 : [0, 1] \rightarrow [0, 1]$ be defined by $f_2(u) = e^{-u}$ and D be a distance measure on LDFSs. Then, we obtain

$$S_D^{f_2}(L_1, L_2) = \frac{f_2(D(L_1, L_2)) - f_2(1)}{f_2(0) - f_2(1)} = \frac{e^{-D(L_1,L_2)} - e^{-1}}{e^0 - e^{-1}} = \frac{e^{-D(L_1,L_2)} - e^{-1}}{1 - e^{-1}}.$$

(3) Let $f_3 : [0, 1] \to [0, 1]$ be defined by $f_3(u) = \frac{1}{1+u}$ and D be a distance measure on LDFSs. Then, we obtain

$$S_D^{f_3}(L_1, L_2) = \frac{f_3(D(L_1, L_2)) - f_3(1)}{f_3(0) - f_3(1)} = \frac{\frac{1}{1+D(L_1,L_2)} - \frac{1}{2}}{1 - \frac{1}{2}} = \frac{1 - D(L_1, L_2)}{1 + D(L_1, L_2)}.$$

Using the distance measures introduced and described above, we can obtain similarity measures as follows;

Corollary 2 *Let $U = \{u_1, u_2, .., u_n\}$ and $L_1, L_2 \in \mathbf{L}(U)$. Then the similarity measures $S_{D_H}^{f_1}$, $S_{D_E}^{f_1}$, $S_{D_M}^{f_1}$, $S_{D_H}^{f_2}$, $S_{D_E}^{f_2}$, $S_{D_M}^{f_2}$, $S_{D_H}^{f_3}$, $S_{D_E}^{f_3}$ and $S_{D_M}^{f_3}$ are obtained as follows:*

$$\begin{aligned} S_{D_H}^{f_1}(L_1, L_2) &= 1 - D_H(L_1, L_2) \\ &= 1 - \Big(\frac{1}{4n}\sum_{i=1}^{n}[|\mu_{L_1}(u_i) - \mu_{L_2}(u_i)| + |\nu_{L_1}(u_i) - \nu_{L_2}(u_i)| \\ &\quad + |\alpha_{L_1}(u_i) - \alpha_{L_2}(u_i)| + |\beta_{L_1}(u_i) - \beta_{L_2}(u_i)|]\Big) \end{aligned}$$

$$\begin{aligned} S_{D_E}^{f_1}(L_1, L_2) &= 1 - D_E(L_1, L_2) \\ &= 1 - \Big(\frac{1}{4n}\sum_{i=1}^{n}[|\mu_{L_1}(u_i) - \mu_{L_2}(u_i)|^2 + |\nu_{L_1}(u_i) - \nu_{L_2}(u_i)|^2 \\ &\quad + |\alpha_{L_1}(u_i) - \alpha_{L_2}(u_i)|^2 + |\beta_{L_1}(u_i) - \beta_{L_2}(u_i)|^2]\Big)^{1/2} \end{aligned}$$

$$\begin{aligned} S_{D_M}^{f_1}(L_1, L_2) &= 1 - D_M(L_1, L_2) \\ &= 1 - \Big(\frac{1}{4n}\sum_{i=1}^{n}[|\mu_{L_1}(u_i) - \mu_{L_2}(u_i)|^a + |\nu_{L_1}(u_i) - \nu_{L_2}(u_i)|^a \\ &\quad + |\alpha_{L_1}(u_i) - \alpha_{L_2}(u_i)|^a + |\beta_{L_1}(u_i) - \beta_{L_2}(u_i)|^a]\Big)^{1/a} \end{aligned}$$

$$\begin{aligned} S_{D_H}^{f_2}(L_1, L_2) &= \frac{e^{-D_H(L_1,L_2)} - e^{-1}}{1 - e^{-1}} \\ &= \Big(exp(1 - 1/4n\sum_{i=1}^{n}[|\mu_{L_1}(u_i) - \mu_{L_2}(u_i)| + |\nu_{L_1}(u_i) - \nu_{L_2}(u_i)| \end{aligned}$$

$$+|\alpha_{L_1}(u_i) - \alpha_{L_2}(u_i)| + |\beta_{L_1}(u_i) - \beta_{L_2}(u_i)|]\Big)\Big) - 1/(e-1)$$

$$S^{f_2}_{D_E}(L_1, L_2) = \frac{e^{-D_E(L_1,L_2)} - e^{-1}}{1 - e^{-1}}$$

$$= \Big(exp(1 - \frac{1}{4n}\sum_{i=1}^{n}[|\mu_{L_1}(u_i) - \mu_{L_2}(u_i)|^2 + |\nu_{L_1}(u_i) - \nu_{L_2}(u_i)|^2$$

$$+|\alpha_{L_1}(u_i) - \alpha_{L_2}(u_i)|^2 + |\beta_{L_1}(u_i) - \beta_{L_2}(u_i)|^2])^{1/2}\Big) - 1/(e-1)$$

$$S^{f_2}_{D_M}(L_1, L_2) = \frac{e^{-D_M(L_1,L_2)} - e^{-1}}{1 - e^{-1}}$$

$$= \Big(exp(1 - \frac{1}{4n}\sum_{i=1}^{n}[|\mu_{L_1}(u_i) - \mu_{L_2}(u_i)|^a + |\nu_{L_1}(u_i) - \nu_{L_2}(u_i)|^a$$

$$+|\alpha_{L_1}(u_i) - \alpha_{L_2}(u_i)|^a + |\beta_{L_1}(u_i) - \beta_{L_2}(u_i)|^a])^{1/a}\Big) - 1/(e-1)$$

$$S^{f_3}_{D_H}(L_1, L_2) = \frac{1 - D_H(L_1, L_2)}{1 + D_H(L_1, L_2)}$$

$$= \Big(1 - \frac{1}{4n}\sum_{i=1}^{n}[|\mu_{L_1}(u_i) - \mu_{L_2}(u_i)| + |\nu_{L_1}(u_i) - \nu_{L_2}(u_i)|$$

$$+ |\alpha_{L_1}(u_i) - \alpha_{L_2}(u_i)| + |\beta_{L_1}(u_i) - \beta_{L_2}(u_i)|]\Big)$$

$$/\Big(1 + \frac{1}{4n}\sum_{i=1}^{n}[|\mu_{L_1}(u_i) - \mu_{L_2}(u_i)| + |\nu_{L_1}(u_i) - \nu_{L_2}(u_i)|$$

$$+|\alpha_{L_1}(u_i) - \alpha_{L_2}(u_i)| + |\beta_{L_1}(u_i) - \beta_{L_2}(u_i)|]\Big)$$

$$S^{f_3}_{D_E}(L_1, L_2) = \frac{1 - D_E(L_1, L_2)}{1 + D_E(L_1, L_2)}$$

$$= \Big(1 - \frac{1}{4n}\sum_{i=1}^{n}[|\mu_{L_1}(u_i) - \mu_{L_2}(u_i)|^2 + |\nu_{L_1}(u_i) - \nu_{L_2}(u_i)|^2$$

$$+|\alpha_{L_1}(u_i) - \alpha_{L_2}(u_i)|^2 + |\beta_{L_1}(u_i) - \beta_{L_2}(u_i)|^2]^{1/2}\Big)$$

$$/\Big(1 + \frac{1}{4n}\sum_{i=1}^{n}[|\mu_{L_1}(u_i) - \mu_{L_2}(u_i)|^2 + |\nu_{L_1}(u_i) - \nu_{L_2}(u_i)|^2$$

$$+|\alpha_{L_1}(u_i) - \alpha_{L_2}(u_i)|^2 + |\beta_{L_1}(u_i) - \beta_{L_2}(u_i)|^2]^{1/2}\Big)$$

$$S^{f_3}_{D_M}(L_1, L_2) = \frac{1 - D_M(L_1, L_2)}{1 + D_M(L_1, L_2)}$$

$$= \Big(1 - \frac{1}{4n}\sum_{i=1}^{n}[|\mu_{L_1}(u_i) - \mu_{L_2}(u_i)|^a + |\nu_{L_1}(u_i) - \nu_{L_2}(u_i)|^a$$

$$+|\alpha_{L_1}(u_i) - \alpha_{L_2}(u_i)|^a + |\beta_{L_1}(u_i) - \beta_{L_2}(u_i)|^a]^{1/a}\Big)$$

$$/\Big(1 + \frac{1}{4n}\sum_{i=1}^{n}[|\mu_{L_1}(u_i) - \mu_{L_2}(u_i)|^a + |\nu_{L_1}(u_i) - \nu_{L_2}(u_i)|^a$$

$$+|\alpha_{L_1}(u_i) - \alpha_{L_2}(u_i)|^a + |\beta_{L_1}(u_i) - \beta_{L_2}(u_i)|^a]^{1/a}\Big)$$

2.2 *Basic Edge Detection Techniques*

Edge detection plays a crucial role in various image-related tasks such as image processing, image analysis, image pattern recognition, and computer vision techniques. The outcome of the edge detection process on an image provides a collection of connected curves that represent object boundaries, surface markings, and variations in surface orientation. By applying an edge detection algorithm to an image, it becomes possible to significantly reduce the data volume for processing, effectively filtering out less relevant information while preserving the essential structural characteristics of the image. Successful execution of the edge detection step allows for a simplified interpretation of the information contained in the original image. The widely employed edge detection methods encompass Sobel, Prewitt, LoG (Laplacian of Gaussian), Canny, and Roberts edge detection techniques.

In the Sobel technique for edge detection, edges are identified by employing a 3×3 image filter in the local neighborhood.

The Prewitt technique estimates edge detection by utilizing a simplified 3×3 image filter in the local neighborhood.

The Laplacian technique detects edges by searching for points where the second derivative crosses zero. The second derivative is more effective in capturing fine details compared to the first derivative. However, it has the disadvantage of being highly sensitive to noise. Therefore, to mitigate this, a Gaussian filter is applied beforehand to remove noise.

The Canny technique involves multiple steps and is designed to reduce noise, accentuate actual edges, and enhance the precision of edge detection. It consists of several steps for determining the edges, firstly it finds the edge candidates in the image and then selects the optimal edges.

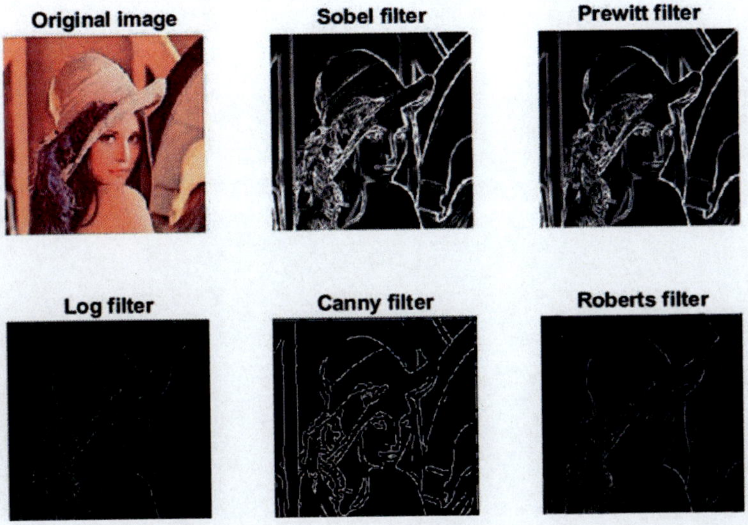

Fig. 1 Edge detection for original image

The Roberts technique utilizes a pair of 2×2 convolution masks to calculate the gradient magnitude of the image. These masks are applied to the image by convolving them with the corresponding pixel neighborhood. The edge detection methods applied to the original image are illustrated in Fig. 1.

Fuzzification of an image is particularly useful when working with algorithms or techniques that require a specific range of inputs, such as image processing algorithms based on normalized values. Generally, fuzzification of an image involves rescaling pixel values to the range [0, 1], this can be made in various ways using different membership functions. This process allows for easier analysis and processing of image data while ensuring consistency and compatibility with image processing techniques and algorithms. The above-mentioned techniques are employed on the image after fuzzification, and the resulting outputs are displayed in Fig. 2.

3 An Application of the LDFSs Sımılarıty for Edge Detection

The similarities between each edge detection filter were calculated based on the similarity measures obtained from the mentioned above distance measures and by using random α *and* β parameters. Different outcomes are obtained when employing various α *and* β parameters, which are randomly chosen. To illustrate, specified α *and* β values for each filter given in Table 1 are used and the corresponding outcomes are presented in Tables 2, 3, and 4.

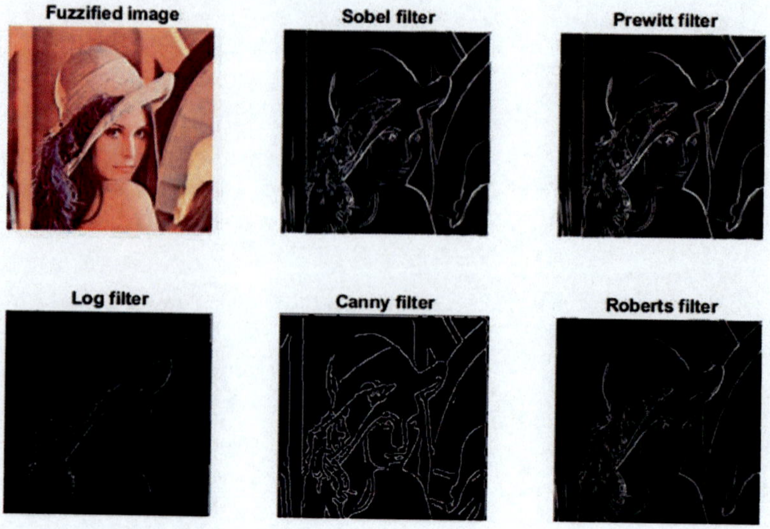

Fig. 2 Edge detection for fuzzified image

Table 1 Parametres for filtres

Filters	α	β
Sobel	0.7447	0.1890
Prewitt	0.6868	0.1835
LoG	0.3685	0.6256
Canny	0.7802	0.0811
Roberts	0.4868	0.4359

Table 2 Similarity results under $S^{f_1}_{D_H}$, $S^{f_1}_{D_E}$ and $S^{f_1}_{D_M}$

		$S^{f_1}_{D_H}$	$S^{f_1}_{D_E}$	$S^{f_1}_{D_M}$ $(a = 3)$
Sobel	Prewitt	0.9998	0.9966	0.9913
Sobel	LoG	0.9991	0.9804	0.9460
Sobel	Canny	0.9975	0.9518	0.8679
Sobel	Roberts	0.9987	0.9814	0.9544
Prewitt	LoG	0.9990	0.9813	0.9484
Prewitt	Canny	0.9976	0.9545	0.8754
Prewitt	Roberts	0.9989	0.9843	0.9613
LoG	Canny	0.9966	0.9467	0.8630
LoG	Roberts	0.9990	0.9840	0.9593
Canny	Roberts	0.9976	0.9626	0.9036

The similarities between the fuzzified images obtained from each edge detection technique were calculated with the defined similarity measures. According to the results obtained, it is seen that the Sobel and Prewitt filters are largely similar to each other. And this similarity gives close results even if the alpha and beta parameters

Table 3 Similarity results under $S_{D_H}^{f_2}$, $S_{D_E}^{f_2}$ and $S_{D_M}^{f_2}$

		$S_{D_H}^{f_2}$	$S_{D_E}^{f_2}$	$S_{D_M}^{f_2}(a=3)$
Sobel	Prewitt	0.9997	0.9947	0.9863
Sobel	LoG	0.9985	0.9694	0.9169
Sobel	Canny	0.9961	0.9256	0.8042
Sobel	Roberts	0.9980	0.9709	0.9294
Prewitt	LoG	0.9984	0.9707	0.9205
Prewitt	Canny	0.9962	0.9296	0.8147
Prewitt	Roberts	0.9983	0.9753	0.9400
LoG	Canny	0.9946	0.9179	0.7975
LoG	Roberts	0.9984	0.9749	0.9368
Canny	Roberts	0.9962	0.9420	0.8547

Table 4 Similarity results under $S_{D_H}^{f_3}$, $S_{D_E}^{f_3}$ and $S_{D_M}^{f_3}$

		$S_{D_H}^{f_3}$	$S_{D_E}^{f_3}$	$S_{D_M}^{f_3}(a=3)$
Sobel	Prewitt	1	0.9933	0.9828
Sobel	LoG	0.9981	0.9616	0.8976
Sobel	Canny	0.9951	0.9081	0.7666
Sobel	Roberts	0.9975	0.9636	0.9127
Prewitt	LoG	0.9980	0.9633	0.9019
Prewitt	Canny	0.9952	0.9130	0.7785
Prewitt	Roberts	0.9979	0.9690	0.9255
LoG	Canny	0.9932	0.8988	0.7590
LoG	Roberts	0.9979	0.9685	0.9217
Canny	Roberts	0.9953	0.9280	0.8242

change. Both filters are popular image processing methods used for edge detection. Both perform a gradient calculation to detect the horizontal and vertical edges in the image and use the differences between pixel values to determine the boundaries. However, there are also some differences. For example, the Sobel filter does more computation than the Prewitt filter and may therefore have a higher computational cost. Also, the Sobel filter may have a better edge redirection ability than the Prewitt filter. In general, however, Sobel and Prewitt filters give similar results and are interchangeable.

4 Conclusion

Overall, fuzzy logic provides a powerful framework for handling ambiguity and uncertainty in image processing. Image processing algorithms can achieve more robust, adaptive and accurate results using fuzzy techniques and facilitate image analysis, understanding and interpretation. Hence, it becomes imperative and crucial to incorporate fuzzy sets and their generalizations into every stage of image processing.

References

1. Aldring, J., Santhoshkumar, S., Ajay, D.: A decision-making approach using linear diophantine fuzzy sets with Dombi operations. In: International Conference on Intelligent and Fuzzy Systems, pp. 684–692. Springer International Publishing, Cham (2022). https://doi.org/10.1007/978-3-031-09176-6-76
2. Atanassov, K.: Intuitionistic fuzzy sets. Fuzzy Sets Syst. **20**(1), 87–96 (1986). https://doi.org/10.1016/s0165-0114(86)80034-3
3. Aydoğdu, E., Güner, E., Aldemir, B., Aygün, H.: Complex spherical fuzzy TOPSIS based on entropy. Expert Syst. Appl. **215**, 119331 (2023). https://doi.org/10.1016/j.eswa.2022.119331
4. Bustince, H., Barrenechea, E., Pagola, M., Orduna, R.: Image thresholding computation using Atanassov's intuitionistic fuzzy sets. J. Adv. Comput. Intell. Intell. Inf. **11**(2), 187–194 (2007). https://doi.org/10.20965/jaciii.2007.p0187
5. Cuong, B.C., Kreinovich, V.: Picture fuzzy sets-a new concept for computational intelligence problems. In: 2013 Third World Congress on Information and Communication Technologies (WICT 2013), pp. 1–6 (2013). https://doi.org/10.1109/WICT.2013.7113099
6. De, S.K., Biswas, R., Roy, A.R.: An application of intuitionistic fuzzy sets in medical diagnosis. Fuzzy Sets Syst. **117**(2), 209–213 (2001). https://doi.org/10.1016/S0165-0114(98)00235-8
7. Ejegwa P.A.: Distance and similarity measures for Pythagorean fuzzy sets. Granular Comput. **5**(2), 225–238 (2020). https://doi.org/10.1007/s41066-018-00149-z
8. Gül, S., Aydoğdu, A.: Novel distance and entropy definitions for linear Diophantine fuzzy sets and an extension of TOPSIS (LDF-TOPSIS). Expert Syst. **40**(1), e13104 (2023). https://doi.org/10.1111/exsy.13104
9. Jurio, A., Paternain, D., Bustince, H., Guerra, C., Beliakov, G.: A construction method of Atanassov's intuitionistic fuzzy sets for image processing. In: 2010 5th IEEE International Conference Intelligent Systems, pp. 337–342 (2010). https://doi.org/10.1109/IS.2010.5548390
10. Kamacı, H.: Linear Diophantine fuzzy algebraic structures. J. Ambient Intell. Hum. Comput. **12**(11), 10353–10373 (2021). https://doi.org/10.1007/s12652-020-02826-x
11. Liu, S., Zhang, J., Niu, B., Liu, L., He, X.: A novel hybrid multi-criteria group decision-making approach with intuitionistic fuzzy sets to design reverse supply chains for COVID-19 medical waste recycling channels. Comput. Ind. Eng. **169**, 108228 (2022). https://doi.org/10.1016/j.cie.2022.108228
12. Mahmood, T., Rehman, U.U.: A novel approach towards bipolar complex fuzzy sets and their applications in generalized similarity measures. Int. J. Intell. Syst. **37**(1), 535–567 (2022). https://doi.org/10.1002/int.22639
13. Mohammad, M.M.S., Abdullah, S., Al-Shomrani, M.M.: Some linear Diophantine fuzzy similarity measures and their application in decision-making problem. IEEE Access **10**, 29859–29877 (2022). https://doi.org/10.1109/ACCESS.2022.3151684
14. Riaz, M., Hashmi, M.R.: Linear Diophantine fuzzy set and its applications towards multi-attribute decision-making problems. J. Intell. Fuzzy Syst. **37**(4), 5417–5439 (2019). https://doi.org/10.3233/JIFS-190550
15. Shanthi, S.A., Valarmathi, R.: Edge detection on fuzzy near sets. Mater. Today Proc. **51**, 2504–2511 (2022). https://doi.org/10.1016/j.matpr.2021.12.120
16. Xu, Z.S., Chen, J.: An overview of distance and similarity measures of intuitionistic fuzzy sets. Int. J. Uncertain. Fuzz. Knowl. Based Syst. **16**(04), 529–555 (2008). https://doi.org/10.1142/S0218488508005406

17. Xuecheng, L.: Entropy, distance measure and similarity measure of fuzzy sets and their relations. Fuzzy Sets Syst. **52**(3), 305–318 (1992). https://doi.org/10.1016/0165-0114(92)90239-Z
18. Zadeh, L.A.: Fuzzy sets. Inf. Control **8**(3), 338–353 (1965). https://doi.org/10.1016/S0019-9958(65)90241-X
19. Zhang, X., Xu, Z.: Extension of TOPSIS to multiple criteria decision-making- with Pythagorean fuzzy sets. Int. J. Intell. Syst. **29**(12), 1061–1078 (2014). https://doi.org/10.1002/int.21676

Delamination Resistance of Laminated Glass Plates Having Ethyl Vinyl Acetate, Polyvinyl Butyral and Sentryglas Plus Interlayers

Ebru Dural

Abstract Laminated glass is produced by laminating glass layers with an interlayer at specified pressure and temperatures using an autoclave. The characteristic properties of laminated glass can be affected considerably by the used interlayers. Polyvinyl Butyral, Ethyl Vinyl Acetate and SentryGlass are three common interlayers used by glass manufacturers. In order to select the appropriate interlayer parameters such as blast performance, availability, cost, durability, optical property and manufacturing equipment should be considered. Usage of laminated glass is expanding due to its safety and security property as well as comfort and design. As a consequence of possess superiority of laminated glass for structural applications in various industries, they strengthen their position between building materials. Nevertheless delamination which may be result of manufacturing process and service is regarded the primary concern and most undesirable failure mode for the analysis of laminated glass unit. For that cause it is hot research area of composite industry. Delamination strength of laminated glass with the mentioned three types of interlayers are presented in the current research. A mathematical model is presented to analyze delamination behavior of laminated glass plate. Five nonlinear equations are written in matrix form. Solution of matrix system is obtained by using special matrix solvers and successive over relaxation method is used to overcome the convergence problems. Deflection and stress values which represent mechanical behavior of unit are presented in figures. In order to carry out validation assumptions of model the finite element model was constituted. A good fit between the results is observed.

E. Dural (✉)
Department of Civil Engineering, Adnan Menderes University, Aydin, Turkey

Department of Technical Programs, Kyrgyz-Turkish Manas University, Bishkek, Kyrgyz Republic
e-mail: ebru.dural@manas.edu.kg; ebru.dural@adu.edu.tr

V. Gayoso Martínez et al. (eds.), *Mathematical Methods for Engineering Applications*, Springer Proceedings in Mathematics & Statistics 439,
https://doi.org/10.1007/978-3-031-49218-1_20

1 Introduction

Laminated glasses which consist of glass layers and a polymeric interlayer become widely preferred materials in modern structures. Since laminated glass units can handle large displacements, behavior of them is complex and need to brought to light to manufacturers and engineers. For the purpose of obtain true behavior of laminated glass unit large deflection theory must be taken into account in the analysis.

Wide application of laminated glass in construction, automotive, aerospace industries lead to increasing studies on the laminated glasses during the last decade. Firstly, Hooper [1] carried out experimental and theoretical researches concerning the bending behavior of laminated glass beam. It is concluded that interlayer shear modulus and thickness has a considerable effect on the bending resistance of laminated glass beam. In 1983 Phillip Davies, Robert Davis and Cadwallader [2] conducted studies on reasons of delamination and the prevention strategies of it.

Vallabhan [3] used Von Karman equations to symbolize the attitude of glass plate units subjected to pressure in lateral direction. Equations were solved using iterative finite difference technique.

Vallabhan et al. [4] conducted experimental studies to compare the stresses of laminated glass and monolithic glass with the results of their theoretical stress analyzes.

In 1986 Behr et al. [5] investigated attitude of laminated glass under impact loading. They have noted that interlayer thickness effects the impact attitude of the laminated glass unit.

Vallabhan et al. [6] developed a model on laminated glass units using calculus of variations and the minimum potential energy theorem and performed experiments to verify the mathematical model.

Aşık [7] conducted a study to reveal the nonlinear analysis of laminated glass plates. By using the finite difference, the nonlinear behavior of laminated glass unit was symbolized by five nonlinear partial differential equations.

Aşık and Tezcan [8] modelled laminated glass beams subjected to different boundary conditions. They detected that laminated glass beams can behave linearly or nonlinearly under different boundary conditions and constraints.

Effect of geometric and mechanical properties on the structural response of laminated glass beams are analyzed analytically by Foraboschi [9]. Their model is able to presages strength and stress of unit.

Foraboschi [10] modelled analytical behavior of laminated glass rectangular plate subjected to simply supported boundary conditions under uniform pressure. He presented an analytical model to explain mechanical behavior by explicit systems of equations.

In 2012 Laura Galuppi and Gianni Royer-Carfagni [11] developed time-dependent behavior of simply supported laminated glass beam consists of a viscoelastic interlayer by using series of Prony's Maxwell elements.

Hooper et al. [12] conducted studies on delamination behavior of laminated glass windows subjected to blast loading by performing tensile tests on cracked laminated glass units. They observed that the delamination process absorbs a significant amount of energy. Researches in the field of laminated glass with delamination are increasing day by day [13–16]. The analysis by Dural and Oyar [13] showed the effect of boundary conditions on the behavior of laminated glass subjected to delamination at different regions of unit. Behavior of laminated composite plate represented by five nonlinear partial differential equations. Four of the equations are used to represent axial displacements and the last one for the lateral deflection of unit.

In the present study, Asik's model [7] is modified to examine the effect of interlayer type on the response of delaminated glass plate with initial delamination. Fixed supported boundary conditions are assumed in the analyses. Equations are obtained through variational principle in terms of displacements and solved numerically by employing iterative solution procedure. At the end of solution process the lateral, in plane displacements and stresses are obtained. In the derivation of model, interlayers compression strain is neglected and only the shear strain is considered for interlayer.

2 Mathematical Modelling

In this research laminated glass plate shown in Fig. 1 is considered. It is subject to fixed supports and each glass layer has its own neutral axis. For the individual components of unit, it is assumed that plane sections remain plane after deformation. In strain tensor quadratic term in $\frac{dw}{dx}$ is taken part to include the so-called "membrane effect" as in Von Karman plate theory.

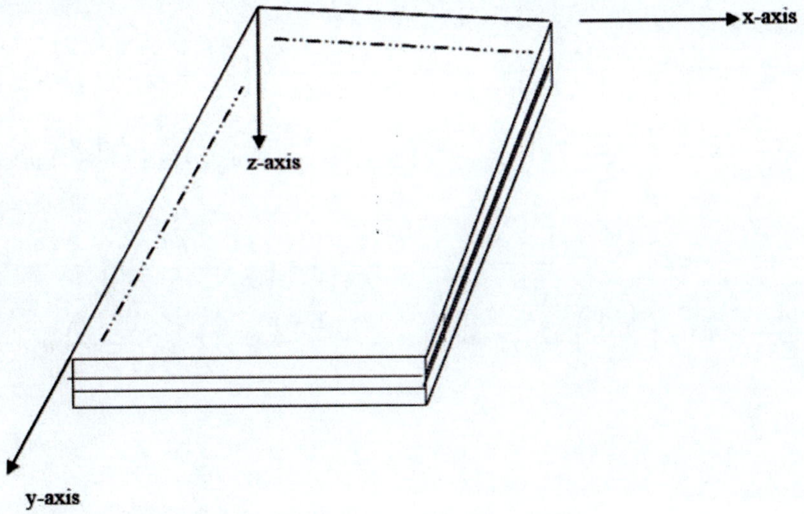

Fig. 1 Laminated glass plate

The complexity of behavior is enhanced due to the large lateral displacement which is greater than thickness of unit under applied lateral forces. Besides to the conventional bending stresses, membrane stresses develops and stiffen the unit, thus the nonlinear behavior which is represented as geometric nonlinearity is involved to the analysis.

Since bending resistance of interlayer negligible the bending stress of interlayer is ignored and only shear stress of interlayer is considered. For this reason while writing the total potential energy of laminated glass plate both membrane and bending strain energies of glass sheets are included, while only shear strain energy of the flexible interlayer is included.

Total potential energy in the unit is $\Pi = U_m^1 + U_b^1 + U_m^2 + U_b^2 + U_I + V$, where U_m^1 and U_m^2 are the membrane strain energies for the top and bottom glass sheets, respectively; and U_b^1 and U_b^2 are the bending strain energies for the top and bottom glass sheets, respectively; U_I is the shear strain energy of the interlayer, and V is the load potential energy function. Applying variational method to the potential energy of plate with regard to in plane and lateral displacements governing equations and boundary conditions are obtained as follows:

$$\begin{aligned}
&\left[(D1 + D2)\,\nabla^4 - \frac{G}{t}\left(\frac{h1}{2} + \frac{h2}{2} + t\right)^2 \nabla^2\right] w \\
&= p - \frac{G}{t}\left(\frac{h1}{2} + \frac{h2}{2} + t\right)\left(u_{1,x} - u_{2,x} + v_{1,y} - v_{2,y}\right) \\
&+ \frac{Eh1}{1-\nu^2}\left[\left[\left(\varepsilon_{1x} + \nu\varepsilon_{1y}\right]w_{,xx} + \left[\left(\varepsilon_{1y} + \nu\varepsilon_{1x}\right]w_{,yy} + (1-\nu)\,\varepsilon_{1xy}w_{,xy}\right]\right. \\
&+ \frac{Eh2}{1-\nu^2}\left[\left[\left(\varepsilon_{2x} + \nu\varepsilon_{2y}\right]w_{,xx} + \left[\left(\varepsilon_{2y} + \nu\varepsilon_{2x}\right]w_{,yy} + (1-\nu)\,\varepsilon_{2xy}w_{,xy}\right]\right.
\end{aligned} \tag{1}$$

$$\begin{aligned}
&u1_{,xx} + \frac{1-\nu}{2}u1_{,yy} - \frac{G\left(1-\nu^2\right)}{Eh1t}u1 = -\frac{G\left(1-\nu^2\right)}{Eh1t}\left(\frac{h1}{2} + \frac{h2}{2} + t\right)w_{,x} \\
&\frac{G\left(1-\nu^2\right)}{Eh1t}u2 - \left(\frac{1+\nu}{2}\right)v1_{,xy} - w_{,x}\left(w_{,xx} + \frac{1-\nu}{2}w_{,yy}\right) - \left(\frac{1+\nu}{2}\right)w_{,y}w_{,xy}
\end{aligned} \tag{2}$$

$$\begin{aligned}
&u2_{,xx} + \frac{1-\nu}{2}u2_{,yy} - \frac{G\left(1-\nu^2\right)}{Eh1t}u2 = \frac{G\left(1-\nu^2\right)}{Eh1t}\left(\frac{h1}{2} + \frac{h2}{2} + t\right)w_{,x} \\
&-\frac{G\left(1-\nu^2\right)}{Eh1t}u1 - \left(\frac{1+\nu}{2}\right)v2_{,xy} - w_{,x}\left(w_{,xx} + \frac{1-\nu}{2}w_{,yy}\right) - \left(\frac{1+\nu}{2}\right)w_{,y}w_{,xy}
\end{aligned} \tag{3}$$

$$v1_{,yy}+\frac{1-\nu}{2}v1_{,xx}-\frac{G\left(1-\nu^2\right)}{Eh1t}v1=-\frac{G\left(1-\nu^2\right)}{Eh1t}\left(\frac{h1}{2}+\frac{h2}{2}+t\right)w_{,y} \\ -\frac{G\left(1-\nu^2\right)}{Eh1t}v2-\left(\frac{1+\nu}{2}\right)u1-w_{,y}\left(w_{,yy}+\frac{1-\nu}{2}w_{,xx}\right)-\left(\frac{1+\nu}{2}\right)w_{,x}w_{,xy} \tag{4}$$

$$v2_{,yy}+\frac{1-\nu}{2}v2_{,xx}-\frac{G\left(1-\nu^2\right)}{Eh2t}v2=\frac{G\left(1-\nu^2\right)}{Eh2t}\left(\frac{h1}{2}+\frac{h2}{2}+t\right)w_{,y} \\ -\frac{G\left(1-\nu^2\right)}{Eh2t}v1-\left(\frac{1+\nu}{2}\right)u2-w_{,y}\left(w_{,yy}+\frac{1-\nu}{2}w_{,xx}\right)-\left(\frac{1+\nu}{2}\right)w_{,x}w_{,xy} \tag{5}$$

Since delamination is a critical failure mode which causes degradation of adhesion between the individual layers of laminated composites, the shear stress transferred by the interlayer will be inefficient in the delaminated regions of plate and the governing equations will be revised as shown in Eqs. (6–10):

$$\left[(D1+D2)\,\nabla^4\right]w=p+\frac{Eh1}{1-\nu^2}\left[\left(\varepsilon_{1x}+\nu\varepsilon_{1y}\right]w_{,xx}+\left[\left(\varepsilon_{1y}+\nu\varepsilon_{1x}\right]w_{,yy} \\ +(1-\nu)\,\varepsilon_{1xy}w_{,xy}+\frac{Eh2}{1-\nu^2}\left[\left(\varepsilon_{2x}+\nu\varepsilon_{2y}\right]w_{,xx}+\left[\left(\varepsilon_{2y}+\nu\varepsilon_{2x}\right]w_{,yy} \\ +(1-\nu)\,\varepsilon_{2xy}w_{,xy}+ \tag{6}$$

$$u1_{,xx}+\frac{1-\nu}{2}u1_{,yy} \\ =-\left(\frac{1+\nu}{2}\right)v1_{,xy}-w_{,x}\left(w_{,xx}+\frac{1-\nu}{2}w_{,yy}\right)-\left(\frac{1+\nu}{2}\right)w_{,y}w_{,xy} \tag{7}$$

$$u2_{,xx}+\frac{1-\nu}{2}u2_{,yy} \\ =-\left(\frac{1+\nu}{2}\right)v2_{,xy}-w_{,x}\left(w_{,xx}+\frac{1-\nu}{2}w_{,yy}\right)-\left(\frac{1+\nu}{2}\right)w_{,y}w_{,xy} \tag{8}$$

$$v1_{,yy}+\frac{1-\nu}{2}v1_{,xx}=-\left(\frac{1+\nu}{2}\right)u1-w_{,y}\left(w_{,yy}+\frac{1-\nu}{2}w_{,xx}\right)-\left(\frac{1+\nu}{2}\right)w_{,x}w_{,xy} \tag{9}$$

$$v2_{,yy}+\frac{1-\nu}{2}v2_{,xx}=-\left(\frac{1+\nu}{2}\right)u2-w_{,y}\left(w_{,yy}+\frac{1-\nu}{2}w_{,xx}\right)-\left(\frac{1+\nu}{2}\right)w_{,x}w_{,xy} \tag{10}$$

In the above equations q is lateral pressure, D_1 and D_2 are the flexural rigidity of glass sheets, ν is the Poisson's ratio of glass and G is the modulus of rigidity of interlayer, E is the modulus of elasticity of glass. The non-linear membrane strains expressed in terms of the lateral and axial displacements as seen below,

$$\varepsilon_{ix} = ui_{,x} + \frac{1}{2}(w_{,x})^2$$

$$\varepsilon_{iy} = vi_{,y} + \frac{1}{2}(w_{,y})^2$$

$$\varepsilon_{ixy} = ui_{,y} + vi_{,x} + w_{,x}w_{,y}$$

Assuming symmetry with respect to x and y-axes, only a quarter of plate is considered. For the case of a fix supported rectangular plate with no in-plane restraints on the edges subjected to a uniform pressure, the obtained boundary conditions are given in Asik [7]. Since it is difficult and complex to solve five nonlinear equations, the iterative solution method used to solve the obtained equations. Finite difference method is applied as the appropriate tool for the solution of equations.

3 Numerical Results

A numerical model is represented to anticipate the effect of bonding interlayer on strength of delaminated glass. Rectangular laminated glass plate 1 × 1 m is used in this analyze. Because of the symmetric geometry, quarter of the plate is used in the research. In x and y directions of quarter plate the number of subdivisions are equal to 20. The maximum value of applied pressure is 5 kPa. Young's modulus of glass is 70 GPa while its Poission's ratio is 0.22. Thickness of top and bottom glass layers are equal to each other and it is 5 mm while the interlayer thickness is 0.76 mm. The interlayer used in the lamination keeps the glass shard together and remains intact after breakage. Polymeric interayers such as ethylene-vinyl acetate (EVA), polyvinyl butyral (PVB) and Sentry Glas Plus (SGP) are the most common interlayers used in laminated glass industry. Interlayer type used in the glazing has a great influence on the strength of laminated glass unit due to the varying mechanical properties of interlayer material.Three different interlayers considered in the present research as Ethyl Vinyl Acetate (G = 5300 kPa), Polyvinyl Butyral (G = 1287 kPa) and Sentryglas Plus (G = 111000 kPa).

Delamination is mode of failure for laminated composites which may be defined as the reducing the bond between the layers. For layered composites it is an undesirable failure mode however it is the most common type of damage due to low strength between layers of material. Delamination may cause to degrade

Fig. 2 Delaminated specimen (delamination between 0 and 0.25 m)

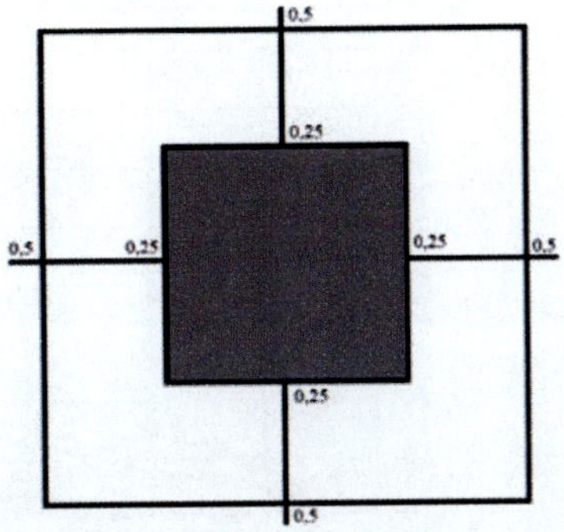

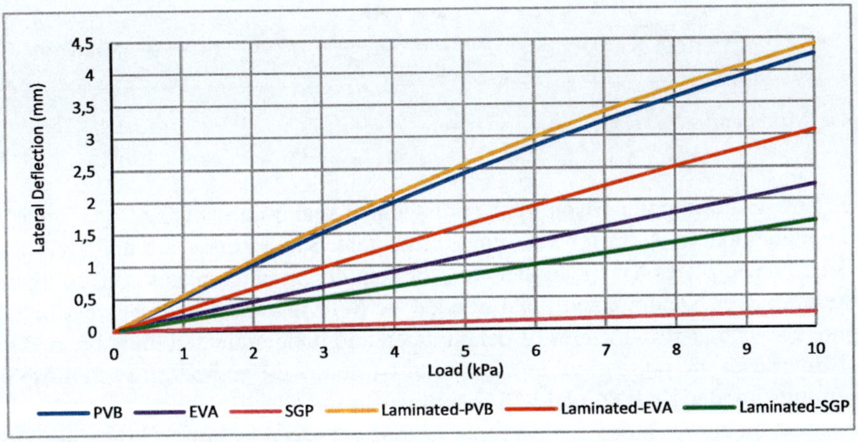

Fig. 3 Lateral deflection versus load

the compression strength, stability and durability and it may affect the physical appearance of the material. Figure 2 shows the laminated glass plate subject to delamination at the plate center.

Figure 3 represents the load versus deflection lines of delaminated and laminated glass plates with PVB, EVA and SGP interlayers. In the below figures Laminated-PVB represents for undelaminated glass plate bonded by PVB interlayer, Laminated-EVA represents for undelaminated glass plate bonded by EVA interlayer and Laminated-SGP represents for undelaminated glass plate bonded by SGP interlayer while PVB, EVA and SGP represents for delaminated glass plates with the Polyvinyl Butyral, Ethyl Vinyl Acetate, and SentryGlas Plus interlayers. Nonlinear behavior of plates are observed since the membrane forces are developing because of fixed end conditions and layering structure. Figure shows that lateral deflection of delaminated glass bonded by SGP interlayer is very small compared to other cases. Deflections of laminated glass unit bonded by PVB interlayer are greater than deflections of laminated units bonded by EVA and SGP interlayers.

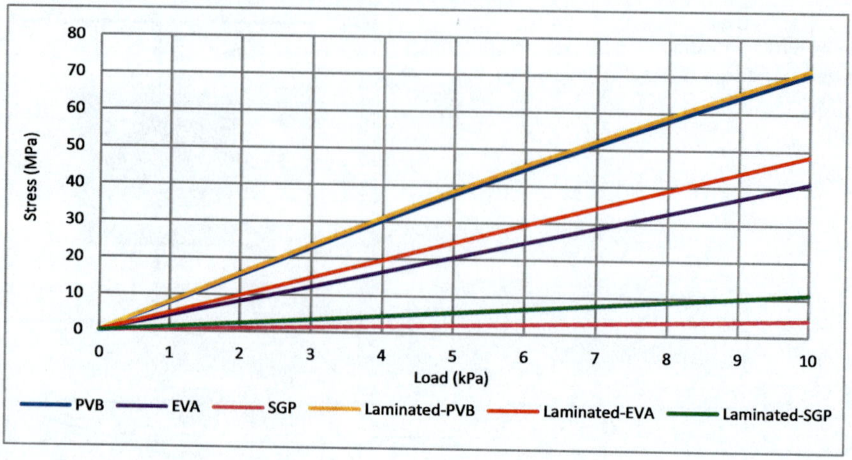

Fig. 4 Maximum stress versus load

Figure 4 shows comparison of the behavior of maximum stress of delaminated and undelaminated plates for 3 varying interlayers. Stress curves are nonlinear and undelaminated plates has the higher stresses than delaminated plates bonded by the same interlayer. Delaminated plate bonded by SGP has the smallest stress value. Figure shows that stress values of delaminated and undelaminated units bonded by PVB interlayer are quite close to each other while the difference between those of other units bonded by EVA and SGP interlayers quite a lot.

Deflection curves along the diagonal of unit are given in Fig. 5. It is noted that the deflection values of undelaminated unit bonded by PVB unit is above that of delaminated unit bonded by PVB, and almost same with it. Deflections for the delaminated system are far smaller than deflections of undelaminated glass units. Lateral deflections are maximum at the center of unit and they are zero at the corner of glass plate.

Figures 6 and 7 present comparison of the maximum principal stresses for applied 10 kPa pressure through the diagonal of the plate unit on the bottom and top surface of unit, respectively. Top surface stresses are in compression and they are maximum at the center of plate. As seen in figure they take both positive and negative values. Maximum principal stress on the bottom surface are in tension through the diagonal line. Top surface stresses change sign through the diagonal and their values reduce to zero at a point close to the middle of the line. Minimum principal stress curves of top and bottom glass surfaces are presented in Figs. 8 and 9. Figures present the minimum principal stress distribution on the surfaces along the diagonal of plate. It is seen that minimum principal stress on top surface are both in tension and compression along the plate diagonal. They take their highest value at the center of plate. They are zero at the corner of unit and at another point on the diagonal line. Stress curve of laminated glass bonded by PVB is quite similar

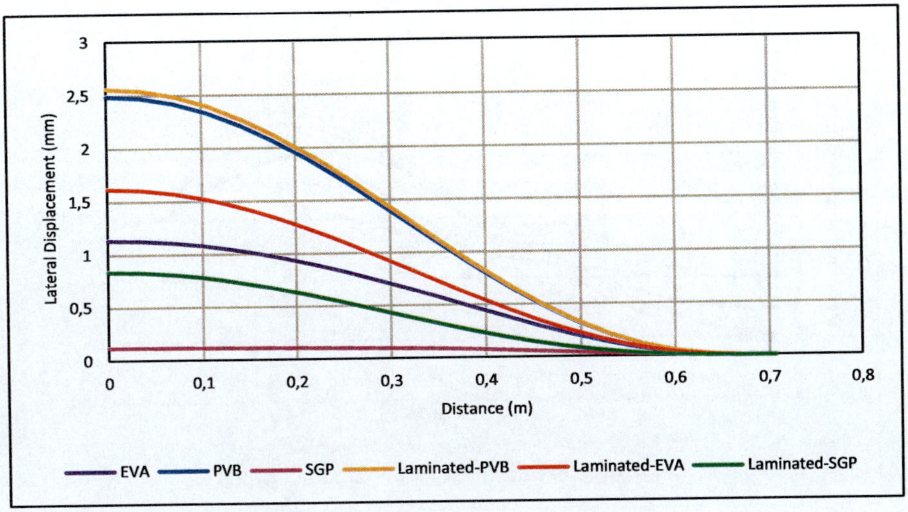

Fig. 5 Lateral Deflection along the diagonal of unit

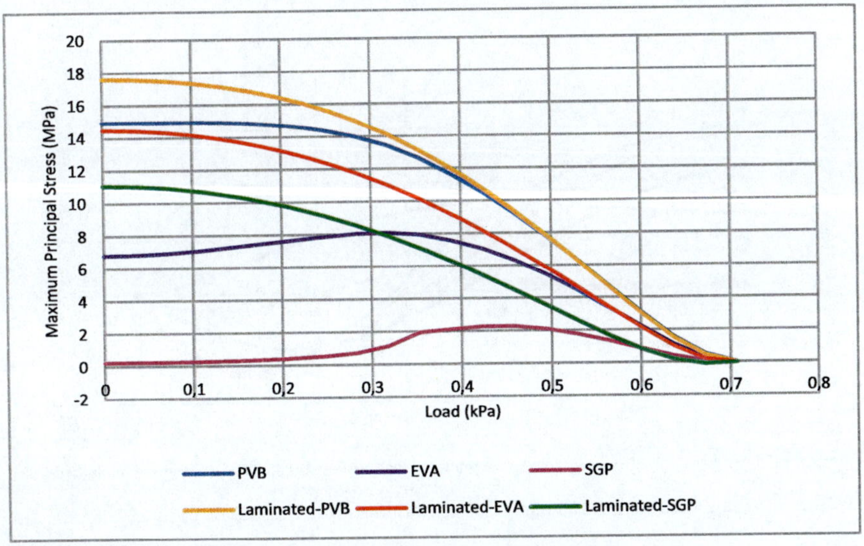

Fig. 6 Maximum principal stress on the bottom surface along the diagonal of unit

to that of Laminated-PVB unit. Top surfaces minimum stresses are compressive through the diagonal and they are maximum at a point close to the midpoint of the diagonal line.

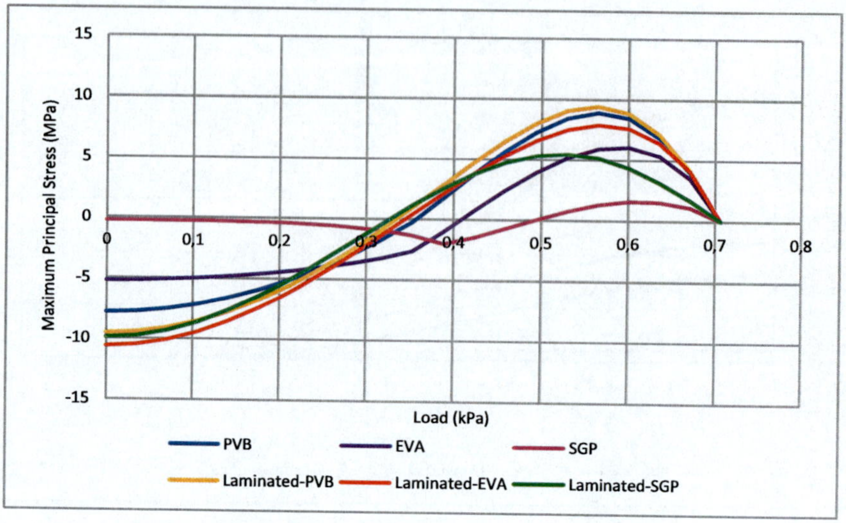

Fig. 7 Maximum principal stress on the top surface along the diagonal of unit

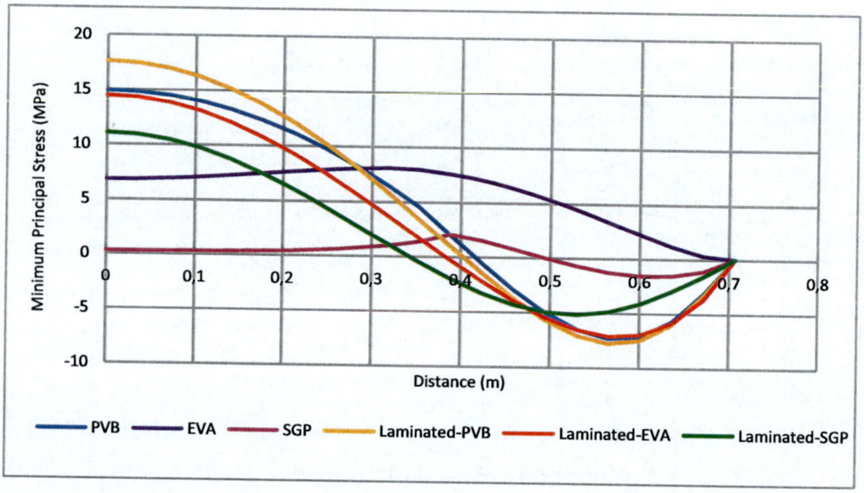

Fig. 8 Minimum principal stress on the bottom surface along the diagonal of unit

4 Conclusion

In this research study, delamination strength of laminated glass with three different types of interlayers are evaluated. We focused on a mathematical model that describe nonlinear response of delaminated glass plates. The principle of minimum potential energy and variational method are attained in the derivation. Two different sets

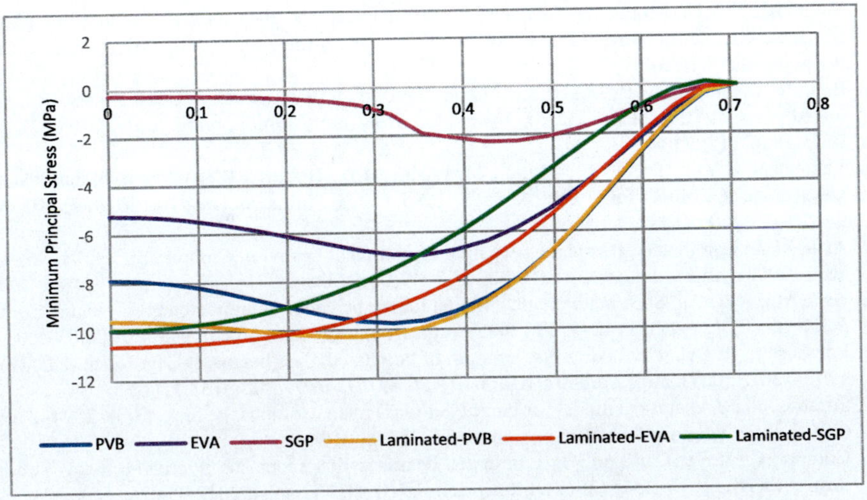

Fig. 9 Minimum principal stress on the top surface along the diagonal of unit

of equations are derived for delaminated and undelaminated regions of unit. In each set there are five coupled nonlinear partial differential equations. The discrete forms of equations are transformed to discrete forms applying numerical finite difference method. Special solvers are applied to figure out the solution of derived matrix equations. Difficulties related to convergence have been overcome by using successive over relaxation method for w and increasing the load gradually for a convergent solution. A computer program based on the obtained formulation has been used in the analysis and it has also been modified for the analysis of laminated glass units composed of glass plates with different interlayers. It can be inferred that, the interlayer type has a great influence on the behavior of delaminated glass units. The maximum load carrying capacity of laminated glasses bonded with SGP is greater than that for the ones bonded with EVA and PVB under the same loading conditions. Higher stress values, either compression or tension, occur on the undelaminated glass plate while the value of stresses are lower for delaminated units.

References

1. Asik, M.Z., Hooper, J.A.: On the bending of architectural laminated glass. Int. J. Mech. Sci. **15**(4), 309–323 (1973). https://doi.org/10.1016/0020-7403(73)90012-x
2. Davies, P., Cadwallader, R.: Delamination issues with laminated glass–causes and prevention. Glass Proces. Days. Vol Session **16**, 427–430 (1984)
3. Vallabhan, C.V.G.: Iterative analysis of rectangular glass plates. J. Struct. Eng. **109**(2), 489–502 (1983). https://doi.org/10.1061/(ASCE)0733-9445(1983)109:2(489)

4. Vallabhan, C.V.G., Minor, J.E., Nagalla, S.: Stress in layered glass units and monolithic glass plates. ASCE J. Struct. Eng. **111**(11), 2416–2426 (1985). https://doi.org/10.1061/(ASCE)0733-9445(1987)113:1(36)
5. Behr, R.A., Linden, M.P., Minor, J.E.: Load duration and interlayer thickness effects on laminated glass. J. Struct. Eng. **112**(6), 1441–1453 (1986). https://doi.org/10.1061/(ASCE)0733-9445(1986)112:6(1441)
6. Vallabhan, C.V.G., Das, Y.C., Magdi, M., Asik, M.Z., Bailey, J.R.: Analysis of laminated glass units. J. Struct. Eng. **119**(5), 1572–1585 (1993). https://doi.org/10.1061/(asce)0733-9445(1993)119:5(1572)
7. Asik, M.Z.: Laminated glass plate: revealing of nonlinear behavior. Comput. Struct. **81**, 2659–2671 (2003). https://doi.org/10.1016/S0045-7949(03)00325-0
8. Asik, M.Z., Tezcan, S.: A mathematical model for the behavior of laminated glass beams. Comput. Struct. **83**(21–22), 1742–1753 (2005). https://doi.org/10.1016/j.compstruc.2005.02.020
9. Foraboschi, P.: Behavior and failure strength of laminated glass beams. J. Eng. Mech. **133**(12), 1290–1301 (2007). https://doi.org/10.1061/(ASCE)0733-9399(2007)133:12(1290)
10. Foraboschi, P.: Analytical model for laminated-glass plate. Compos. B. Eng. **43**(5), 2094–2106 (2012). https://doi.org/10.1016/j.compositesb.2012.03.010
11. Galuppi, L., Royer-Carfagni, G.: Laminated beams with viscoelastic interlayer. Int. J. Solids Struct. **49**(18), 2637–2645 (2012). https://doi.org/10.1016/j.ijsolstr.2012.05.028
12. Del Linz, P., Hooper, P.A., Arora, H., Wang, Y., Smith, D., Blackman, B.R., Dear, J.P.: Delamination properties of laminated glass windows subject to blast loading. Int. J. Impact Eng. **105**, 39–53 (2017). https://doi.org/10.1016/j.ijimpeng.2016.05.015
13. Dural, E., Oyar, F.: Effect of delamination size, location and boundary conditions on the behavior of a laminated glass plate. Structures. **47**(2023), 121–133 (2022). https://doi.org/10.1016/j.istruc.2022.11.034
14. Dural, E.: Experimental and numerical treatment of delamination in laminated glass plate structures. J.Reinf.Plast.Comp. **35**(1), 56–70 (2016). https://doi.org/10.1177/0731684415613163
15. Dural, E.: Analysis of delaminated glass beams subjected to different boundary conditions. Compos. B. Eng. **101**, 132–146 (2016). https://doi.org/10.1016/j.compositesb.2016.07.002
16. Vedrtnam, A.: Experimental and simulation studies on delamination strength of laminated glass composites having polyvinyl butyral and ethyl vinyl acetate inter-layers of different critical thicknesses. Def. Technol. **14**, 313–317 (2018). https://doi.org/10.1016/j.dt.2018.02.002

Algebraic and Quantum Mechanical Approach to Spinors

Tahir Manzoor and S. N. Hasan

Abstract Clifford algebras and quantum mechanics are closely associated to each other. We are familiar with some popular terms, such as algebra, light polarization, and quantum spin. Dirac spinors, Majorana spinors, and Weyl spinors are discussed in particular as subspaces of Clifford algebras with some remarkable algebraic features. Furthermore, we are interested in demonstrating how the quantum spin state and classical polarization of light waves can be derived from one another along with the Bloch sphere and Poincare sphere representations.

1 Introduction

Ehrenfest [1] coined the term spinor in quantum mechanics. Rotational transformations distinguish spinors from tensors and vectors. Representations of the rotation group and their transformation features bridge tensors and spinors. The two-component complex vector structure of a spinor in Hamilton's quaternions was recognized by Cartan [2] which is why both the spin matrices and Pauli spinors are represented by quaternions and the algebraic properties and geometric significance of both are equivalent [3]. Rotating an object by 2π if a negative sign is generated is a spinor such as $\psi(\theta)$ in $\psi(2\pi + \theta) = -\psi(\theta)$. A negative sign is observed while rotating a vector by 2π along the Mobius band; otherwise, visualization of the spinors is not simple. Particles such as spin-half-possessing state functions that undergo a sign change are spinors, and the probability amplitude of the spin state is given by the former. In 1927, for the basic interpretation of 2×2 complex matrices σ_k, Pauli introduced spinors. Pauli spin algebra in the description of electron spin (non-zero) is invariant under an external force or field and has variant orientation [4], and is the key component for discussing various spinor types.

T. Manzoor (✉) · S. N. Hasan
Department of Mathematics, Maulana Azad National Urdu University, Hyderabad, India

V. Gayoso Martínez et al. (eds.), *Mathematical Methods for Engineering Applications*, Springer Proceedings in Mathematics & Statistics 439,
https://doi.org/10.1007/978-3-031-49218-1_21

Clifford algebras not only possessing quantum theory of spin and its natural formulism, but the geometrical information as well as graded nature of spinors is encoded in them. The dedicated Clifford algebra for spinors is even subalgebra Cl_3^+ of Cl_3. Dirac's primitive purpose in 1928 for the relativistic theory of electron spin was because of the connectivity between spinors and Lorentz group [5]. Without using a relativity reference, the evolution of the spinors occurs automatically in the Lorentz group. For all finite-order Lorentz group representations, van der Waerden [6] in 1929 introduced the term spinor. Space–time geometry is beyond the matter and quantum mechanics for spinors, as discussed in [7, 8]. In rotational dynamics, conjugate column vectors (spinor kets) defined by Cayley-Klein parameters similar to those of the spinors were first used by Klein [9]. Dirac formalism is necessary to elucidate the Majorana spinors. Vector fields such as photons A, Z boson Z^o, $W^\dagger$ boson $W^\dagger$ gluon G are formed by spin-1 force-carrying particles. Spin-$\frac{1}{2}$ matter particles form electron field Ψ_e, up quark U, down quark D, and neutrino V_e fields.

This paper is organized into five sections, excluding conclusions. Part 2 constitutes tabular information of particles in terms of spin values, rotation matrices, commutation relations, and exponentiation of spin generators. Section 3 presents a flag pole demonstration of spinors, and Sect. 4 concerns the Weyl equations and spinors derived from the Schrödinger wave equation and Lorentz transformations. Section 5 is enriched with the Clifford algebra and quantum spin representations of different spinors.

2 Particles with Spin and Rotation Matrices

Before commencing rotation matrix algebra for atomic and subatomic particles, we are interested in having a glimpse of salient features. Spin [10–12] is a quantum mechanical phenomenon without a classical analog, and its intrinsic nature is clearly explained by its zero connection with any rotating internal part of a particle. The standard model comprises fundamental particles and natural forces. Higgs field is considered to be a hypothetical ubiquitous quantum field supposed to be accountable for the masses of particles, and the particle related to such a field is the Higgs boson [13]. Einstein's phenomenon of redirecting the compass needle is the best way to discover bosons [14]. Quarks and leptons are major subatomic particles containing electrons, protons, neutrons, and neutrinos as their ingredients. Neutrinos can hold the key to the disruption of matter anti-matter symmetry in the early universe, as mentioned in [15–17]. Three rotation matrices, $R_{xy}\theta$, $R_{yz}\phi$, $R_{zx}\psi$ are in $Su(3)$ group of 3D rotations. Such generator matrices are also known as quantum angular momentum operators that satisfy the commutation relations $[R_{yz}\phi, R_{zx}\psi] = R_{xy}\theta$, $[R_{zx}\psi, R_{xy}\theta] = R_{yz}\phi$, $[R_{xy}\theta, R_{yz}\phi] = R_{zx}\psi$ of $SO(3)$ shown in Table 3. Different particles with their spin values and corresponding rotation matrices along x direction are listed in Table 1. Similarly, rotation matrices in y and z directions can be derived.

Table 1 Spin value of particles and generator matrix S_x

Particles	Higgs boson	Electron, neutrino, quarks	Photon	(Δ^{++} Baryon)	Graviton
Spin value	0	$\frac{1}{2}$	1	$\frac{3}{2}$	2
x-rotation generator matrix S_x	(0)	$\frac{i}{2}\begin{pmatrix} 0 & 1 \\ 1 & 0 \end{pmatrix}$	$\frac{i}{\sqrt{2}}\begin{pmatrix} 0 & 1 & 0 \\ 1 & 0 & 1 \\ 0 & 1 & 0 \end{pmatrix}$	$\frac{i}{2}\begin{pmatrix} 0 & \sqrt{3} & 0 & 0 \\ \sqrt{3} & 0 & 2 & 0 \\ 0 & 2 & 0 & \sqrt{3} \\ 0 & 0 & \sqrt{3} & 0 \end{pmatrix}$	$\frac{i}{2}\begin{pmatrix} 0 & \sqrt{2} & 0 & 0 & 0 \\ \sqrt{2} & 0 & \sqrt{3} & 0 & 0 \\ 0 & \sqrt{3} & 0 & \sqrt{3} & 0 \\ 0 & 0 & \sqrt{3} & 0 & \sqrt{2} \\ 0 & 0 & 0 & \sqrt{2} & 0 \end{pmatrix}$

Table 2 Exponentiation of generator matrices

R_{1xy}	R_{2yz}	R_{3zx}
Spin and $e^{\theta R_{3zx}}$	$e^{\theta R_{2yz}}$	$e^{\theta R_{1xy}}$
0 and (1)	(1)	(1)
$\frac{1}{2}$ and $\begin{pmatrix} e^{i\frac{\theta}{2}} & 0 \\ 0 & e^{-i\frac{\theta}{2}} \end{pmatrix}$	$\begin{pmatrix} \cos\frac{\theta}{2} & -\sin\frac{\theta}{2} \\ \sin\frac{\theta}{2} & \cos\frac{\theta}{2} \end{pmatrix}$	$\begin{pmatrix} \cos\frac{\theta}{2} & i\sin\frac{\theta}{2} \\ i\sin\frac{\theta}{2} & \cos\frac{\theta}{2} \end{pmatrix}$
1 and $\begin{pmatrix} \cos\theta & 0 & \sin\theta \\ 0 & 0 & 0 \\ -\sin\theta & 0 & \cos\theta \end{pmatrix}$	$\begin{pmatrix} \cos\theta & 0 & \sin\theta \\ 0 & \cos\theta & -\sin\theta \\ 0 & \sin\theta & \cos\theta \end{pmatrix}$	$\begin{pmatrix} \cos\theta & -\sin\theta & 0 \\ \sin\theta & \cos\theta & 0 \\ 0 & 0 & 0 \end{pmatrix}$

Table 3 Generator matrices corresponding to the commutation relations

Spin	$R_{xy}\theta$	$R_{yz}\phi$	$R_{zx}\psi$
0	(0)	(0)	(0)
$\frac{1}{2}$	$\frac{1}{2}\begin{pmatrix} 0 & i \\ i & 0 \end{pmatrix}$	$\frac{1}{2}\begin{pmatrix} 0 & -1 \\ 1 & 0 \end{pmatrix}$	$\frac{1}{2}\begin{pmatrix} i & 0 \\ 0 & -i \end{pmatrix}$
1	$\begin{pmatrix} 0 & -1 & 0 \\ 1 & 0 & 0 \\ 0 & 0 & 0 \end{pmatrix}$	$\begin{pmatrix} 0 & 0 & 0 \\ 0 & 0 & -1 \\ 0 & 1 & 0 \end{pmatrix}$	$\begin{pmatrix} 0 & 0 & 1 \\ 0 & 0 & 0 \\ -1 & 0 & 0 \end{pmatrix}$
$\frac{3}{2}$	$\frac{i}{2}\begin{pmatrix} 0 & \sqrt{3} & 0 & 0 \\ \sqrt{3} & 0 & 2 & 0 \\ 0 & 2 & 0 & \sqrt{3} \\ 0 & 0 & \sqrt{3} & 0 \end{pmatrix}$	$\frac{1}{2}\begin{pmatrix} 0 & -\sqrt{3} & 0 & 0 \\ \sqrt{3} & 0 & -2 & 0 \\ 0 & 2 & 0 & -\sqrt{3} \\ 0 & 0 & \sqrt{3} & 0 \end{pmatrix}$	$\frac{i}{2}\begin{pmatrix} 3 & 0 & 0 & 0 \\ 0 & 1 & 0 & 0 \\ 0 & 0 & -1 & 0 \\ 0 & 0 & 0 & -3 \end{pmatrix}$

From [18] and [19], baryons contains triplets of quarks and various baryons in particle accelerators are Λ, Σ^-, $\Sigma^+, \cdots, \Delta^+$, Δ^{++}. The exponentiation of the rotation generators, which are also matrices, are listed in Table 2. If graviton exists, it must be massless, implied by an unlimited range of gravitational force (Table 3).

3 Spinors as Flag Poles

Topologically S^3 sphere determines the compactness of $SU(2)$.

$$x \in SU(2) \implies xx^{\dagger} = I = \begin{pmatrix} pp^* + qq^* & pr^* + qs^* \\ rp^* + sq^* & rr^* + ss^* \end{pmatrix} \tag{1}$$

$x = \begin{pmatrix} p & q \\ -q^* & p^* \end{pmatrix} \implies |p|^2 + |q|^2 = 1 = x_1^2 + x_2^2 + x_3^2 + x_4^2$ is bounded parameterization for $p = x_1 + ix_2$ and $q = x_3 + ix_4$.

Thus $p = e^{i\theta_1}\cos\theta_2, \quad q = e^{i\theta_3}\sin\theta_2$ with $0 \leq (\theta_1, \theta_2, \theta_3) \leq 2\pi$. $S^3 = \partial B^4$ three dimensional manifold is embedded in $\mathbb{R}^4$. Axes of rotations in $3D$ or a point on the sphere are demonstrated by flag poles which are actually $SO(3)$ elements [20].

Behavior of spinors is unaffected under the complex number multiplication. $(\xi_1\ \xi_2)^T \equiv \left(\frac{\xi_1}{\xi_2}\ 1\right)^T = \begin{pmatrix} \xi_1^{\sim} \\ 1 \end{pmatrix}$ or $(1\ \xi_2^{\sim})^T$. Overall phase and scaling degrees of freedom are determined by flag pole visualization. Flag pole intersection at a point on a sphere governs a specific spinor. Real projective line $\mathbb{R}P^1$ and plane $\mathbb{R}P^2$ are sufficient to learn about the complex projective line $\mathbb{C}P^1$ which has spinors as members. For the arbitrary points $\begin{pmatrix} a_1 \\ a_2 \end{pmatrix} \equiv \alpha \begin{pmatrix} a_1 \\ a_2 \end{pmatrix}, \beta \begin{pmatrix} a_1 \\ a_2 \\ a_3 \end{pmatrix}, \ldots \gamma \begin{pmatrix} a_1 \\ a_2 \\ \cdot \\ \cdot \\ \cdot \\ a_n \end{pmatrix} \implies$

$\begin{pmatrix} 1 \\ \frac{a_2}{a_1} \end{pmatrix} \in \mathbb{R}P^1, \quad \begin{pmatrix} 1 \\ \frac{a_2}{a_1} \\ \frac{a_3}{a_1} \end{pmatrix} \in \mathbb{R}P^2, \quad \begin{pmatrix} 1 \\ \frac{a_2}{a_1} \\ \frac{a_3}{a_1} \\ \cdot \\ \cdot \\ \cdot \\ \frac{a_n}{a_1} \end{pmatrix} \in \mathbb{R}P^n$. Description of $\mathbb{C}P^1$ elements is same as that for the $\mathbb{R}P^1$. Points with $a_1 = 0$ gets projected to the point a infinity like $\left(1\ \frac{a_2}{a_1}\right)^T = \begin{pmatrix} 1 \\ \infty \end{pmatrix}$. With the orientation λ of flag pole and scaling μ, the factor $\frac{a_2}{a_1}$ is same in the spinor

$$|\psi\rangle \equiv \mu e^{i\lambda} \begin{pmatrix} \cos\left(\frac{\theta}{2}\right) \\ \sin\left(\frac{\theta}{2}\right) e^{i\phi} \end{pmatrix} \equiv \begin{pmatrix} \mu e^{i\lambda} a_1 \\ \mu e^{i\lambda} a_2 \end{pmatrix}$$

Flagpole constructed on the Poincare sphere is representation of a spinor given in Fig. 1.

Direction of the flag pole on the sphere is $\frac{a_2}{a_1} = \tan\left(\frac{\theta}{2}\right)e^{i\phi}$. General matrices acting on a spinor are Mobius transformations [21] $\left\{M^{\sim} = \frac{ax^{\sim}+b}{cx^{\sim}+d}\right\}$, since $\frac{\alpha}{\beta} =$

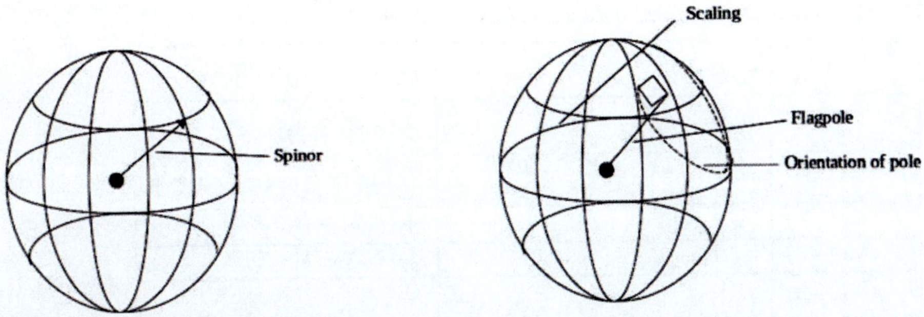

Fig. 1 Flagpole representation of spinors

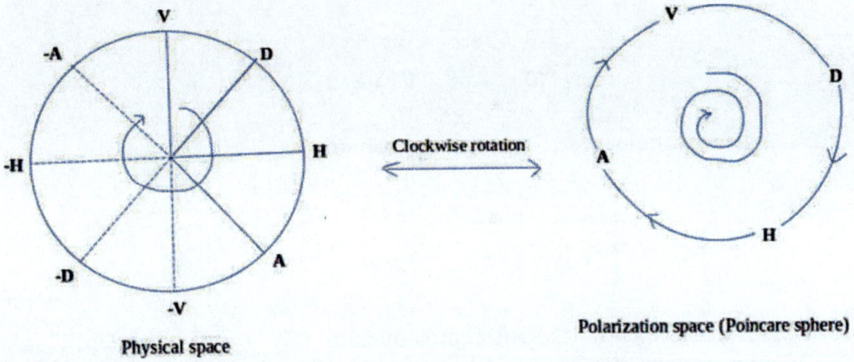

Fig. 2 Rotation in physical space by 360° and 720° in polarization space

$\frac{a(\frac{x}{y})+b}{c\frac{x}{y}+d}$ in $\begin{pmatrix}\alpha\\ \beta\end{pmatrix} = \left(ax + by\ cx + dy\right)^T$. Therefore, a spinor is member of $\mathbb{C}p^1$ and Bloch sphere in the this context is known by Riemann sphere [22].

Rotation of Jones vectors $\left(A^x e^{i\phi_x}\ A^y e^{i\phi_y}\right)^T$ by a twofold turn in polarization space explains their spinor nature as explained in Fig. 2. Jones vectors in physical state and their corresponding counterparts in Poincare sphere are in Table 4. Angle θ in Poincare sphere corresponds to the $\frac{\theta}{2}$ in physical space. $SU(2)$ Lie group $\subseteq U(2)$ group composed of unitary unimodular 2×2 anti hermitian matrices is used for rotations in different polarizations such as in anti-diagonal, anti-horizontal,..., and vertical.

4 Weyl Equations and Spinors

Equation $-\frac{h^2}{2m}\nabla^2\psi = \overrightarrow{E}\psi$ in [23] as the solution of Schrodinger's wave eqn. $ih\partial_t\Psi = H\Psi$ is significant for the success of quantum mechanics. Hermann Weyl [24] proposed a pair of decoupled eqns.

Table 4 Comparison of Jones vectors and quantum states

$\vec{A} \longleftrightarrow \|-X\rangle$	$\frac{1}{\sqrt{2}}\vec{H} - \frac{1}{2}\vec{V}$	$\frac{1}{\sqrt{2}}\|+Z\rangle - \frac{1}{\sqrt{2}}\|-Z\rangle$
$\vec{D} \longleftrightarrow \|+X\rangle$	$\frac{1}{\sqrt{2}}\vec{H} + \frac{1}{2}\vec{V}$	$\frac{1}{\sqrt{2}}\|+Z\rangle + \frac{1}{\sqrt{2}}\|-Z\rangle$
$\vec{H} \longleftrightarrow \|+Z\rangle$	$\vec{H} \longmapsto \begin{pmatrix}1\\0\end{pmatrix}, \quad \vec{V} \longmapsto \begin{pmatrix}0\\1\end{pmatrix}$	$\|+Z\rangle \longmapsto \vec{H}, \quad \|-Z\rangle \longmapsto \vec{V}$
$\vec{L} \longleftrightarrow \|+Y\rangle$	$\frac{1}{\sqrt{2}}\vec{H} + \frac{i}{2}\vec{V}$	$\frac{1}{\sqrt{2}}\|+Z\rangle + \frac{i}{2}\|-Z\rangle$
$\vec{R} \longleftrightarrow \|-Y\rangle$	$\frac{1}{\sqrt{2}}\vec{H} - \frac{i}{2}\vec{V}$	$\frac{1}{\sqrt{2}}\|+Z\rangle - \frac{i}{2}\|-Z\rangle$
$\vec{V} \longleftrightarrow \|-Z\rangle$	$\begin{pmatrix}0\\1\end{pmatrix}$	$\begin{pmatrix}0\\1\end{pmatrix}$

$$\begin{cases} i(\partial_t\phi - \vec{\sigma}\cdot\nabla\phi) = 0 \\ i(\partial_t\tau + \vec{\sigma}\cdot\nabla\tau) = 0 \end{cases} \tag{2}$$

for the massless particles with 2-component spinors ϕ and τ. Coupled pair

$$\begin{cases} i\partial_t\phi + m\phi = i\vec{\sigma}\cdot\nabla\phi \\ \left(i(\partial_t\tau + \vec{\sigma}\cdot\nabla\tau) - m\tau\right) = 0 \end{cases} \tag{3}$$

is defined for massive particles. Differential operator $i(\partial_t - \vec{\sigma}\cdot\nabla)$ in Klein-Gordon [25–27] (relativistic wave) eqn.

$$(m^2 - \Box)\phi = 0 \text{ or } \left((m + \partial_t)^2 - (2m\partial_t - \nabla^2)\right)\phi = 0 \tag{4}$$

with Minkowski metric is same as ϕ. This implies coupled pair (3).

Weyl spinors τ and ϕ as helicity eigen states with ± 1 eigen values are depicted by canonical solutions

$$E_\phi = -\vec{\sigma}\cdot\vec{p}\phi \quad , \quad E_\tau = \vec{\sigma}\cdot\vec{p}\tau \implies \tau = \vec{\sigma}\cdot\hat{p}\tau \text{ and } \phi = -\vec{\sigma}\cdot\hat{p}\phi$$

using relativity relationship $E^2 - \vec{p}^{\,2} = m^2$, $E = \vec{p}\,\hat{p}^{-1}$. Therefore, left and right Weyl eqns. are;

$$\left(E_\phi^{1*}\ E_\phi^{2*}\right)^T = \begin{pmatrix} x_1 + ix_2 & -x_3 \\ x_3 & x_1 - ix_2 \end{pmatrix}\begin{pmatrix}\phi^{1*}\\ \phi^{2*}\end{pmatrix} \tag{5a}$$

$$\left(E_\tau^1\ E_\tau^2\right)^T = \begin{pmatrix} x_3 & x_1 - ix_2 \\ x_1 + ix_2 & -x_3 \end{pmatrix}\begin{pmatrix}\tau^1\\ \tau^2\end{pmatrix} \tag{5b}$$

Under Lorentz rotations $\{J_x, \ J_y, \ J_z\} \in SO(4)$ and boosts $\{K_x, \ K_y, \ K_z\}$, which anti-commute $[K_x, K_y] = -iJ_z$ and two boosts in different directions led to a rotation. Whileas, commutator $[\pm i K_x, \pm i K_y] = i J_z$ is obtained on annexation of $\pm i$ to any K_i such that

$$J_i = \frac{\sigma_i}{2} \implies \pm i K_i = \pm\frac{\sigma_i}{2} \implies e^{K_i} = e^{\pm i \frac{\sigma_i}{2}}.$$

Then ψ_L andψ_R transforms into;

$$\begin{pmatrix} \psi'_L \\ \psi'_R \end{pmatrix} = \begin{pmatrix} 0 & e^{\frac{1}{2}i(\sigma\cdot\theta-\sigma\cdot\phi)} \\ e^{\frac{1}{2}i(\sigma\cdot\theta+\sigma\cdot\phi)} & 0 \end{pmatrix} \begin{pmatrix} \psi_L \ \psi_R \end{pmatrix}^T \quad (6)$$

Let ζ_i is the ith direction of $\phi \implies \zeta_{\pm i} = \zeta_1 \pm i\zeta_2$. Then

$$\begin{pmatrix} \psi'_{L_1} \\ \psi'_{R_1} \\ \psi'_{L_2} \\ \psi'_{R_1} \end{pmatrix} = e^{\frac{1}{2}i\sigma\cdot\theta} \begin{pmatrix} 0 & 0 & h'_3 & h'_1 \\ h_3 & h_1 & 0 & 0 \\ 0 & 0 & h_2 & h_3 \\ h'_2 & h'_3 & 0 & 0 \end{pmatrix} \begin{pmatrix} \psi_{L_1} \\ \psi_{R_1} \\ \psi_{L_2} \\ \psi_{R_12} \end{pmatrix} \quad (7)$$

Where $h_1 = \zeta - \sinh\frac{\phi}{2}$, $h'_1 = -\zeta - \sinh\frac{\phi}{2}$, $h_2 = -\zeta + \sinh\frac{\phi}{2}$, $h'_2 = \zeta + \sinh\frac{\phi}{2}$, $h_3 = \cosh\frac{\phi}{2} + \zeta_3 \sinh\frac{\phi}{2}$, $h'_3 = \cosh\frac{\phi}{2} - \zeta_3 \sinh\frac{\phi}{2}$

Weyl transformation 4×4 matrix can also be written as

$$\Omega_T = \gamma^o\gamma\cdot\zeta \sinh\frac{\phi}{2} + \cosh\frac{\phi}{2} I = e^{\frac{1}{2}(\gamma^o\gamma\cdot\zeta)}$$

5 Spinors in Clifford Algebras Cl_2 and $\mathbb{C}l_6$

To define generalized ideals (subspaces) intuitively from some arbitrary algebra, a specific multiplication operation is needed that we denote by $(\diamond)$. Let the algebra of our choice be V with its subalgebra V_1, then $\diamond(x, y) \in V_1, \quad \forall x \in V, \ y \in V_1$ is a generalized ideal which is distinguished from the ubiquitous left ideals [28]. Dirac, left and right Weyl, and Majorana spinor types can be demonstrated with complex $(\diamond_c)$, Hermitian $(\diamond_h)$ and quaternionic $(\diamond_q)$ invariant operations. In particular, spinors are provided in $(\diamond_c)$, scalars are field strength tensor ϕ and $\mathcal{F}^{\mu\nu}$ by $(\diamond_q)$, and $(\diamond_h)$ gives rise to four vectors P_{mu} [29–31]. Spinors or left spinors are the elements of minimal left ideals, while as minimal right ideals behave as conjugate spinors. The concept of spinors in Clifford algebras Cl_2 and Cl_6 can be extended to other relevant algebras. Cl_6 algebra of complex octonions lacks associativity, so to bring the octonions under Clifford formalism we have applied the idea of left action mapping generating chain structure. Generating spaces for both the algebras are discussed apart from the minimal left ideals construction.

Spinor construction into even and odd dimensions can be generally decided on γ-matrices. Chirality operator $\hat{\gamma}$ in even dimension D of $SO(D)$ spinor representation $\hat{\gamma} = i^{\pm\frac{D}{2}} \Pi_{m=1}^{D} \gamma^m$ is Hermitian, satisfies $\hat{\gamma}^2 = 1$ with ± 1 eigen values and commutes with all γ matrices. $\hat{\gamma}$ commutes with γ matrices,so no Weyl spinors can exist for odd dimensions. Dirac's eqn. is the foundation for quantum field theory. Weyl representation of γ-matrices;

$$\gamma^0 = \begin{pmatrix} \mathbf{0}_{2\times 2} & \mathbf{1}_{2\times 2} \\ \mathbf{1}_{2\times 2} & \mathbf{0}_{2\times 2} \end{pmatrix}, \ \gamma^1 = \begin{pmatrix} 0 & 0 & 1 & -1 \\ 0 & 0 & -1 & 0 \\ 0 & 1 & 0 & 0 \\ 1 & 0 & 0 & 0 \end{pmatrix}$$

$$\gamma^2 = \begin{pmatrix} 0 & 0 & 0 & i \\ 0 & 0 & -i & 0 \\ 0 & -i & 0 & 0 \\ i & 1 & 0 & 0 \end{pmatrix}, \quad \gamma^3 = \begin{pmatrix} 0 & 0 & -1 & 0 \\ 0 & 0 & 0 & 1 \\ 1 & 0 & 0 & 0 \\ 0 & -1 & 0 & 0 \end{pmatrix}$$

with $(\gamma^o)^2 = I, \quad (\gamma^i)^2 = -I, \quad \gamma^\mu\gamma^\nu + \gamma^\nu\gamma^\mu = 2\eta^{\mu\nu}$.

We will derive chirality operators in some dimensions provided the nature of spinor types.

- Chirality operator for $SO(2)$ is $\hat{\gamma} = i\gamma^{12} \begin{pmatrix} 0 & -i \\ i & 0 \end{pmatrix}$ along with spin generator $\sigma^{12} = \begin{pmatrix} 0 & -1 \\ 1 & 0 \end{pmatrix}$ or $\sigma^{12} = \frac{1}{2}\begin{pmatrix} -i & 0 \\ 0 & i \end{pmatrix}$ and $\hat{\gamma} = i\gamma^{12}\begin{pmatrix} 1 & 0 \\ 0 & -1 \end{pmatrix}$.

  $SO(2) \simeq U(1)$ and $Spin(2) \simeq SO(2)$ as two components of spinor transform under $Spin(2)$.
- $su(2)$ algebra and $Spin(3) \simeq SU(2)$ have no real representations $\Longrightarrow$ neither Majorana nor Weyl spinors exist in 3D case.
- $D = 4 \Longrightarrow$ negative and positive chirality Weyl spinors are in

$$\hat{\gamma} = -\gamma^{1234} = \begin{pmatrix} -i\sigma^{123} & 0 \\ 0 & i\sigma^{123} \end{pmatrix} = \begin{pmatrix} \mathbf{1}_{2\times 2} & \mathbf{0} \\ \mathbf{0} & \mathbf{1}_{2\times 2} \end{pmatrix}$$

- Chirality operator in 6D is $\hat{\gamma} = i\gamma^{12345} = \begin{pmatrix} \mathbf{1}_{4\times 4} & \mathbf{0} \\ \mathbf{0} & -\mathbf{1}_{4\times 4} \end{pmatrix}$ with

$$\gamma^\alpha = \begin{pmatrix} \mathbf{0} & i\gamma^\alpha \\ -i\gamma^\alpha & \mathbf{0} \end{pmatrix} \text{ and } \gamma^6 = \begin{pmatrix} \mathbf{0} & \mathbf{1}_{4\times 4} \\ \mathbf{1}_{4\times 4} & \mathbf{0} \end{pmatrix}, \quad \alpha = 1,\ 2,\ 3,\ 4,\ 5.$$

Table 5 Spinors in various Clifford algebras

Spinor types			
Pinor representation	Clifford algebra	Spinors	Dim.
Real 2-component Majorana	$Cl^{+}_{1,1} \simeq \mathbb{R} \otimes \mathbb{R}$	1-component Majorana Weyl	
	$Cl^{+}_{2} \simeq \mathbb{C}$	1-component Weyl	2
2-component complex (quaternionic)		complex Weyl	
Unique quaternionic	$Cl_{1,3} \simeq Mat_4(\mathbb{R})$	Majorana	
	$Cl^{+}_{1,3} \simeq Mat_2(\mathbb{C})$		
	$Cl_{3,1} \simeq Mat_2(\mathbb{H})$	Dirac	
	$Cl^{+}_{3,1} \simeq Mat_2(\mathbb{C})$	pseudo-Majorana	
	$Cl_{4,0} \simeq Mat_2(\mathbb{H})$		4
	$Cl_{2,2} \simeq Mat_4(\mathbb{R})$		
	$Cl^{+}_{2,2} \simeq Mat_2(\mathbb{R})$	2-component Majorana Weyl	
Unique quaternionic	$Cl_{5,1} \simeq Mat_4(\mathbb{H})$	8-component Dirac	
	$Cl_{5,1} \simeq Cl_{4,0} \otimes Cl_{1,1}$		
	$Cl_{6,0} \simeq Mat_8(\mathbb{R})$	8-component Majorana	6
	$Cl^{+}_{6,0} \simeq Mat_4(\mathbb{C})$	4-component complex	
32-component real	$Cl_{9,1} \simeq Cl_{1,9} \simeq Mat_{32}(\mathbb{R})$	Majorana	
	$Cl_{9,1} \simeq Cl_{4,0} \otimes Cl_{0,2} \otimes Cl_{2,0} \otimes Cl_{1,1}$	16-component Majorana Weyl	10
32-component complex	$Cl_{10,1} \simeq Mat_{32}(\mathbb{C})$	Majorana in -ve metric and pseudo-Majorana in +ve metric	
	$Cl_{10,1} \simeq Mat_{32}(\mathbb{R}) \otimes Mat_{32}(\mathbb{R})$		11
64-dimensional real	$Cl_{10,2} \simeq Mat_{64}(\mathbb{R})$	32-component Majorana Weyl	12

Weyl representation of $SU(4)$ is generated by a set of 15 anti-Hermitian matrices projected by $\sigma^{\alpha\beta}$. γ^{α} is real for $\alpha = 1,\ 3,\ 6$ and purely imaginary in 2, 4, 5.

$$\gamma^{\alpha} = \begin{pmatrix} \mathbf{0} & i\gamma^{\alpha} \\ -i\gamma^{\alpha} & \mathbf{0} \end{pmatrix} \text{ and } \gamma^{8} = \begin{pmatrix} \mathbf{0} & \mathbf{1}_{8\times 8} \\ \mathbf{1}_{8\times 8} & \mathbf{0} \end{pmatrix}, \quad \alpha = 1,\ 2,\ \ldots,7.$$

Purely imaginary γ^{α} and γ matrices as real ones provides 8 dimensional Majorana representations which splits into Weyl. Thus, we have two Majorana-Weyl representations in $8D$. Spinors with corresponding Clifford algebra types are is discussed in detail in Table 5.

In an algebra $\mathcal{A}$, left action is defined as for any $f,\ g \in \mathcal{A}$

$$\Longrightarrow f\,:\,g \longrightarrow g\prime \Longrightarrow\; fg = g', \quad \forall f,\, g,\, h \in \mathcal{A} \Longrightarrow\; h(fg) = g''$$
$$\Longrightarrow \overleftarrow{hf}\,:\,g \longrightarrow g''.$$

Since such maps are associative and can be annexed for finite elements. We have

$$\left\{\alpha_o g = \binom{1}{0} + \alpha_n \overleftarrow{e_n}\, g = \binom{7}{1} + \alpha_{lm}\overleftarrow{e_l e_m} g = \binom{7}{2} + \alpha_{ijk}\overleftarrow{e_i e_j e_k} g = \binom{7}{3}\right\} = 64$$

degrees of freedom in octonion chains.

$$Lg = \alpha_{ijk}\overleftarrow{e_i e_j e_k} g + \alpha_{lm}\overleftarrow{e_l e_m} g + \alpha_n \overleftarrow{e_n}\, g + \alpha_o g \tag{8}$$

$\overleftarrow{e_i e_j e_k} g$, $\overleftarrow{e_l e_m} g$, $\overleftarrow{e_n}\, g$ are octonion chains. Their Clifford algebra structure $\forall g \in \mathbb{C} \otimes \mathbb{O}$ is;

$$\overleftarrow{e_l e_m} g = \begin{cases} -\overleftarrow{e_m e_l} g & l \neq m \\ -g & l = m \end{cases} \tag{9}$$

This can be extended to the complex sedenions $\mathbb{C} \otimes \mathbb{S} \simeq (\mathbb{C} \otimes \mathbb{O}) \oplus (\mathbb{C} \otimes \mathbb{O})$ with 128 freedom degrees.

Generating space of Cl_2 in terms of basis $\left\{\alpha = \frac{i\hat{q_x} - \hat{q_y}}{2}\quad, \quad \alpha^\dagger = \frac{i\hat{q_x} + \hat{q_y}}{2}\right\}$ satisfies $\{\alpha, \quad \alpha^\dagger\} = 0 = \{\alpha^\dagger, \quad \alpha^\dagger\}$ and $\{\alpha, \quad \alpha^\dagger\} = 1$

In Cl_6, generating space under the basis $\left\{\alpha_1 = \frac{ie_1 - e_3}{2},\ \alpha_2 = \frac{ie_4 - e_5}{2},\ \alpha_3 = \frac{ie_2 - e_6}{2},\ \alpha_1^\dagger = \frac{ie_1 + e_3}{2},\ \alpha_2^\dagger = \frac{ie_4 + e_5}{2},\ \alpha_3^\dagger = \frac{ie_2 + e_3}{2}\right\}$ holds $\{\alpha_i, \quad \alpha_j\}g = 0 = \{\alpha_i^\dagger, \quad \alpha_j^\dagger\}g \quad \forall\, g \in Cl_6$ and $\{\alpha_i, \quad \alpha^\dagger\}g = \delta_{ij} g$.

α_i and $\alpha_i^\dagger$ behave as fermionic lowering and raising operators, and have $U(3) \simeq SU(3) \times U(1)/\mathbb{Z}_3$ symmetry.

Minimal left ideals $\mathcal{I}_1$ and $\mathcal{I}_2 \in Cl_2$, $\mathcal{I}^O$ and $\mathcal{I}^{O'} \in Cl_6$ are;

Let $\mathcal{P}$ be the projector [32] and idempotent in Cl_2, then $\mathcal{P} = \alpha\alpha^\dagger$, left ($\psi_1$) and right ($\psi_2$) Weyl spinors are

$$Cl_2\mathcal{P} = \psi_1^\uparrow \mathcal{P} + \psi_1^\downarrow \alpha^\dagger \mathcal{P} \equiv \mathcal{I}_1 \tag{10a}$$

$$Cl_2\mathcal{P}^* = \psi_2^\downarrow \mathcal{P}^* + \psi_2^\uparrow \alpha^\dagger \mathcal{P}^* \equiv \mathcal{I}_2 \tag{10b}$$

Spinors in Cl_6 are composed of $\bar{\zeta}$, ν, ω, $\bar{D}^r$, $\bar{D}^g$, $\bar{D}^b$, ν^r, ν^g, ν^b, and $\zeta^\dagger$ (electrons, neutrinos, anti-down quarks, up quarks and positrons...).

$\mathcal{I}^O \equiv Cl_6\nu$ and $\mathcal{I}^{O'} \equiv Cl_6\nu^\dagger$ are two minimal ideals spanned by $\{\alpha_i\ ,\ \alpha_i^\dagger\}$ basis vectors. Where α_i, $\alpha_i^\dagger$ are here termed as lowering and raising operators which transform into

Table 6 Minimal left ideals as Spinors

$Cl_6\nu \equiv \mathcal{I}^O$	$SU(3)_6$	Q	$Cl_6\nu^\dagger \equiv \mathcal{I}^{O'}$	$SU(3)_6$	$-Q^*$
$\nu\mathcal{P}$	1	0	$\bar{\nu}\mathcal{P}^*$	1	0
$D^r\alpha_1^\dagger\mathcal{P}\ D^g\alpha_2^\dagger\mathcal{P}\ D^b\alpha_3^\dagger\mathcal{P}$	3^*	$\frac{1}{3}$	$D^r\alpha_1\mathcal{P}^*\ D^g\alpha_2\mathcal{P}^*\ D^b\alpha_3\mathcal{P}^*$	3	$-\frac{1}{3}$
$\eta^r\alpha_{23}^\dagger\nu\ \eta^g\alpha_{31}^\dagger\nu\ \eta^b\alpha_{12}^\dagger\nu$	3	$\frac{2}{3}$	$\bar{\eta}^r\alpha_{23}\nu^*\ \bar{\eta}^g\alpha_{31}\nu^*\ \bar{\eta}^b\alpha_{12}\nu^*$	3^*	$-\frac{2}{3}$
$\zeta^\dagger\alpha_{123}^\dagger\nu$	1	1	$\zeta\alpha_{123}\nu^*$	1	-1

$$U(3) = SU(3) \times \frac{U(1)}{\mathbb{Z}_3}$$

Number of eigen values of the number operator $U(1) \equiv N$ are $\{0\ ,\ 1\ ,\ 1\ ,\ 1\ ,\ 2\ ,\ 2\ ,\ 2\ ,\ 3\}$ which gives rise to the electric charges $\equiv \{0\ ,\ \frac{1}{3}\ ,\ \frac{1}{3}\ ,\ \frac{1}{3}\ ,\ \frac{2}{3}\ ,\ \frac{2}{3}\ ,\ \frac{2}{3}\ ,\ 1\}$.

Therefore, Table 6 contains the information regarding the action of electric charge operator $Q \equiv \frac{N}{3}$ on various states. Multiplying any arbitrary element ω in the right side of any element of $\mathcal{I}^O$ gives the corresponding basis vector for $\mathcal{I}^{O'}$ elements. Boson behavior of ω and $\omega^\dagger$ is produced by applying the relations

$$[Q\ ,\ \omega] = -\omega \tag{11a}$$

$$[Q\ ,\ \omega^\dagger] = \omega^\dagger \tag{11b}$$

Thus, we can say that electric charges are quantized as N operator takes only integer values. Therefore from (11a) and (11b),$-\omega\ ,\ \omega^\dagger$ are negatively and positively charged. From Table 6, the values $\{1\ ,\ 3^*\ ,\ 3\ ,\ 1\}$ and $\{1\ ,\ 3\ ,\ 3^*\ ,\ 1\}$ under Q are $SU(3)_c$ representations for one full generation of quarks and leptons [33–35].

Theorem 1 *Let S_1 and S_2 be the subspaces of Clifford algebra $\mathbb{C} \times \mathbb{H} \simeq Cl_2 \simeq Mat_2(\mathbb{R})$. From these subspaces, Weyl, Dirac and Majorana spinors can be deduced.*

Proof Let $\mathcal{W}_1,\ \mathcal{W}_2,\ \mathcal{D},\ \mathcal{M}_1$ and $\mathcal{M}_2$ be the notions for Weyl, Dirac and Majorana spinors.

$$\mathcal{W}_1 + \mathcal{W}_2 = w_1^\uparrow q_{\uparrow\uparrow} + w_1^\downarrow q_{\downarrow\uparrow} + w_2^\uparrow q_{\uparrow\downarrow} + w_2^\downarrow q_{\downarrow\downarrow} \tag{12a}$$

$$\mathcal{W}_1 - \mathcal{W}_2 = w_1^\uparrow q_{\uparrow\uparrow} - w_2^\uparrow q_{\uparrow\downarrow} + w_1^\downarrow q_{\downarrow\uparrow} - w_2^\downarrow q_{\downarrow\downarrow} \tag{12b}$$

Since $\{1,\ iq_x,\ iq_y,\ iq_z\}$ and $\{q_{\uparrow\uparrow},\ q_{\uparrow\downarrow},\ q_{\downarrow\uparrow},\ q_{\downarrow\downarrow}\}$ are the bases which can be transformed into each other.

$q_{\uparrow\uparrow} \equiv \frac{1+iq_z}{2},\ q_{\uparrow\downarrow} \equiv \frac{iq_x-1}{2},\ q_{\downarrow\uparrow} \equiv \frac{q_y+iq_x}{2},\ q_{\downarrow\downarrow} \equiv \frac{1-iq_z}{2}$ have $q_x,\ q_y,\ q_z$ quaternionic units and $i \in \mathbb{C}$.

$w_1^{\uparrow},\quad w_1^{\downarrow},\quad w_2^{\uparrow},\quad w_2^{\downarrow}\quad \in \mathbb{C}$

$$(12a) + (12b) \implies \frac{1}{2}\left\{w_1^{\uparrow} q_{\downarrow\uparrow} + w_2^{\uparrow} q_{\uparrow\downarrow} - w_1^{\downarrow} q_{\uparrow\downarrow} - w_2^{\uparrow} q_{\downarrow\uparrow}\right\} \tag{13}$$

$$\mathcal{M}_1 = \frac{1}{2}\left\{\mathcal{W}_1 + \mathcal{W}_2 - w_1^{\downarrow}(q_{\uparrow\downarrow} + q_{\downarrow\uparrow}\Omega^{\dagger}\Omega) - (w_2^{\uparrow} q_{\downarrow\uparrow} + w_2^{\downarrow}\Omega\Omega^{\dagger})\right\} \tag{14}$$

$$\mathcal{M}_2 = w_1^{\uparrow} + w_2^{\downarrow} + (2w_1^{\downarrow} q_{\downarrow\uparrow} + w_2^{\uparrow} q_{\uparrow\downarrow})\Omega^{\dagger}\Omega + w_2^{\downarrow}\Omega\omega^{\dagger} + \cdots - w_2^{\downarrow} q_{\uparrow\uparrow} \tag{15}$$

Therefore, $\mathcal{M}_1$ and $\mathcal{M}_2$ are two Majorana spinors, and Majorana-Weyl spinors can also be obtained from this.

6 Conclusion

In this paper, spinors are discussed in terms of the Clifford algebras and quantum physics. Numeric and symbolic computations of spinors in the Clifford algebra approach by using some packages like LUCY [36] for MAPLE was created. Our future goal is to derive different spinor denominations in higher Clifford algebras, find their various applications in multiple domains, and finally to develop their computational packages for PYTHON.

References

1. Veblen, O.: Spinors. Science **80**(2080), 415–419 (1934)
2. Cartan, É.: Les groupes projectifs qui ne laissent invariante aucune multiplicité plane. Bull. Soc. Math. France **41**, 53–96 (1913)
3. De Sabbata, V., Datta, B.K.: Geometric Algebra and Applications to Physics. CRC Press (2006)
4. Ficek, Z.: Quantum Physics for Beginners. CRC Press (2017)
5. Held, A., Newman, E.T., Posadas, R.: The Lorentz group and the sphere. J. Math. Phys. **11**(11), 3145–3154 (1970)
6. van der Waerden, B.L.: Group Theory and Quantum Mechanics, vol. 214. Springer Science & Business Media (2012)
7. Penrose, R., Rindler, W.: Spinors and Space-time: Volume 1, Two-spinor Calculus and Relativistic Fields, vol. 1. Cambridge University Press (1984)
8. Penrose, R., Rindler, W.: Spinors and Space-time. Volume 2: Spinor and Twistor Methods in Space-time Geometry (1986)
9. Klein, F., Sommerfeld, A.: The Theory of the Top. Volume IV: Technical Applications of the Theory of the Top, vol. 4. Springer Science & Business Media (2014)
10. Eaton, G.R., Eaton, S.S.: Foundations of Modern EPR. World Scientific (1998)

11. Sebens, C.T.: How electrons spin. Stud. Hist. Philos. Sci. B Stud. Hist. Philos. Mod. Phys. **68**, 40–50 (2019)
12. Uhlenbeck, G.E.: Personal reminiscences. Phys. Today **29**(6), 43–48 (1976)
13. Howard Haber Stephen Reucroft and Chris Quigg Michael Dine. What exactly is the Higgs Boson? Scientific American.
14. Greene, B.: How the Higgs Boson was Found. Smithsonian Magazine (2013)
15. Abe, K., Abgrall, N., Aihara, H., Ajima, Y., Albert, J. B., Allan, D., et al.: The T2K experiment. Nucl. Instrum. Methods Phys. Res. Sect. A Acceler. Spectrom. Detect. Assoc. Equip. **659**(1), 106–135 (2011)
16. Kaneyuki, K., T2K Collaboration: T2K experiment. Nucl. Phys. B Proc. Suppl. **145**, 178–181 (2005)
17. Valish, L.: Why is the Universe Made Up Almost Exclusively of Matter? neutrinos may hold the key. University of Rochester News Center (2020)
18. Sacchi, R., H1 Collaboration, & ZEUS Collaboration: Leading baryons at HERA. Nucl. Phys. B (Proc. Suppl.) **191**, 214–220 (2009)
19. Trimble, V.: Existence and nature of dark matter in the universe. Annu. Rev. Astron. Astrophys. **25**(1), 425–472 (1987)
20. Penrose, R., Rindler, W.: Spinors and Space-time: Volume 1, Two-spinor Calculus and Relativistic Fields, vol. 1. Cambridge University Press (1984)
21. González-Miret Zaragoza, L.: Möbius Transformations (TrabajoFindeGradoInédito). Universidad de Sevilla, Sevilla (2019)
22. Yur'ev, D.V.E.: Complex projective geometry and quantum projective field theory. Theoret. Math. Phys. **101**(3), 1387–1403 (1994)
23. Griffiths, D.J., Schroeter, D.F.: Introduction to Quantum Mechanics. Cambridge University Press (2018)
24. Newman, M.H.A.: Hermann Weyl, 1885–1955 (1957)
25. Gravel, P., Gauthier, C.: Classical applications of the Klein–Gordon equation. Am. J. Phys. **79**(5), 447–453 (2011)
26. Kaup, D.J.: Klein-gordon geon. Phys. Rev. **172**(5), 1331 (1968)
27. Shatah, J.: Normal forms and quadratic nonlinear Klein-Gordon equations. Commun. Pure Appl. Math. **38**(5), 685–696 (1985)
28. Furey, C.: Unified theory of ideals. Phys. Rev. D **86**(2), 025024 (2012)
29. Bongaarts, P., Bongaarts, P.: Quantum Field Theory and Particle Physics: An Introduction. Quantum Theory: A Mathematical Approach, pp. 247–264 (2015)
30. Carmeli, M., Malin, S.: Theory of Spinors: An Introduction. World Scientific Publishing Company (2000)
31. Cottingham, W.N., Greenwood, D.A.: An Introduction to the Standard Model of Particle Physics. Cambridge University Press (2023)
32. Abłamowicz, R.: Construction of spinors via Witt decomposition and primitive idempotents: A review. In: Clifford Algebras and Spinor Structures: A Special Volume Dedicated to the Memory of Albert Crumeyrolle (1919–1992), pp. 113–123 (1995)
33. Harari, H.: Quarks and leptons. Phys. Rep. **42**(4), 235–309 (1978)
34. Shupe, M.A.: A composite model of leptons and quarks. Phys. Lett. B **86**(1), 87–92 (1979)
35. Raitio, R.: A model of lepton and quark structure. Phys. Scr. **22**(3), 197 (1980)
36. Schray, J., Tucker, R.W., Wang, C.H.T.: LUCY: a Clifford algebra approach to spinor calculus approach to spinor calculus, pp. 121–143 (1996)

New G-Closed Sets with Related to an Ideal

Aynur Keskin Kaymakci

Abstract Ideal is a family which has two properties as known heredity and finite additive. Besides it is a partially ordered collection of sets. This using this concept with a topology was obtained a new topology finer than old. Elements of topology are open sets and their complements are closed sets. It is known that generalized closed set is weaker than closed. Then, several authors studied on this subject both topological spaces and ideal topological spaces. In this paper, we recall some concepts and their basic properties. Then, we introduce new g-closed set called $*Ig$-closed set by using ideal and obtained some properties of it. We have obtain a diagram showing the relationships between them and other types of g-closed sets in the literature. Besides, we define a concept of $*Ig$-open set and investigate it. We give that a family of consisting of this sets in any I-space forms a minimal structure. Finally, we mention when $*Ig$-closed is conserved.

1 Introduction and Preliminaries

Let W be a set and $\mathcal{P}(W)$ be the power set of W. A nonempty family $I \subseteq P(W)$ is called to be an ideal [11, 12, 17] on W if $D \subset C \in I$ requires $D \in I$ and $C, D \in I$ requires $(C \cup D) \in I$. If (W, υ) is a topological space with the ideal I, the triplet (W, υ, I) is called an ideal topological space [7]. An ideal topological space is renamed to I-space throughout this paper. Since I-space is related to both υ and I of course $(\upsilon \cap J) = \{\emptyset\}$ is an important equality. It is termed as Hayashi-Samuel space by Jankovic and Hamlett [7]. Actually Hayashi [5] and Samuels [16] defined two different concepts as Hayashi-space and Samuels-space, respectively independently of each other. However, Dontchev et al. [7] found that these two concepts were equivalent to each other and named it Hayashi-Samuels space. It should not go without mentioning that; ideal is a generalization of important concepts in topology

A. K. Kaymakci (✉)
Selcuk University, Faculty of Science, Department of Mathematics, Konya, Turkey
e-mail: akeskin@selcuk.edu.tr

V. Gayoso Martínez et al. (eds.), *Mathematical Methods for Engineering Applications*, Springer Proceedings in Mathematics & Statistics 439,
https://doi.org/10.1007/978-3-031-49218-1_22

such as condensation point, accumulation point and closure point [7]. The simplest ideals are $\{\emptyset\}$ and $\mathcal{P}(W) = \{C : C \subseteq W\}$. It is known that most important ideals in any I-space (W, υ, I) I_f (i.e., the ideal of finite subsets of W), I_c (i.e., the ideal of countable subsets of W), I_n (i.e., the ideal of nowhere dense sets in (W, υ)). An ideal is sometimes called a dual filter which is nonempty collection of nonempty sets closed under the operations of superset and finite intersection. The interior (resp. closure) of the subset C in topological space (W, υ) is denoted by $i(C)$ (resp. $c(C)$). On the other hand since the notion of localization is important in topology as well as analysis, Kuratowski [11, 12] introduced a new definition is called local function using the concepts topology and ideal in what follows. For a subset $C \subset W$, $C^*(I, \upsilon) = \{w \in W : (P \cap C) \notin I$ for every $P \in \upsilon(w)\}$ is called to be the local function of C with respect to I and υ. $\upsilon(w)$ is a family of open neighborhoods of $w \in W$. Sometimes, $C^*(I)$ or C^* are used instead of the notion $C^*(I, \upsilon)$. It must state that W^* is often a proper subset of W. $C^*(\{\emptyset\}) = c(C)$ and $C^*(\mathcal{P}(W)) = \emptyset$ for every $C \subset W$. The ideal has been studied by many topologists for years because it is an important concept. Therefore, the reader can read the following articles for the main and interesting parts of the subject [1, 3–10, 15] and [16]. It is well-known that $c^*(C) = C \cup C^*$ is a Kuratowski closure operator for a topology $\upsilon^*(I)$. We use υ^* instead of $\upsilon^*(I)$. υ^* is finer than υ. If $C \in \upsilon^*$(resp. $(W - C) \in \upsilon^*$), then C is called to be υ^*-open (resp. υ^*-closed) [7]. A subset C of an I-space $(W, \upsilon$,I) is υ^*-closed [7](resp. $*$-perfect [5], *I-open* [1]) iff $C^* \subset C$ (resp. $C^* = C$, $C \subset Int(C^*)$). We use υ^*-c set instead of υ^*-closed set. Besides, $\psi(C) \subset Int^*(C)$ is hold. $Int^*(C)$ is an interior operator related to υ^* topology for every $C \subset W$ [15].

In [13] Levine introduced the concept of generalized closed sets. He obtained that it is weaker than closed set. Later, this topic was studied extensively by many topologists. Besides, several generalization of it using the notion of ideal are obtained. Indeed in [4] and [6] authors both introduced the concept of I-g-closed sets and I-g-closed sets, respectively and obtained some properties of them. On the other hand, a new type of g-closed set is defined in I-spaces by using I-open set instead of open set in [10]. Let's talk about the fact that openness and I-openness are independent concepts [1].

Let's consider some definitions, lemmas and one conclusion that will be needed in this article

Lemma 1 ([7]) *Let (W, υ) be a topological space with I and J ideals on W and let C, D two subsets of W. Then, the following features are valid:*

(1) If $C \subset D$, then $C^* \subset D^*$,
(2) $C^* = c(C^*) \subset c(C)$,
(3) $(C^*)^* \subset C^*$,
(4) $(C \cup D)^* = C^* \cup D^*$,
(5) If $I \subset J$, then $C^*(J) \subset C^*(I)$,
(6) If $S \in I$, then $S^* = \emptyset$.
(7) $(I \cap J)$ is an ideal on W.

Definition 1 Let (W, υ, I) be an I-space and $C \subset W$. Then C is called to be

(1) generalized closed [13] (shortly g-c) if $c(C) \subset P$, whenever $C \subset P \in \upsilon$,
(2) I-g-closed [4] (shortly I-g-c) if $C^* \subset P$, whenever $C \subset P \in \upsilon$.

Using above definitions, one can obtain next diagram and find counter examples in [13] and [4].

Diagram I

$$\begin{array}{ccc} closed & \longrightarrow & \tau^* - c \\ \downarrow & & \downarrow \\ g - c & \longrightarrow & I - g - c \end{array}$$

We must state that to avoid confusion we use Ig-c as gI-c in this paper.
Now, we recall another one generalization of g-c sets.

Definition 2 Let (W, υ, I) be an I-space and $C \subset W$. Then C is called to be Ig-closed [6] (shortly, Ig-c) if $(c(C) - P) \in I$, whenever $C \subset P \in \upsilon$.

We must state that to avoid confusion we use Ig-c as gI-c in this paper.

Remark 1 ([6]) Although every g-c is gI-c in any I-space, the converse of this requirement is not generally true.

Another type of g-c set is generalized-I-closed. A subset C of (W, υ, I) is called to be generalized I-closed [10] if $c(C) \subset P$, whenever $C \subset P$ and P is an I-open set of W.

Finally, we recall the following result related to the concepts of ideal.

Lemma 2 ([11]) *Let* $f : (W, \upsilon, I) \to (R, \varphi)$ *be a function, then* $f(I) = \{f(S) : S \in I\}$ *is an ideal on* R.

This study consists of three main parts as "$*Ig$-Closed Sets", "$*Ig$-Open Sets" and "Conservation of $*Ig$-Closed Sets". Now let's take it one by one.

2 $*Ig$-Closed Sets

Firstly, we defined a new type of closed set called $*Ig$-closed. We obtain that it is weaker than both $I - g$-c and gI-c. Thus, we gave a diagram of its relations with some types of g-c sets in the literature and determined its location. We investigate properties of it and give a characterization of it using local function operator.

Definition 3 Let (W, υ, I) be an I-space and $C \subset W$. Then C is call to be $*I$-generalized-closed (shortly, $*Ig$-c) if $(C^* - P) \in I$, whenever $C \subset P$ and $P \in \upsilon$.

Example 1 Let $(\mathcal{R}, \upsilon)$ be the real numbers with usual topology υ and I_f the ideal of all finite sets of $\mathcal{R}$. Let $\mathcal{Q}$ be the set of all rationals. Since $\mathcal{R}$ is only open set containing Q and $Q^* = \mathcal{R}$, we have $(Q^* - \mathcal{R}) = \emptyset \in I$ and hence $\mathcal{Q}$ is $*Ig$-c.

Remark 2 Let (W, υ, I) be an I-space and $C \subset W$. Then, the next features are valid:

(1) $W, \emptyset$ and C^* are always $*Ig$-c.
(2) If $C \in I$, C is $*Ig$-c.
(3) If (W, υ) be an indiscrete space, then every subset of W is $*Ig$-c.

Proposition 1 *Let (W, υ, I) be an I-space. Then, the following features are hold:*

(1) Every I-g-c set is $*Ig$-c,
(2) Every gI-c set is $*Ig$-c.

Proof

(1) Let $C \subset W$ is I-g-c set. Then, $C^* \subset P$ whenever $C \subset P$ and $P \in \upsilon$. Since $(C^* - P) \subset (P - P) = \emptyset$ and $\emptyset \in I$, we have $(C^* - P) \in I$. Therefore we have $(C^* - P) \in I$, whenever $C \subset P$ and $P \in \upsilon$. This shows that C is $*Ig$-c.
(2) Let $C \subset W$ is gI-c set. Then $(c(C) - P) \in I$, whenever $C \subset P$ and $P \in \upsilon$. Using Lemma 1(2), $(C^* - P) \in I$. Therefore we have $(C^* - P) \in I$, whenever $C \subset P$ and $P \in \upsilon$. This shows that C is $*Ig$.

We have the next diagram using Diagram I, Remark 1, and Proposition 1.

Diagram II

$$\begin{array}{ccc} closed & \longrightarrow & \tau^* - c \\ \downarrow & & \downarrow \\ g - c & \longrightarrow & I - g - c \\ \downarrow & & \downarrow \\ gI - c & \longrightarrow & *Ig - c \end{array}$$

Remark 3 The reverses of each of the requirements in Diagram II are generally not true as given in the related papers [4, 6, 13] and the examples below.

Example 2 Let $W = \{u, v, w\}$, $\tau = \{\emptyset, W, \{u\}, \{u, w\}\}$ and $I = \{\emptyset, \{v\}\}$. Set $C = \{w\}$ is $*Ig$-c but it is not I-g-c in I-space (W, υ, I). For $C = \{w\}$, $C^* = \{v, w\}$. Since $(C^* - \{u, w\}) = \{v\} \in I$ and $(C^* - W) = \emptyset \in I$, where $C \subset \{u, w\} \in \upsilon$ and $C \subset W \in \upsilon$, respectively. Hence, C is $*Ig$-c. Besides since $C = \{w\} \subset \{u, w\} \in \upsilon$, we have $C^* = \{v, w\} \not\subset \{u, w\}$. This shows that C is not I-g-c. Besides since $C^* = \{v, w\} \not\subset \{w\} = C$, C is not also υ^*-c.

Example 3 Let $W = \{u, v, w\}$, $\upsilon = \{\emptyset, W, \{v\}, \{u, v\}\}$ and $I = \{\emptyset, \{w\}, \{v\}, \{v, w\}\}$. Set $C = \{v\}$ is $*Ig$-c but it is not gI-c in (W, υ, I). For $C = \{v\}$ since $\{v\} \in I$, we have $C^* = \emptyset$ from Lemma 1(6) and hence $(C^* - P) = \emptyset \in I$ is valid for every $P \in \upsilon(C)$. Therefore $(C^* - P) \in I$, whenever $C \subset P$ and $P \in \upsilon$ and hence C is $*Ig$-c. On the other hand for $C = \{v\}$, $c(C) = X$ and $(c(C) - \{v\}) = \{u, v\} \notin I$. This shows that C is not gI-c. ($\upsilon(C)$ is a family of open neighborhoods of $C \in W$.)

Example 4 Let (W, υ, I) is an I-space in Example 3, i.e., $W = \{u, v, w\}$, $\upsilon = \{\emptyset, W, \{v\}, \{u, v\}\}$ and $I = \{\emptyset, \{w\}, \{v\}, \{v, w\}\}$. Set $C = \{v\}$ is Ig-c but it is not gI-c in (W, υ, I). For $C = \{v\}$ since $\{v\} \in I$, we have $C^* = \emptyset$ from Lemma 1(6) and hence $C^* \subset P$ is valid for every $P \in \upsilon(C)$. Therefore, we have $C^* \subset P$, whenever $C \subset P$ and $P \in \upsilon$ and hence C is Ig-c. On the other hand C is not gI-c.

Example 5 Let $W = \{u, v, w, t\}$, $\upsilon = \{\emptyset, W, \{u, w\}, \{t\}, \{u, w, t\}\}$ and $I = \{\emptyset, \{v\}\}$. Set $C = \{u, w\}$ is gI-c but it is not Ig-c in (W, υ, I). For $C = \{u, w\}$ since $\upsilon(C) = \{\{u, w\}, \{u, w, t\}, W\}$ and $c(C) = \{u, v, w\}$, we have $(c(C) - \{u, v\}) = (c(C) - \{u, w, t\}) = \{v\} in I$ and $(c(C) - W) = \emptyset \in I$. Therefore, C is gI-c. On the other hand for $C = \{u, w\}$, since $C^* = \{u, v, w\}$ and $\{u, v, w\} \not\subset \{u, w\} \in \upsilon(C)$. This shows that C is not Ig-c.

Remark 4 Examples 4 and 5 show that a gI-c set and an I-g-c set are independent notions.

Theorem 1 *Let (W, υ, I) be an I-space such that $I = \{\emptyset\}$. Then, gI-closedness and $*Ig$-closedness are coincide in (W, υ, I).*

Proof Since we give in Proposition 1(2) that every gI-c is $*Ig$-c, we will only prove the opposite of this implication. Since $I = \{\emptyset\}$, $(C^* = c(C))$ for every $(C \subset W)$. Let C is $*Ig$-c. Then we have $(C^* - P) \in J$, whenever $C \subset P \in \upsilon$ and hence $(c(C) - P) \in J$. Therefore $(c(C) - P) \in J$, whenever $C \subset P \in \upsilon$. This shows that C is gI-c.

We give the next proposition directly from Lemmas 1(5) and 1(7).

Proposition 2 *Let I and J be two ideals on W, (W, υ) be a topological space and $C \subset W$. Then, the following features are valid:*

(1) If $I \subset J$, then every $*Ig$-c set is $*Jg$-c.
(2) If C is $*(I \cap J)g$-c set, then it is both $*Ig$-c and $*Jg$-c.

Now, we give a characterization of $*Ig$-c set.

Theorem 2 *A set C is $*Ig$-c set in (W, υ, I) iff $L \subset (C^* - C)$ and L is a closed in W requires $L \in I$.*

Proof Sufficiency. Suppose that C is $*Ig$-c, $L \subset (C^* - C)$ and L is a closed. Then, $C \subset (W - L) \in \upsilon$. Since C is $*Ig$-c, we have $(C^* - (W - L)) \in I$ and hence $L \in I$.

Necessity. Suppose that $L \subset (C^* - C)$ and L is a closed in (W, υ, I) requires that $L \in I$. Let $C \subset P \in \upsilon$. Then $(C^* - P) = C^* \cap (W - P)$ is a closed set in W, that is $(W - P) \subset (C^* - C)$. By hypothesis, we have $(C^* - C) \in I$. Hence, C is $*Ig$-c.

Theorem 3 *If C and D are two $*Ig$-c sets in (W, υ, I), then $(C \cup D)$ is $*Ig$-c.*

Proof Let C and D are two $*Ig$-c sets in (W, υ, I). If $(C \cup D) \subset P$ and $P \in \upsilon$, we have $C \subset P$ and $D \subset P$. Since C and D are two $*Ig$-c sets, $(C^* - P) \in I$ and

$(D^* - P) \in I$. Hence, $((C \cup D)^* - P) = ((C^* \cup D^*) - P) = (C^* - P) \cup (D^* - P)$ and hence $((C \cup D)^* - P) \in I$. This shows that $(C \cup D)$ is $*Ig$-c.

Remark 5 The intersection of two $*Ig$-c sets need not be a $*Ig$-c set as shown by the next example.

Example 6 Let (W, υ, I) be an I-space such that $W = \{u, v, w\}$, $\upsilon = \{\emptyset, W, \{v\}\}$ and $I = \{\emptyset\}$. Then $C = \{u, v\}$ and $D = \{v, w\}$ are $*Ig$-c, but $C \cap D = \{v\}$ isn't $*Ig$-c.

Theorem 4 *Let C and D be two subsets in an I-space (W, υ, I) such that $C \subset D \subset C^*$. If C is $*Ig$-c, then D is $*Ig$-c.*

Proof Assume that C is $*Ig$-c and $C \subset D \subset C^*$. Let $D \subset P$ and $P \in \upsilon$. Since C is $*Ig$-c, we have $C \subset P$ and $(C^* - P) \in I$. Besides since $D \subset C^*$ we have $D^* \subset (C^*)^* \subset C^*$ using Lemma 1(3). Hence $(D^* - P) \subset (C^* - P)$ and we have $(D^* - P) \in I$. Thus, D is $*Ig$-c.

The next fact is obtained from Theorem 4 and definition of $*$-perfect set.

Corollary 1 *Every superset of a set which $*$-perfect $*Ig$-c is also $*Ig$-c in any I-space.*

Definition 4 ([2]) A subset of D in an I-space (W, υ, I) is call to be an I-subspace and denoted by (D, υ_D, I_D). ($I_D = \{S \cap D : S \in I\}$ is an ideal on D.)

Theorem 5 *Let $C \subset D \subset W$ and C is $*Ig$-c set in (W, υ, I), then C is a $*Ig$-c set relative to D.*

Proof Assume that $C \subset (D \cap P)$ and $P \in \upsilon$. Since C is $*Ig$-c set in (W, υ, I), $(C^* - P) \in I$. Hence $[(C^* \cap D) - (P \cap D)] = [(C^* - P) \cap D] \in I_D$ using Definition 4, whenever $C \subset (D \cap P)$ and $(D \cap P) \in \upsilon_D$. This shows that C is a $*Ig$-c set relative to the subspace D.

Theorem 6 *Let (W, υ, I) be an I-space and C, L be two subsets of W. If C is a $*Ig$-c and L is a closed, then $(C \cap L)$ is a $*Ig$-c.*

Proof Suppose that $(C \cap L) \subset P$ and $P \in \upsilon$. Then, $[C \subset (P \cup (W - L))]$ and $(P \cup (W - L)) \in \upsilon$. Since C is $*Ig$-c, we have $[C^* - (P \cup (W - L))] \in I$. Then, using Lemma 1(2) we have $(C \cap L)^* \subset (C^* \cap (L^*) \subset (C^* \cap c(L)) = (C^* \cap L) = (C^* \cap L) - (W - L)$ and $[(C \cap L)^* - P] \subset [(C^* \cap L) - (P \cup (W - L))] \subset [C^* - (P \cup (W - L))]$. Therefore, we have $[(C \cap L)^* - P] \subset [C^* - (P \cup (W - L))]$ and so we obtain that $[(C \cap L)^* - P] \in I$ using hypothesis. This shows that $(C \cap L) \subset W$ is a $*Ig$-c.

3 $*Ig$-Open Sets

In here, we define the concept of $*Ig$-open set related to $*Ig$-c and obtain some of its properties. Its most important feature is: A family of consisting of $*Ig$-open sets in any I-space forms a minimal structure. Also, we have given a characterization of $*Ig$-open set with the help of the ψ operator. Finally, we have obtained an equivalent of $*Ig$-c set for $*Ig$-open using local function.

Definition 5 A subset C of an I-space (W, υ, I) is call to be $*I$-generalized-open (shortly, $*Ig$-o) if $(W - C)$ is $*Ig$-c.

We will denote by $*IGO(W, \upsilon)$ the collection of all $*Ig$-o subsets of an I-space (W, υ, I).

One can obtain the next theorem using Remark 2(1), Definition 5.

Theorem 7 *$*IGO(W, \upsilon)$ forms a minimal structure.(A family $\mu \subseteq P(W)$ is said to be minimal structure on W [2] if $\emptyset$, W $\in \mu$.)*

Theorem 8 *$C \in *IGO(W, \upsilon)$ iff $(L - S) \subset \psi(C)$ for some $S \in I$, whenever $L \subset C$ and L is closed.*

Proof Necessity. Assume that C is a $*Ig$-o set, $L \subset C$ and L is a closed. Then, we have $(W - C) \subset (W - L) \in \upsilon$. By assumption, $(W - C)^* \subset (W - L) \cup S$, for some $S \in I$. This implies that $[W - ((W - L) \cup S)] \subset [W - (W - C)^*]$ and hence $(L - S) \subset \psi(C)$.

Sufficiency. Let $L \subset C$ and L is a closed set imply $(L - S) \subset \psi(C)$. Let $P \in \upsilon$ such that $(W - C) \subset P$. Then, $(W - P) \subset C$. By assumption, we have $[(W - P) - S] \subset \psi(C)$ and $[W - (P \cup S)] \subset \psi(C)$. Then, we have $(W - C)^* \subset (P \cup S)$, for some $S \in I$. This shows that $[(W - C)^* - P] \in I$ and hence $(W - C)$ is $*Ig$-c.

We have the following three facts from previous section.

Corollary 2 *If C and D are two $*Ig$-o sets in (W, υ, I), then $(C \cap D)$ is $*Ig$-o.*

Proof Let C and D are two $*Ig$-o sets in (W, υ, I). Then, $(W - C)$ and $(W - D)$ are $*Ig$-c. According to Theorem 2, we have $[W-(C\cap D)] = [(W-C)\cup(W-D)]$ is $I * g$-c and hence $(C \cap D)$ is $*Ig$-o.

Remark 6 The union of two $*Ig$-o sets is usually not $*Ig$-o set as shown in the following.

Example 7 Let (W, υ, I) be an I-space as Example 2, i.e.,$W = \{u, v, w\}$, $\upsilon = \{\emptyset, W, \{v\}\}$ and $I = \{\emptyset\}$. Then $C = \{w\}$ and $D = \{u\}$ are $*Ig$-o sets, but their union $(C \cup D) = \{u, w\}$ is not $*Ig$-o.

From now on, one can obtain the next conclusion.

Corollary 3 *In any I-space, $I * GO(W, \upsilon)$ forms neither topology nor supra topology [14] which is defined as discarding the finite intersection property from topology.*

Definition 6 Let C, D be two subsets of W and (W, υ, I) be an I-space. The sets C and D are called $*I$-separated if $(C^* \cap D) = \emptyset$ and $(C \cap D^*) = \emptyset$.

Theorem 9 *Let C, D be two subsets of W and (W, υ, I) be an I-space. If C and D are $*I$-separated $*Ig$-o sets, then $(C \cup D)$ is $*Ig$-o.*

Proof Let C and D are $*I$-separated $*Ig$-o sets in (W, υ, I) and $(W-L) \in \upsilon_{(C \cup D)}$. Since L is a closed, we have $(L \cap C^*) \subset C$ and $(L \cap D^*) \subset D$. By assumption $[(L \cap C^*) - S] \subset \psi(C)$ and $[(L \cap D^*) - T] \subset \psi(D)$, for some $S, T \in I$. This state that $[(L \cap C^*) - \psi(C)] \in I$ and $[(L \cap D^*) - \psi(D)] \in I$. Then, we have $([(L \cap C^*) - \psi(C)] \cup [(L \cap D^*) - \psi(D)]) \in I$. Hence $([(L \cap (C^* \cup D^*)) - (\psi(C) \cup \psi(D))] \in I$. Since $L \subset (C \cup D)$, $L = L \cap (C \cup D) \subset L \cap (C \cup D)^*$ and we have $[L - \psi(C \cup D)] \subset ([L \cap (C \cup D)^*] - \psi(C \cup D)) \subset ([L \cap (C \cup D)^*] - [\psi(C) \cup \psi(D)]) \in I$. Therefore $(L - M) \subset \psi(C \cup D)$, for some $M \in I$. Thus, we obtain that $(C \cup D)$ is $*Ig$-o according to Theorem 6.

Corollary 4 *Let C, D are two $*Ig$-c sets in (W, υ, I). If $(W - C)$ and $(W - D)$ are $*$-separated, then $(C \cap D)$ is $*Ig$-c.*

Theorem 10 *Let $C \subset D \subset W$ and C is $*Ig$-o set relative to D and D is $*Ig$-o set in (W, υ, I), then C is a $*Ig$-o set in (W, υ, I).*

Proof Let $C \subset D \subset X$, C is $*Ig$-o set relative to D and D is $*Ig$-o set in (W, υ, I). Assume that $L \subset C$ and L is closed. Since C is $*Ig$-o set relative to D, we have $(L - S) \subset \psi_D(C)$, for some $S \in I_D$ by Theorem 6. This implies there exists an open set P such that $(L - S) \subset (P \cap D \cap C)$, for some $S \in I$. Since D is $*Ig$-o $L \subset D$ and L is closed; $(L - T) \subset \psi(D)$, for some $T \in I$. This implies that there exists an open set N such that $(L - T) \subset N \subset D$, for some $T \in I$. We have $(L - (S \cup T)) \subset ((L - S) \cap (L - T)) \subset (P \cap N) \subset (P \cap D) \subset C$. This implies that $(L - (S \cup T)) \subset \psi(C)$, for some $(S \cup T) \in I$ and hence $C \in *IGO(W, \upsilon)$.

Theorem 11 *Let C, D are subsets of an I-space (W, υ, I). If $\psi(C) \subset D \subset C$ and C is $*Ig$-o, then D is $*Ig$-o.*

Proof Since $\psi(C) \subset D \subset C$, we have $(W - C) \subset (W - D) \subset (W - \psi(C)) = (W - C)^*$ from definition of ψ operator. Besides since C is $*Ig$-o, then $(W - C)$ is $*Ig$-c. According to Theorem 4, we obtain that $(W - D)$ is $*Ig$-c and hence $D \in *IGO(W, \upsilon)$.

Theorem 12 *Let (W, υ, I) be an I-space and C be a subset of W. C is $*Ig$-c in (W, τ, I) iff $(C^* - C) \in *IGO(W, \upsilon)$.*

Proof Necessity. Let $(L - (C^* - C))$ and L is closed. Assume that C is $*Ig$-c. According to Theorem 1, $L \in I$. This implies that $(L - S) = \emptyset$, for some $S \in I$. Obvious that $(L - S) \subset \psi(C^* - C)$ and therefore $(C^* - C)$ is $*Ig$-o according to Theorem 6.

Sufficiency. Let $C \subset P \in \upsilon$. $(C^* \cap (W - P)) \subset (C^* \cap (W - C)) = (C^* - C) = \emptyset$, for some $S \in I$ are hold. By assumption since $(C^* - C)$ is $I * g$-o we have $[(C^* \cap (W - P)) - S] \subset \psi(C^* - C) = \emptyset$, for some $S \in I$. Therefore, we have $(C^* \cap (W - P)) \subset S$ and hence $(C^* - P) \in I$. Consequently, we obtain that C is $*Ig$-c.

4 Conservation of $*Ig$-Closed Sets

In this section, we have obtained under which conditions the image of $*Ig$-c set is also $*Ig$-c sets according to a function. So we have expressed when being $*Ig$-c is conserved.

Theorem 13 *Let $f : (W, \upsilon, I) \to (R, \varphi)$ be a continuous function and $f(C)^* \subset f(C^*)$ for every subset C of W. If C is $*Ig$-c in (W, υ, I), then $f(C)$ is $*f(I)g$-c in $(R, \varphi, f(I))$.*

Proof Let $C \subset W$ and C is $*Ig$-c set. Assume that $f(C) \subset N$ and $N \in \varphi$. Since f is continuous function, then we have $C \subset f^{-1}(N)$ and $f^{-1}(N) \in \tau$. By assumption since C is $*Ig$-c set, according to Definition 3 we have $(C^* - f^{-1}(N)) \in I$. Using Lemma 2, we have $f(C^* - f^{-1}(N)) \in f(I)$. By hypothesis since $f(C)^* \subset f(C^*)$ for every subset C of W, $(f(C)^* - N) \subset (f(C^*) - N) \subset (f(C^* - f^{-1}(N))$ and $(f(C)^* - N) \in f(I)$ from definition of ideal. This shows that hence $f(C)$ is $*f(I)g$-c.

One can obtain the next fact using Theorems 3 and 13.

Corollary 5 *Let $f : (W, \tau, I) \to (R, \varphi)$ be a continuous, surjection function and $f(C)^* \subset f(C^*)$ for every subset C of W. If $(C_\alpha)_{\alpha\in\Delta}$ be a family consisting of $*Ig$-c sets and Δ is finite set, then $(f(C_\alpha))_{\alpha\in\Delta}$ be a family consisting of $*f(I)g$-c. Besides $\cup(C_\alpha)_{\alpha\in\Delta}$ and $\cup(f(C)_\alpha)_{\alpha\in\Delta}$ is a $*Ig$-c and a $*f(I)g$-c, respectively.*

References

1. Abd El-Monsef, M.E., Lashien, E.F., Nasef, A.A.: On I−open sets and I−continuous functions. Kyungpook Math. J. **32**(1), 21–30 (1992)
2. Alimohammady, M., Roohi, M.: Fixed point in minimal spaces. Nonlinear Anal. Model. Control **10**, 305–314 (2005)
3. Dontchev, J.: On Hausdorff spaces via topological ideals and I-irresolute functions. Ann. N. Y. Acad. Sci. **767**, 28–38 (1995). Papers on General Topology and Applications
4. Dontchev, J., Ganster, M., Noiri, T.: Unified operation approach of generalized closed sets via topological ideals. Math. Japon. **49**, 395–401 (1999)
5. Hayashi, E.: Topologies defined by local properties. Math. Ann. **156**, 205–215 (1964)
6. Jafari, S., Rajesh, N.: Generalized closed sets with respect to an ideal. Eur. J. Pure Appl. Math. **4**(2), 147–152 (2011)
7. Jankovic, D., Hamlett, T.R.: New topologies from old ideals. Amer. Math. Mon. **97**, 295–310 (1990)
8. Jankovic, D., Hamlett, T.R.: Compatible extensions of ideals. Boll. Un. Mat. Ital. **7**(6-B), 453–465 (1992)
9. Keskin, A., Yüksel, S., Noiri, T.: On I-extremally disconnected spaces. Commun. Fac. Sci. Univ. Ank. Ser. A1 Math. Stat. **5**(1), 33–40 (2007)
10. Keskin Kaymakci, A.: g-I-closed sets. Conf. Proc. Sci. Technol. **5**(2), 241–245 (2022)
11. Kuratowski, K.: Topologies. Vol. I. Warszawa (1933)
12. Kuratowski, K.: Topology. Vol. I. Academic Press, New York (1966)

13. Levine, N.: Generalized closed sets in topological spaces. Rend. Circ. Mat. Palermo **19**, 89–96 (1970)
14. Mashhour, A.S., Allam, A.A., Mahmoud, F.S., Khedr, F.H.: On supra topological spaces. Indian J. Pure Appl. Math. **14**(4), 502–510 (1983)
15. Natkaniec, T.: On I-continuity and I-semicontinuity points. Math. Slovaca **36**(3), 297–312 (1986)
16. Samuels, P.: A topology formed from a given topological space. J. Lond. Math. Soc. **2**(10), 409–416 (1975)
17. Vaidyanathasvamy, R.: Set Topology. Chelsa Publishing Company, New York (1960)

Service-Learning Activity in a Statistics Course

Araceli Queiruga-Dios, Deolinda M. L. D. Rasteiro, Beatriz Sánchez Barbero, Ángel Martín-del Rey, and Ion Mierlus-Mazilu

Abstract Learning through service, a combination of community-based activities with learning objectives, or experiential education, are some features of service-learning (SL) methodology, which improves the realities where the service is performed, and which considers who receives the service as a central element. A SL activity was proposed to undergraduate sophomore students from Industrial Engineering degrees. Students developed a teamwork as part of their statistics course collaborating with a local entity. To make this possible, teachers identified a community need in line with the course contents and after that students were invited to participate. A teamwork was proposed to all students and one group decided to collaborate with an association that helps families in need. The work consisted of creating a database with the data gathered by volunteers and after that present a basic descriptive analysis. The built database and the analysis performed were grounded on the association needs to better help the families that it serves. This SL activity was assessed as part of the course activities.

1 Introduction

The current educational paradigm, and teaching and learning strategies are changing according to students, degrees and current situations. However teachers from all educational levels use to be under pressure due to time constraints and the need of

A. Queiruga-Dios (✉) · B. Sánchez Barbero · Á. Martín-del Rey
Institute of Fundamental Physics and Mathematics, Universidad de Salamanca, Salamanca, Spain
e-mail: queirugadios@usal.es; beatrizsanchezb@usal.es; delrey@usal.es

D. M. L. D. Rasteiro
Polytechnic University of Coimbra, Coimbra Engineering Institute, Coimbra, Portugal
e-mail: dml@isec.pt

I. Mierlus-Mazilu
Technical University of Civil Engineering Bucharest, Bucharest, Romania
e-mail: ion.mierlusmazilu@utcb.ro

V. Gayoso Martínez et al. (eds.), *Mathematical Methods for Engineering Applications*, Springer Proceedings in Mathematics & Statistics 439,
https://doi.org/10.1007/978-3-031-49218-1_23

complete the contents and courses syllabi. Most universities undergo these problems and they try to solve them and find suitable solutions to address them. In some cases they decided to reduce the content of the syllabus [17].

A service-learning proposal includes a set of activities that provide impact of curricular development, it could be part of a specific course, and it takes place within a community service framework [13]. Although SL is not a set of volunteer activities neither an internships, it is considered a experiential education that allow students engage in social, human and community activities and provides a different form of learning [15].

SL combines service with academic learning in such a way that students learn while they engage in society and collaborate with the community. The targets of the SL activities are the central element although SL is a course-based learning experience with defined learning objectives [11]. In fact, SL is included into the contents of courses curriculum, because community and social service is connected to the learning outcomes and the curricular objectives. In this educational approach students are motivated because they may apply their knowledge and acquired competencies to real-world problems and challenges [4, 10]. In that way, students participate in solving community problems and needs, and they are the protagonists of their own learning process [16].

To develop a service-learning project, the following stages can be considered (see Fig. 1):

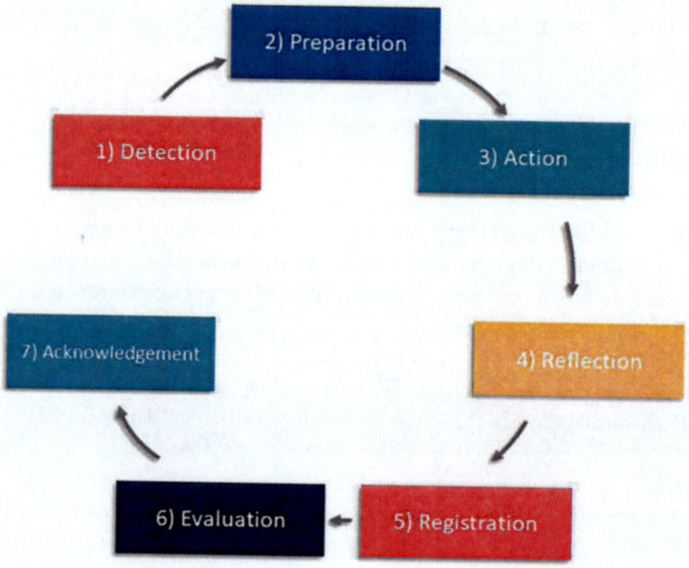

Fig. 1 Diagram of a service-learning project

1. Detection, research or asking the community about its needs. Community members, faculty and students share their ideas providing feedback to each other. A meeting with community members could be useful.
2. Preparation: includes different aspects such as looking for other partners or associations to address the need, to identify learning objectives related to the curriculum. The design of the activity is also part of this stage.
3. Action or implementation. Once the community need is detected is time to work on it. Sometimes, this phase is separated in preparation and development.
4. Reflection: After concluding the service, a reflection is needed to be sure that the need was covered. In some cases, a new project could be proposed to improve community life.
5. Registration: Evidences of the activity must be stored in any format or medium. This is useful to make known the activities.
6. Evaluation: As every learning process, an evaluation step is needed.
7. Acknowledge: Not all projects include this phase, but it is appropriate to recognize the work done and the benefit it brings to the community.

Universities play a key role in service-learning. First, campuses are plenty of students, teachers, spaces (classrooms), technology, research, etc., and, on the other hand, all of this could provide an important service (cultural, social, educational or economic development, among others) to society and the community [6].

This paper is distributed as follows: Sect. 2 includes specific concepts and some projects related to engineering. Section 3 is related to different possibilities of suing active methodologies as part of a mathematics lesson. After that, the specific case study related to the statistics course in detailed in Sect. 4, and finally, some conclusions are stated.

2 Service-Learning in Engineering Degrees

Service-learning methodology started in the USA. It originated to serve the community in a process of social evolution. The government implemented a SL program for K-12 and university students. Since 1950 several studies have been published related to SL proposals and projects. This pedagogy is a kind of experiential education that allow students to be involved in their own learning and make them reflect about their process of learning. Several proposals were implemented in American universities as colleges, some examples could me mentioned such as the EPICS (Engineering Projects in Community Service) at the Purdue University, or collaborations with Engineers Without Borders, Engineers for a Sustainable World, Engineering World Health, Engineering for Developing Communities, in this case integrating social needs into several engineering courses [20].

Several types of SL project are being implementing in engineering degrees. Some of them could be aimed to children (activities in the school), or outreach proposals, or projects related to sustainability or environmental care, etc. These are some examples of projects developed by engineering degree students: service learning project at a science museum [2], Filtrón drinking water filter [3], hazards, vulnerability, and risk projects [4], genetically modified crops and their use in agriculture for high school students [8], design of adapted devices for persons with disabilities [24], etc.

One of the key learning outcomes of a SL activity is the civil learning. With academic learning and personal growth, civil learning is the third pillar of service learning. Moreover, academic contents and competencies, community-engaged activities, and critical reflection provide civil learning to students [5].

3 Mathematics and Active Methodologies

Engineering and science degrees include a set of mathematics courses. Depending on the degree, mathematics courses duration is different. To propose a different teaching and learning methodology is difficult, because of time constraints and the need to cover all contents and competencies. Most of the teachers like to innovate and to change the way of teaching and the way how students learn. Although this is usual in primary and secondary education, it is not very widespread in higher education. However, several active methodologies are used in different degrees. Specifically, when paying attention to mathematics, problem or project-based learning [7, 19], game based learning [21], the use of specific software [12, 14], etc. are some of the methodologies that are used among science teachers.

Authors of this paper collaborate together to promote different methodologies among university students. As part of the Erasmus+ GIRLS project (Generation for Innovation, Resilience, Leadership and Sustainability. The game is on!), a team made up of eight partners participate in it. This partnership includes four universities (Universidad de Salamanca, Spain; Instituto Superior de Engenharia de Coimbra, Portugal; Universidad Francisco de Vitoria, Spain; and the Universidad Vasco de Quiroga, Mexico), the Spanish National Research Council (CSIC), a school from Salamanca (Spain) and a non-profit organization (AELCLÉS). The objectives of the GIRLS project are the following:

1. Train in digital skills and innovative pedagogies.
2. Bring the university closer to society.
3. Promote sustainable development in higher education.
4. Promote transformation and individual change in organizations, promoting improvements, new approaches and institutional changes.

Furthermore, the proposed project is focused on 4 methodologies: research-based learning, game-based learning, competency-based learning, and service-learning. These educational approaches are being developed in higher education, starting by partners institutions (https://girlsproject.eu/).

The case study detailed in this study was developed as part of a mathematics course that is part of some industrial engineering degrees, and its duration is around

60 on-site classes. In this case, the course is aimed to sophomore students and it is divided into two blocks: Numerical Methods and Statistics.

During the statistics block a team work is proposed to students. They can choose the topic of their work and one option is to participate in a service-learning project. The teacher with the help of the university's Social Affairs Service, contacted some non-governmental organizations to check different possibilities to help the community. There is no many options to learn while learning numerical methods or statistics.

A organization which is based in the same city as the engineering school need to manage a large amount of data. To develop any type of project, particularly a service-learning project, it must be organized, prepared and implement with the best quality. This will provide positive outcomes, increase commitment and impact learners [9].

In mathematics, service-learning may be considered as the bridge between the application of mathematics and its knowledge, i.e., as the applications connect the theory with the real world, this could be somehow and in some cases, considered as service-learning. Moreover, apart from research and teaching, university life is also made of service. These three elements make up the SL [13].

4 Case Study: Service-Learning in a Statistics Course

As as authors know, all undergraduate degree programs include a statistics course. This course could be a basic one or an advance course. In the case of this study, the statistics course belong to Industrial Engineering degrees, i.e., Mechanical, Electronic and Automatic, and Electricity.

This second-year statistics course does from descriptive Statistics to confidence intervals, hypothesis testing and inference, through other topics such as probability, discrete and continuous random variables, sampling and simulation. It is a short course that takes less than a semester, usually 30 hours. There is not too much time to go deeper into Statistics.

As part of this course, students participate int some practical classes with the computer, using MS excel or SPSS (this depend on the teacher) and the assessment process is separated in the following tasks:

- Some problems, which are solve by students and give back to the teacher to get her feedback.
- Practical exercises with a statistic software.
- Team work about a course topic.
- Some written exams about course topics.

As part of the team work, some students were invited to participate in a service-learning activity. In this case, as this was the firs time, the work was to solve a need from a local non-profit organization (NPO) that welcomes people coming from other countries, with several needs. This people received different help from the

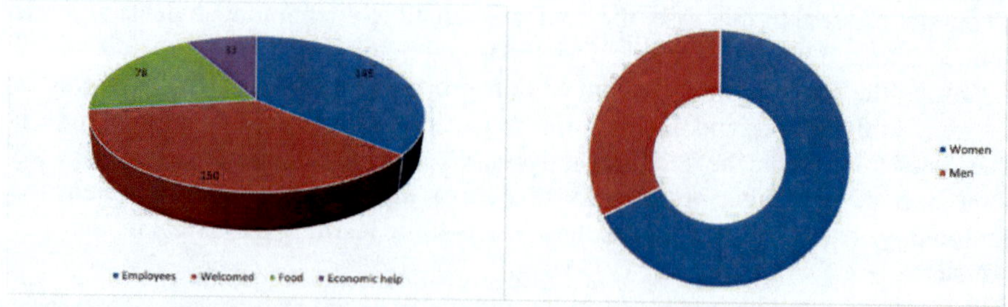

Fig. 2 Some results from the SL activity: Graphs with the help provided by the NPO and the distribution of women and men

association, such as food, money for paying bills (electricity, water, etc.), etc. This NPO had several excel files with information related to the people they help and they need to join them in a unique file, to get a database including all the data.

The NPO received the group of students that wanted to participate in the SL activity and showed them how they work and send them the files with data. These files contained the following information: name, surname, identity number, gender, birthdate, country of origin, members of the family unit, date of arrival to the NPO, results of the evaluation, and other information related to their social problems. Apart from this, several information related to the help was also included in the files: food, employment (if they have), economic help, and social, health, economic or unemployment benefits.

The group of students prepared a file with all the information, they developed a descriptive statistical study with the information with some graphs (see Fig. 2 as an example).

Surveys and several questionnaires are the more common procedures to assess SL projects. These techniques allow to get opinions from all people involved in the activities and they provide a lot of useful information to improve both, the service and the learning [20].

In the case study provided here, students presented their work to the teacher and she provided the information to the NPO. As there were not enough time to make a final presentation all together, the SL project team proceed as the rest of students, which developed a more "classical" work about statistics.

Students that participated in the SL project increased their self-confidence in mathematics and improved their mathematics competencies [22, 23]. Moreover, they improve almost all the 8 competencies established by the Mathematics working group of the European Society for Engineering Education (SEFI), i.e., (1) thinking mathematically, (2) posing and solving mathematical problems, (3) modeling mathematically, (4) reasoning mathematically, (5) representing mathematical entities, (6) handling mathematical symbols and formalism, (7) communicating in, with, and about mathematics, and (8) making use of aids and tools [1, 18].

5 Conclusions

This paper shows the case study of a service-learning project developed by undergraduate degree teachers and students from a second-year statistics course in engineering degrees. The proposed team work consisted in the improvement of the database of a nonprofit organization, as well to provide it with a basic descriptive statistic and the corresponding graphs. This work was developed using MS Excel, which was considered suitable for the expected results.

The service-learning project engaged students and made them be involved in mathematics classes. The use of this methodology motivate them and increased their interest.

Acknowledgments This research is part of the Erasmus+ project GIRLS—Generation for Innovation, Resilience, Leadership and Sustainability. The game is on! co-funded by the European Union.

References

1. Alpers, B.A., Demlova, M., Fant, C.H., Gustafsson, T., Lawson, D., Mustoe, L., Olsen-Lehtonen, B., Robinson, C., Velichova, D.: A framework for mathematics curricula in engineering education: a report of the mathematics working group (2013)
2. Ansari, F., Wang, J., Shelby, R., Patten, E., Pruitt, L.A.: A follow-up study of a first-year leadership and service learning module. In: 2013 ASEE Annual Conference & Exposition, pp. 23–42 (2013)
3. Bielefeldt, A.: Attracting women to engineering that serves developing communities. In: 2006 Annual Conference & Exposition, pp. 11–261 (2006)
4. Brand, B.D., Brascia, K., Sass, M.: The community outreach model of service-learning: a case study of active learning and service-learning in a natural hazards, vulnerability, and risk class. High. Learn. Res. Commun. (2019). https://doi.org/10.18870/hlrc.v9i2.452
5. Bringle, R.G., Clayton, P.H.: Civic learning: A sine qua non of service learning. In: Frontiers in Education, vol. 6, p. 606443. Frontiers Media SA (2021)
6. Bringle, R.G., Hatcher, J.A.: Implementing service learning in higher education. J. High. Educ. **67**(2), 221–239 (1996)
7. Caridade, C.M., Encinas, A.H., Martín-Vaquero, J., Queiruga-Dios, A., Rasteiro, D.M.: Project-based teaching in calculus courses: Estimation of the surface and perimeter of the Iberian Peninsula. Comput. Appl. Eng. Educ. **26**(5), 1350–1361 (2018)
8. Chrispeels, H.E., Chapman, J.M., Gibson, C.L., Muday, G.K.: Peer teaching increases knowledge and changes perceptions about genetically modified crops in non-science major undergraduates. CBE—Life Sci. Educ. **18**(2), ar14 (2019)
9. Conner, J., Erickson, J.: When does service-learning work? Contact theory and service-learning courses in higher education. Michigan J. Community Serv. Learn. **23**(2), 53–65 (2017)
10. Cooper, L.A., Kotys-Schwartz, D.A.: Designing the design experience-identifying factors of student motivation in project-based learning and project-based service-learning. In: 2013 ASEE Annual Conference & Exposition, pp. 23–382 (2013)
11. Derreth, R.T., Wear, M.P.: Critical online service-learning pedagogy: justice in science education. J. Microbiol. Biol. Educ. **22**(1), ev22i1–2537 (2021). https://doi.org/10.1128/jmbe.v22i1.2537

12. Gocheva-Ilieva, S., Kulina, H., Ivanov, A.: Assessment of students' achievements and competencies in mathematics using CART and CART Ensembles and Bagging with combined model improvement by MARS. Mathematics **9**(1), 62 (2020)
13. Hadlock, C.R.: Mathematics in Service to the Community: Concepts and Models for Service-learning in the Mathematical Sciences, vol. 66. Mathematical Association of America (2005)
14. Izquierdo, A.A., León, M.Á.G., Martín-Vaquero, J., Rasteiro, D.M.D., Kováčová, M., Richtáriková, D., Rodríguez-Gonzálvez, P., Rodríguez-Martín, M., Queiruga-Dios, A.: Specific mathematical software to solve some problems. In: Calculus for Engineering Students, pp. 327–347. Elsevier (2020)
15. Jacoby, B., et al.: Service-Learning in Higher Education: Concepts and Practices. The Jossey-Bass Higher and Adult Education Series. ERIC (1996)
16. Lorenzo Moledo, M.d.M., Sáez-Gambín, D., Ferraces Otero, M.J., Varela Portela, C.: Reflection and quality assessment in service-learning projects. when, with whom, and why. In: Frontiers in Education, vol. 5, p. 605099. Frontiers Media SA (2021). https://doi.org/10.3389/feduc.2020.605099
17. Mustoe, L.: Mathematics in engineering education. Eur. J. Eng. Educ. **27**(3), 237–240 (2002)
18. Queiruga-Dios, A., Sánchez, M.J.S., Pérez, J.J.B., Martín-Vaquero, J., Encinas, A.H., Gocheva-Ilieva, S., Demlova, M., Rasteiro, D.D., Caridade, C., Gayoso-Martínez, V.: Evaluating engineering competencies: A new paradigm. In: 2018 IEEE Global Engineering Education Conference (EDUCON), pp. 2052–2055. IEEE (2018)
19. Queiruga-Dios, M.Á., Sáiz-Manzanares, M.C., Montero-García, E.: Adaptive and creative problem-projects in the teaching of science. Description of the methodology and appreciation of the students involved. Res. Educ. Learn. Innov. Arch. (23), 1–23 (2019)
20. Queiruga-Dios, M., Santos Sánchez, M.J., Queiruga-Dios, M.Á., Acosta Castellanos, P.M., Queiruga-Dios, A.: Assessment methods for service-learning projects in engineering in higher education: A systematic review. Front. Psychol. **12**, 629231 (2021). https://doi.org/10.3389/fpsyg.2021.629231
21. Queiruga-Dios, M., Bullón Pérez, J.J., Queiruga-Dios, A.: An escape game to find the owner of a latent fingerprint while learning biometry. In: Breakthroughs in Digital Biometrics and Forensics, pp. 327–344. Springer (2022)
22. Rasteiro, D.M., Caridade, C.M.: Probability and statistical methods: Assessing knowledge and competencies (pre and during covid19 pandemic)–case study at ISEC. The 20th SEFI Special Interest Group in Mathematics-SIG in Mathematics, p. 6 (2020)
23. Rasteiro, D.M., Caridade, C.M.: Statistical methods–what is the problem here? In: International Joint Conference: 12th International Conference on Computational Intelligence in Security for Information Systems (CISIS 2019) and 10th International Conference on EUropean Transnational Education (ICEUTE 2019) Seville, Spain, May 13th–15th, 2019 Proceedings 12, pp. 252–262. Springer (2020)
24. Self, B.P., Widmann, J.M., Slivovsky, L.A., Taylor, J.K., Hey, D.W.: Adapted physical activity design projects: A collaboration between kinesiology and engineering. In: 2013 ASEE Annual Conference & Exposition, pp. 23–139 (2013)

Fermatean Fuzzy Type a Three-Way Correlation Coefficients

Murat Kirişci

Abstract In challenging decision-making circumstances, tools like aggregation operators and information measures are routinely used. Using correlation coefficient (KK)'s estimate of the amount of reliance between two sets is another way to decide which choice is better. There are a few methods for figuring out the KK of Fermatean fuzzy sets (FFSs). These previous methodologies, which are specified between [0, 1], simply evaluate the strength of the link between FFSs. This study, using the notions of variance and covariance, respectively, provides a three-way method for computing the KKs between FFSs. The strength of the association between the considered FFSs is demonstrated by this novel approach, which is defined inside the interval $[-1, 1]$ similar to standard statistics, and it identifies whether the FFSs are positively or negatively associated. The potential of inaccuracy owing to information leakage is reasonably mitigated by the suggested technique's inclusion of the three traditional FFSs parameters. To demonstrate the new technique's usefulness as an accurate information measure, several theoretical results will be used to validate it. The advantages of the new methods over comparable methods will be demonstrated using a few numerical examples. With the help of the novel approach, several decision-making issues involving pattern recognition and diagnostic medicine will be handled. Multi-attribute decision-making issues can be resolved using the three-way technique for determining the correlation coefficient between FFSs.

M. Kirişci (✉)
Istanbul University-Cerrahpaşa, Istanbul, Turkey
e-mail: murat.kirisci@iuc.edu.tr, https://avesis.iuc.edu.tr/murat.kirisci

V. Gayoso Martínez et al. (eds.), *Mathematical Methods for Engineering Applications*, Springer Proceedings in Mathematics & Statistics 439,
https://doi.org/10.1007/978-3-031-49218-1_24

1 Introduction

1.1 Research Motivation

These days, due to the expanding and growing challenges, it is getting harder and harder to select the best solution from a variety of plausible ones during the decision-making (DM) process. In the context of current DM challenges, numerous professionals have rated objects like crisp and interval in various ways, which may make decisions tough. Researchers regularly use the theory of Fuzzy set (FS) [35] and its extensions (intuitionistic FSs (IFSs), Pythagorean FSs (PFSs), and FFSs), to resolve data uncertainties. To study the numerous applications for these sets, researchers have developed aggregation operators and information measurements. The importance of information measures in fusing the recipients' various preferences together has been extensively researched. One of these numerous concepts is determining the optimal alternative by using correlation measures.

Many applications are used in the literature research to deal with DM issues. The KKs, which are used to determine the degree of dependency between two sets, are one way for selecting the best alternative. A correlation is any statistical association (causal or not) between two random variables or bivariate data in statistics. A precise metric known as the KK is employed in correlation studies to demonstrate the strength of the linear relationship between two variables. The formula shows how well the relationship between the variables might fit to an example line formed across the data for two variables by measuring the distance of each data point from the variable mean.

A common task in statistics is determining the connection between two parameters or variables. Pearson's KK has been used in DM, pattern recognition, clustering, medical diagnosis, and data analysis and classification research. Traditional correlation has been shown to be insensitive to ambiguous data. Quantifying human perception and cognitive uncertainty is the focus of fuzzy logic. Binary logic is frequently used by researchers to analyze data. Human reasoning is complex and flawed, therefore applying binary logic to processes in the human brain can produce some distortion. The foundation of fuzzy logic is human cognition. As an example, fuzzy logic is "the modeling of thinking and decision mechanisms that enable humans to make consistent and correct judgments in the face of extensive and imprecise information." Mathematical statistics, fuzzy logic, and statistical KKs were used to expand the fuzzy-type KKs. The KK produced for fuzzy data displays both the intensity of the relationship between FSs and whether those FSs are positively or negatively associated.

Our study enhances the Fermatean fuzzy number (FFN)-based KK to address DM issues because FFSs have a very high capacity to replicate uncertain and imprecise information in real-world implementations.

1.2 Literature

The idea of an FS proposed by Zadeh [35] was used to highlight the ambiguity and irrationality of a membership degree (MD). Then, Atanassov [5] finds the IFS, which, by associating the non-membership degree (ND) of an component with an item, may articulate evaluation information more thoroughly. IFSs, however, have a design that makes it challenging for judgement specialists to make the right assessments because of their considerable limitations in providing preference information. IFS specifies the ND in addition to the MD. MD and ND are both in the [0, 1] range, according to IFS theory.

Any DM process must account for imprecision. To deal with the ambiguous environment of collective DM, many tools and strategies have been proposed. FFSs [28] are one of the newest techniques for coping with uncertainty. Compared to the IFS [5] and PFS [32, 33], which are extensions of FS [35], these sets offer a larger range of applications. In [12], a new extension of the fuzzy sets, entitled q-rung orthopair fuzzy set (q-ROFS), which is an extension of the crisp set, intuitionistic fuzzy set, and the Pythagorean fuzzy sets, is introduced. q-ROFS is PFS for $q = 1,2,3$ q-ROFS is IFS, PFS, and FFS, respectively. Recently, FFSs have inspired many studies [1–4, 13, 18–20, 22, 26, 27, 29].

In challenging DM circumstances, instruments such as aggregation operators and information measures are all the time used. The KKs evaluation of the level of dependency between two groups is one such instrument for determining the best selection. Utilizing KKs, one may define how strongly two variables are related. Several researchers have composed the KKs in fuzzy contexts since information is typically unclear, ambiguous, and partial in various situations. Chiang and Lin [8] have presented a solution for KK based on FSs while also fulfilling the correlation for fuzzy information in terms of conventional statistics. By Liu and Kao [24], the KK of fuzzy information has been investigated using a mathematical programming approximation in line with the conventional notion of KKs. The correlation of interval-valued IFSs (IVIFSs) in finite universal sets was studied in [7]. The correlation of IVIFSs in probability spaces was investigated by Hong [14]. The IFS was represented geometrically by Szmidt and Kacprzyk [30], who also proposed three parameters to characterize the space between IFSs. Based on this type of geometrical background, Zeng and Li [36] considered all three parameters describing IFSs and proposed a novel approach to calculating the correlation and KK between IFSs (KKIFSs), which are comparable to the cosine of the intersectional angle in the finite set and probability space, respectively. KKIFSs were defined by Mitchell [25]. Mitchell demonstrated how to obtain a straightforward, intuitively appealing KK between two IFSs by viewing an IFS as an ensemble of regular FSs. In [31], several operational laws of IVIF numbers, correlation, and KK of IVIFs are introduced with regard to multi-criteria decision-making (MCDM) problems with IVIF information. As a result of PFS, many DM difficulties using PF information have been identified in the literature [15–17, 21, 32–34, 37], which was created to address an IFS problem. Garg [11] pointed out the shortcomings in the existing KKIFSs and proposed a new KK and weighted KK formulation to assess the relationship between two PFSs. Using the KK and entropy measure for the linguistic PFSs (LPFS), Lin et al. [23] provide a novel TOPSIS technique. The correlation coefficient is recommended as a way to gauge how closely LPFSs are related. Ejegwa et al. [10] have presented a three-way approach for KK between PFSs (KKPFSs) by using some statistical concepts.

Recently, new correlation coefficients with FFSs have been defined. Bhatia et al. [6] presented new KKs using defined statistical concepts with FFSs. Further, the weighted KKs also obtained with these KKs. Demir [9] defined four types of KKs for Fermatean hesitant FSs (FHFS). Nevertheless, obtained new KKs have been generalized to interval-valued Fermatean hesitant fuzzy sets (IVFHFSs). The first study on KKs based on FFSs was defined by Kirişci [18] and their fundamental characteristics were studied. This first study was expanded in [19] by redefining it using statistical concepts. Pearson KKs were in turn defined in this study as related to FFSs. In [20], for obtained KKs wit FHFS and IVFHFS, during operations with new KKs, the least common multiple expansion method has been offered to solve the problem when the cardinality of FHF elements and IVFHF elements are different.

1.3 Necessity

The inability of IFS and PFS-based techniques to capture data in FFSs format should go without saying. Additionally, it goes without saying that when experts offer preference values in the form of FFSs, DM methods based on IFS and PFS will fail. However, FFS-based approaches, such as the one presented here, are capable of quickly evaluating data to rank workable alternatives in accordance with criterion values. The FFS ecosystem, which unifies IFSs and PFSs into a single platform, is thought to be able to handle more information and cover a number of themes for processing uncertain data. As a result, there is minimal information loss in this collection.

It is believed that the proposed KKs are a special instance of the existing KKs under the IFS and PFS. Additionally, compared to the current ones, the suggested KKs handle more information.

The provided method contains far more information than the current methodologies utilized in the IFS and PFS contexts for dealing with data uncertainty. It provides more precise and accurate information on the things. As a result, it is an effective instrument for dealing with cloudy and confusing data throughout the DM process.

In these investigations, a new approach to computing defined KK with the help of FFSs based on a three-way statistical method is presented. As can be seen from the outputs of this work, the three-way correlation coefficients obtained using FFSs have a structure that will facilitate the solution of MCDM problems.

1.4 Contribution

This study has made the following contributions:

1. Today's fuzzy or non-standard fuzzy theory commonly uses KKs with values between 0 and 1, which only indicate the intensity of the relationship. A KK with a value in $[-1, 1]$ can be used to easily correlate uncertain notions.
2. For their comparison metrics, the majority of studies on fuzzy and non-standard FSs rely on faked data. The present correlation coefficients based on IF and PF do not meet all or part of these criteria. We therefore suggest some new correlation coefficients for FFSs in this study that are higher than the current correlation coefficients while taking these factors into account.
3. We propose four new correlation coefficients and discuss some of their appealing properties for FFSs with values in $[-1, 1]$.
4. Applications for data mining and medical diagnostics are examined using our proposed FF-correlation coefficients. The suggested KKFFSs are also contrasted with different compatibility metrics that already exist under FF conditions.

2 Preliminaries

U will be used throughout the article as the initial universal set.

For $\alpha_P : U \to [0, 1]$ and $b\beta_P : U \to [0, 1]$, the FFS P is shown by $P = \{(u, \alpha_P(u), \beta_P(u)) : u \in U\}$, where the inequality $0 \le \alpha_P^3(u) + \beta_P^3(u) \le 1$ [28] is valid.

It is defined as $\gamma_P(u) = \sqrt[3]{1 - (\alpha_P^3(u) + \beta_P^3(u))}$ degree of indeterminacy of u to P.

Take three FFSs $P = \{\alpha_P, \beta_P\}$, $P_1 = \{\alpha_{P_1}, \beta_{P_1}\}$ and $P_2 = \{\alpha_{P_2}, \beta_{P_2}\}$. Then, some operations as follows [28]:

(i) $P_1 \cap P_2 = \left(\min\{\alpha_{P_1}, \alpha_{P_2}\}, \max\{\beta_{P_1}, \beta_{P_2}\}\right)$;
(ii) $P_1 \cup P_2 = \left(\max(\alpha_{P_1}, \alpha_{P_2}), \min(\beta_{P_1}, \beta_{P_2})\right)$;
(iii) $P^t = (\beta_P, \alpha_P)$;
(iv) $P_1 \boxplus P_2 = \left(\sqrt[3]{\alpha_{P_1}^3 + \alpha_{P_2}^3 - \alpha_{P_1}^3\alpha_{P_2}^3}, \beta_{P_1}\beta_{P_2}\right)$;
(v) $P_1 \boxtimes P_2 = \left(\alpha_{P_1}^3\alpha_{P_2}^3, \sqrt[3]{\beta_{P_1}^3 + \beta_{P_2}^3 - \beta_{P_1}^3\beta_{P_2}^3}\right)$;
(vi) $\alpha P = \left(\sqrt[3]{1 - (1 - \alpha_P^3)^\lambda}, \alpha_P^\alpha\right), \quad \lambda > 0$;
(vii) $P^\lambda = \left(\alpha_{P_1}^3, \sqrt[3]{1 - (1 - \beta_P^3)^\lambda}\right), \quad \lambda > 0$.

The robust KKs based on PFSs [10] are given as:

Definition 1 For two PFS P, Q in $\mathcal{U}$ and the weight ω,

$$RPFS(P,Q)=\frac{K(P,Q)}{\sqrt{VR(P)VR(Q)}} \tag{1}$$

$$RPFS_{\omega}(P,Q)=\frac{K_{\omega}(P,Q)}{\sqrt{VR_{\omega}(P)VR_{\omega}(Q)}} \tag{2}$$

The KKs defined by Kirişci [18] based on FFSs are defined as follows:

Definition 2 Choose two FFSs P and Q. Then

$$\mathfrak{C}(P,Q)=\frac{C(P,Q)}{[I(P).I(Q)]^{1/2}} \tag{3}$$

$$=\frac{\sum_{i=1}^{n}\left(\alpha_P^3(x_i)\alpha_Q^3(x_i)+\beta_P^3(x_i)\beta_Q^3(x_i)+\gamma_P^3(x_i)\gamma_Q^3(x_i)\right)}{\sqrt{\sum_{i=1}^{n}\left(\alpha_P^6(x_i)+\beta_P^6(x_i)+\gamma_P^6(x_i)\right)}.\sqrt{\sum_{i=1}^{n}\left(\alpha_Q^6(x_i)+\beta_Q^6(x_i)+\gamma_Q^6(x_i)\right)}}$$

$$\mathfrak{D}(P,Q)=\frac{C(P,Q)}{\max\,[I(P).I(Q)]} \tag{4}$$

$$\mathfrak{C}_{\omega}(P,Q)=\frac{C_{\omega}(P,Q)}{[I_{\omega}(P).I_{\omega}(Q)]^{1/2}} \tag{5}$$

$$\mathfrak{D}_{\omega}(P,Q)=\frac{C\omega(P,Q)}{\max\,[I\omega(P).I\omega(Q)]}. \tag{6}$$

are called the KKs between P, Q.

The Pearson Correlation Coefficients defined in [19] are:

Definition 3 For two FFSs P and Q, the Pearson correlation coefficients are:

$$\mathfrak{C}_{pearson}(P,Q)=\frac{1}{2}(Z_1+Z_2) \tag{7}$$

$$\mathfrak{C}_{pearson}(P,Q)=\frac{1}{2}(Z_1+Z_2+Z_3) \tag{8}$$

where

$$Z_1=\frac{\sum_{k=1}^{m}\left\{[\alpha_P^3(u_k)-\overline{\alpha_P}^3]\times[\alpha_Q^3(u_k)-\overline{\alpha_Q}^3]\right\}}{\sqrt{[\alpha_P^3(u_k)-\overline{\alpha_P}^3]}\sqrt{[\alpha_Q^3(u_k)-\overline{\alpha_Q}^3]}}$$

$$Z_2=\frac{\sum_{k=1}^{m}\left\{[\beta_P^3(u_k)-\overline{\beta_P}^3]\times[\beta_Q^3(u_k)-\overline{\beta_Q}^3]\right\}}{\sqrt{[\beta_P^3(u_k)-\overline{\beta_P}^3]}\sqrt{[\beta_Q^3(u_k)-\overline{\beta_Q}^3]}}$$

$$Z_3 = \frac{\sum_{k=1}^{m}\left\{[\gamma_P^3(u_k) - \overline{\gamma_P}^3] \times [\gamma_Q^3(u_k) - \overline{\gamma_Q}^3]\right\}}{\sqrt{[\gamma_P^3(u_k) - \overline{\gamma_P}^3]}\sqrt{[\gamma_Q^3(u_k) - \overline{\gamma_Q}^3]}}$$

3 New Approaches

Definition 4 For two FFSs P, Q and $n > \infty$, the mean values, deviations, variance, and covariance of P, Q defined by

$$\begin{cases} Ort(P) = (\overline{\alpha_P}, \overline{\beta_P}, \overline{\gamma_P}) = \left(\frac{1}{n}\sum_{i=1}^{n}\alpha_P(s_i), \frac{1}{n}\sum_{i=1}^{n}\beta_P(s_i), \frac{1}{n}\sum_{i=1}^{n}\gamma_P(s_i)\right) \\ Ort(Q) = (\overline{\alpha_Q}, \overline{\beta_Q}, \overline{\gamma_Q}) = \left(\frac{1}{n}\sum_{i=1}^{n}\alpha_Q(s_i), \frac{1}{n}\sum_{i=1}^{n}\beta_Q(s_i), \frac{1}{n}\sum_{i=1}^{n}\gamma_Q(s_i)\right), \end{cases} \quad (9)$$

$$\begin{cases} SPM(P) = (\alpha_P^3(s_i) - \overline{\alpha_P}^3) - \left(\beta_P^3(s_i) - \overline{\beta_P}^3\right) - (\gamma_P^3(s_i) - \overline{\gamma_P}^3) \\ SPM(Q) = \left(\alpha_Q^3(s_i) - \overline{\alpha_Q}^3\right) - \left(\beta_Q^3(s_i) - \overline{\beta_Q}^3\right) - \left(\gamma_Q^3(s_i) - \overline{\gamma_Q}^3\right) \end{cases} \quad (10)$$

$$\begin{cases} VR(P) = \frac{1}{n-1}\sum_{i=1}^{n} SPM_i^2(P) \\ VR(Q) = \frac{1}{n-1}\sum_{i=1}^{n} SPM_i^2(Q) \end{cases} \quad (11)$$

$$K(P, Q) = \frac{1}{n-1}\sum_{i=1}^{n} SPM_i(P)SPM_i(Q) \quad (12)$$

Definition 5 For two FFSs P, Q in $\mathcal{U}$, then, using 9, 10, 11, 12,

$$R(P, Q) = \frac{K(P, Q)}{\sqrt{VR(P)VR(Q)}} \quad (13)$$

is said to be the robust KKFFSs.

Proposition 1 *For two FFSs P, Q in $\mathcal{U}$,*

(i) $VR(P) = VR(\overline{P})$,
(ii) $K(P, Q) = K(\overline{P}, \overline{Q})$

Proposition 2 *Let P, Q be two FFSs in $\mathcal{U}$. If $P = Q$, then $K(P, Q) = VR(P) = VR(Q)$.*

Proof Using the definition of covariance, $K(P,Q) = \frac{1}{n-1}\sum_{i=1}^{n} SPM_i(P)$ $SPM_i(Q)$. Assume $P = Q$, then $K(P,P) = \frac{1}{n-1}\sum_{i=1}^{n} SPM_i^2(P) = VR(P)$ and so $K(P,P) = VR(Q)$.

According to the conclusion of this Proposition, Eq. 13 can be written as:

$$R(P,Q) = \frac{K(P,Q)}{\sqrt{K(P,P)K(Q,Q)}}. \tag{14}$$

Proposition 3 *Choose two FFSs P, Q in $\mathcal{U}$. Then $R(P,Q) = R(Q,P)$.*

Proof

$$\begin{aligned} K(P,Q) &= \frac{1}{n-1}\sum_{i=1}^{n} SPM_i(P)SPM_i(Q) \\ &= \frac{1}{n-1}\sum_{i=1}^{n} SPM_i(Q)SPM_i(P) = K(Q,P). \end{aligned}$$

Theorem 1 *For two FFSs P, Q in $\mathcal{U}$, $R(P,Q)$ is a KKFFSs.*

These KKs can be defined in a variety of ways. These new definitions will be based on weight. Because, in many real-world situations, different sets can have different weights. These KKs can be defined in a variety of ways. These new definitions will be based on weight. Because, in many real-world situations, different sets can have different weights, As a result, in new definitions, the weight ω_i of each element $u_i \in \mathcal{U}$ must be considered. Weighted KKs are the KKs that will be specified by weights.

$$R_\omega(P,Q) = \frac{K_\omega(P,Q)}{\sqrt{VR_\omega(P)VR_\omega(Q)}} \tag{15}$$

where

$$K_\omega(P,Q) = \frac{1}{n-1}\sum_{i=1}^{n} \omega_i SPM_i(P)SPM_i(Q) \tag{16}$$

$$VR_\omega(P) = \frac{1}{n-1}\sum_{i=1}^{n} \omega_i SPM_i^2(P), \tag{17}$$

$$VR_\omega(Q) = \frac{1}{n-1}\sum_{i=1}^{n} \omega_i SPM_i^2(Q).$$

Proposition 4 *For the weight vector ω, if P, Q are two FFSs in $\mathcal{U}$ and $P = Q$, $K_\omega(P, Q) = VR_\omega(P) = VR_\omega(Q)$.*

Proposition 5 *For two FFSs P, Q in $\mathcal{U}$ and the weight vector ω,*

(i) $VR_\omega(P) = VR_\omega(\overline{P})$,
(ii) $K_\omega(P, Q) = K_\omega(\overline{P}, \overline{Q})$.

Proposition 6 *For two FFSs P, Q in $\mathcal{U}$ and the weight vector ω, $K_\omega(P, Q) = K_\omega(Q, P)$.*

Theorem 2 *For two FFSs P, Q in $\mathcal{U}$ and the weight vector ω, $R_\omega(P, Q)$ is a KKFFSs.*

Example 1 For $\mathcal{U} = \{u_1, u_2, u_3\}$, FFSs P and Q are

$$P = \{(u_1, 0.6, 0.7), (u_2, 0.8, 0.5), (u_3, 0.9, 0.4)\},$$
$$Q = \{(u_1, 0.9, 0.5), (u_2, 0.8, 0.6), (u_3, 0.7, 0.8)\}.$$

and the weight of $\mathcal{U}$ is $\omega = (0.4, 0.35, 0.25)$.

The mean values: $Ort(P) =$, $Ort(Q) =$ and for ω $Ort_\omega(P) =$, $Ort_\omega(Q) =$.

The variances: $VR(P) =$, $VR(Q) =$ and $VR_\omega(P) =$, $VR_\omega(Q) =$ with weights.

The covariances: $K(P, Q) =$ and $K_\omega(P, Q) =$.

The robust KKFFSs: $R(P, Q) =$ and $R_\omega(P, Q) =$.

4 Applications

4.1 Algorithm

Algorithm 1 The algorithm for the robust KKFFS

Input: The values of the FFSs.
Output: Results of correlation coefficients.
Begin
 1. Calculate mean values using Equations 9 via the values of the FFSs P and Q.
 2. Compute the deviations using Equations 10 via the values of the FFSs P and Q.
 For $i = 1; n$ **do**
  3. Find the variance and covariance using Equations 11 and 12.
  4. Find the variance and covariance with weights using Equations 16 and 17.
 End for
5. Calculate the robust KKFs using Equation 13.
6. Calculate the robust KKFs with weights using Equation 15.
7. Evaluate the values in Step 5 and Step 6 and choose the largest value.
End

4.2 Examples

Example 2 Application of Data Mining: In this example, newly proposed KKF-SSs will be applied to pattern recognition.

When we compare new information with information we have stored in memory, a cognitive process called pattern recognition takes place in our brains. When a mother teaches her child to count, she might say, "One, two, three." When the mother says "one, two...", the child may say "three" after several repetitions. As a result, the child recognizes the pattern.

Pattern-recognition receptors are cells in our bodies that have a specific purpose to identify and combat pathogenic and damage-related chemical patterns. But this is biology, not technology.

The core of the pattern recognition process is the classification of events based on statistical data, historical data, or machine memory. That is, pattern recognition is a procedure that examines available data to see if it has any order to it.

Consider the known patterns $F_i = \{(u_k, \alpha_{F_i}(u_k), \beta_{F_i}(u_k)) : u_k \in \mathcal{U}\}$ and the unknown pattern $G = \{(u_k, \alpha_G(u_k), \beta_G(u_k)) : u_k \in \mathcal{U}\}$ for $i = 1, 2, \cdots, m$ and $k = 1, 2, \cdots, l$. In this process, the unknown model G will be classified with one of the known models F_i.

Take the three known patterns F_1, F_2, F_3 and $U = \{u_1, u_2, u_3\}$.

$$F_1 = \{(u_1, 0.46, 0.68), (u_2, 0.55, 0.73), (u_3, 0.38, 0.84)\},$$

$$F_2 = \{(u_1, 0.81, 0.34), (u_2, 0.24, 0.91), (u_3, 0.27, 0.79)\},$$

$$F_3 = \{(u_1, 0.18, 0.85), (u_2, 0.43, 0.67), (u_3, 0.54, 0.39)\},$$

and the unknown pattern G is:

$$G = \{(u_1, 0.67, 0.33), (u_2, 0.39, 0.44), (u_3, 0.51, 0.55)\}.$$

Now, we computed the KK between G and F_i and obtain the result: F_2.

Example 3 Application of Medical Diagnosis: Kirişci ve Simsek [21] example of infectious illness was modified to illustrate to illustrate how the recommended strategy in MCDM may be applied. The illness state of the patients will be determined by taking into account their symptoms. The ailment from which the patient suffers the most will be identified based on the findings.

The diseases are: A_1 =Influenza(H1N1), A_2 =Hepatitus C, A_3 =sandly fever, A_4 =Crimean Hemorrhagic Fever, and A_5 =norovirus. The symptoms are: s_1 =temperature, s_2 =chest pain, s_3 =headache, s_4 =cough, s_5 =stomach pain. The weights of symptoms: $\omega = (0.3, 0.25, 0.1, 0.25, 0.1)$. The relationship between diseases and symptoms is given in Table 1. Further, the clinical findings of a patient whose disease is to be diagnosed are represented in FF values as: $A = \{(s_1, 0.64, 0.77), (s_2, 0.55, 0.83), (s_3, 0.26, 0.90), (s_4, 0.61, 0.73), (s_5, 0.92, 0.35)\}$.

Table 1 Diseases-Symptoms

	s_1	s_2	s_3	s_4	s_5
A_1	(0.1, 0.9)	(0.2, 0.9)	(0.8, 0.5)	(0.4, 0.5)	(0.9, 0.2)
A_2	(0.2, 0.7)	(0.7, 0.6)	(0.7, 0.4)	(0.8, 0.4)	(0.9, 0.1)
A_3	(0.4, 0.6)	(0.9, 0.2)	(0.1, 0.7)	(0.7, 0.5)	(0.8, 0.4)
A_4	(0.5, 0.7)	(0.2, 0.7)	(0.6, 0.6)	(0.8, 0.3)	(0.9, 0.1)
A_5	(0.3, 0.7)	(0.2, 0.8)	(0.8, 0.5)	(0.9, 0.1)	(0.4, 0.6)

According to the values given in this table, robust KKFFSs are obtained as follows: $R_\omega(A_1, A) = 0.346$, $R_\omega(A_2, A) = 0.054$, $R_\omega(A_3, A) = -0.4856$, $R_\omega(A_4, A) = 0.001$, $R_\omega(A_5, A) = 0.120$. These results indicate that the patient suffered from A_1 disease.

5 Discussion

5.1 Comparison

We applied the pattern recognition and medical DM practices we used for the presented technique for the KKs given in the studies [10, 18, 19]. Comparative analysis of all methods can be seen in Table 2.

5.2 Implications

The FS, IFS, and PFS comprise the classical set, which is generalized by the FFS. The MD and ND satisfaction of the specific option with respect to the requirements, where their total squares are equal to or less than 1, distinguishes one of the most popular extensions, PFS. In some circumstances, the decision-maker may indicate the degree of membership and nonmembership of a specific feature in such a way that the sum of the squares is larger than 1. As a result, the PFS handled this

Table 2 Comparison

Methods	Pattern recognition outputs	Medical DM outputs
Our method	F_2	A_1
KKPFS [10]	F_2	A_5
KKFFS(1) [18]	F_1	A_1
KKFFS(2) [6]	F_1	A_2
KKFFS(3) [19]	F_2	A_1
KKFFS(4) [20]	F_2	A_1

case wrongly. One of the more thorough solutions to this issue is the FFS theory, which can deal with incomplete information as well as unclear and inconsistent information, both of which are frequently present in real-world situations. As a result, Fermatean fuzzy information DM is better suited for real-world applications in science and engineering.

It's difficult and complex to make decisions about real-world issues or come up with answers to them. Therefore, in order to choose the best option or alternatives, it is imperative to lessen uncertainty while making a selection. The connections between the inputs must also be handled effectively for DM to be most beneficial. We put out a fresh approach that combines the benefits of FFS and KK to demonstrate the presence of the association between the variables.

A number of academics have developed an approach that uses the KK for PFSs, as seen from the previous studies. As was already established, some scenarios cannot be captured by PFSs, hence the related algorithm may not produce the desired outcomes.

A specific case of the KK for FFSs is the KK for PFSs. The suggested KK is, therefore, more applicable to a wider range of situations than the current ones, making it better suited to real-world problem-solving. While our new method has yielded very good results, good results have also been obtained with KKs based on predefined FFSs. As can be understood from the new method, the work done with FFSs gives better results.

6 Conclusion

Let's start by discussing the advantages of the suggested technique and how it differs from others. As is well known, FFSs are more able than IFs to analyze situations where there is uncertainty and insufficient information. The sets of Pythagorean and intuitionistic MDs are smaller than the sets of FFSs [28], therefore it stands to reason that FFSs will have many more possibilities than IF and PF for locating and resolving doubt.

There are KKs obtained using IFs and PFs, as well as methods developed using these KKs, in the literature. Because some examples might not be adequately represented by IFs and PFs, as was already mentioned, appropriate results from their respective algorithms might not be obtained. A portion of the KKs obtained with FFSs are the KKs acquired with IFs and PFs. Therefore, compared to existing ones, the proposed KK is more general and better adapted to dealing with actual scenarios.

A statistical technique called correlation exposes the connection between two components. The main effect of a correlation is a KK. The goal of this study is to describe a KK for FFSs. In this study, the KKFFS was added to the list of restrictions that included $MD + ND \leq 1$ for IF and $MD^2 + ND^2 \leq 1$ for PF. In cases where the KKs already present in the IF and PF structures fail, the numerical example that shows that the provided KK may easily operate is applicable. Each FFS has an MD and an ND based on the items in the sample space, and this is one of its fundamental characteristics. Therefore, the link between FFSs has unique characteristics. The results from several researchers' earlier investigations are also analyzed, along with the KKs of FFSs according to both MDs and NDs. The example instances show how the KK in the FFS framework may be used to solve real-world issues with ease. Computed results show the advantages of KKs as described in the FFS as follows: The results produced with the suggested KKs are more accurate. As a result, computing costs are reduced, and the results are more useful in practical situations.

References

1. Akram, M., Shahzadi, G., Ahmadini, A.A.H.: Decision-making framework for an effective sanitizer to reduce COVID-19 under Fermatean fuzzy environment. J. Math. (2020). Article ID 3263407. https://doi.org/10.1155/2020/3263407
2. Akram, M., Ali, G., Alcantud, J.C.R., Riaz, A.: Group decision-making with Fermatean fuzzy soft expert knowledge. Artif. Intell. Rev. (2022). https://doi.org/10.1007/s10462-021-10119-8
3. Akram, M., Amjad, U., Alcantud, J.C.R., Santos-Garcia, G.: Complex Fermatean fuzzy N-soft sets: A new hybrid model with applications. J. Ambient Intell. Human. Comput. (2022). https://doi.org/10.1007/s12652-021-03629-4
4. Ashraf, S., Naeem, M., Khan, A., Rehman, N., Pandit, M.K.: Novel Information Measures for Fermatean Fuzzy Sets and Their Applications to Pattern Recognition and Medical Diagnosis, Volume 2023, Article ID 9273239. https://doi.org/10.1155/2023/9273239
5. Atanassov, K.: Intuitionistic fuzzy sets. Fuzzy Sets Syst. **20**, 87–96 (1986)
6. Bhatia, M., Arora, H.D., Naithani, A.: Some new correlation coefficient measures based on Fermatean fuzzy sets using decision making approach in pattern analysis and supplier selection. Int. J. Math. Eng. Manag. Sci. **8**(2), 245–263 (2023). https://doi.org/10.33889/IJMEMS.2023.8.2.015
7. Bustince, H., Burillo, P.: Correlation of interval-valued intuitionistic fuzzy sets. Fuzzy Sets Syst. **74**, 237–244 (1995)
8. Chiang, D.A., Lin, N.P.: Correlation of fuzzy sets. Fuzzy Sets Syst. **102**(2), 221–226 (1999)
9. Demir, I.: Novel correlation coefficients for interval-valued Fermatean hesitant fuzzy sets with pattern recognition application. Turkish J. Math. **47**(1), 213–233 (2023). https://doi.org/10.55730/1300-0098.3355
10. Ejegwa, P.A., Wen, S., Feng, Y., Zhang, W., Liu, J.: A three-way Pythagorean fuzzy correlation coefficient approach and its applications in deciding some real-life problems. Appl. Intell. **53**, 226–237 (2023). https://doi.org/10.1007/s10489-022-03415-5
11. Garg, H.: A novel correlation coefficients between Pythagorean fuzzy sets and its applications to the decision-making process. Int. J. Intell. Syst. **31**, 1234–1252, (2016). https://doi.org/10.1002/int.21827
12. Garg, H. (ed.): q-Rung Orthopair Fuzzy Sets: Theory and Applications. Springer, Singapore (2022). https://doi.org/10.1007/978-981-19-1449-2
13. Garg, H., Shahzadi, G., Akram, M.: Decision-making analysis based on Fermatean fuzzy Yager aggregation operators with application in COVID-19 testing facility. Math. Probl. Eng. **2020**, Article ID 7279027. https://doi.org/10.1155/2020/7279027
14. Hong, D.H.: A note on correlation of interval-valued intuitionistic fuzzy sets. Fuzzy Sets Syst. **95**, 113–117 (1998)
15. Kirişci, M.: Parameter reduction method for Pythagorean fuzzy soft sets. Univ. J. Math. Appl. **3**(1), 33–37 (2020). https://doi.org/10.32323/ujma.615166

16. Kirişci, M.: Ω-soft sets and medical decision-making application. Int. J. Comput. Math. **98**(4), 690–704 (2021). https://doi.org/10.1080/00207160.2020.1777404
17. Kirişci, M.: Group-based Pythagorean fuzzy soft sets with medical decision-making applications. J. Exp. Theor. Artif. Intell. **36**(1), 27–45 (2024). https://doi.org/10.1080/0952813X.2022.2079006
18. Kirişci, M.: Correlation coefficients of fermatean fuzzy sets with a medical application. J. Math. Sci. Model. **5**(1), 16–23 (2022). https://doi.org/10.33187/jmsm.1039613
19. Kirişci, K.: Fermatean fuzzy type statistical concepts with medical decision-making application. Fuzzy Optim. Model. J. **4**(1), 1–14 (2023). https://doi.org/10.30495/FOMJ.2023.1982256.1082
20. Kirişci, K: Data analysis for panoramic X-ray selection: Fermatean fuzzy type correlation coefficients approach. Eng. Appl. Artif. Intell. **126**, 106824 (2023)
21. Kirişci, M., Şimşek, N.: Decision-making method related to Pythagorean fuzzy soft sets with infectious diseases application, J. King. Saud. Univ. Comput. Inf. Sci. **34**(8), 5968–5978 (2022). https://doi.org/10.1016/j.jksuci.2021.08.010
22. Kirişci, M., Demir, I., Simsek, N.: Fermatean fuzzy ELECTRE multi-criteria group decision-making and most suitable biomedical material selection. Artif. Intell. Med. **127**, 102278 (2022). https://doi.org/10.1016/j.artmed.2022.102278
23. Lin, M.W., Huang, C., Xu, Z.S.: TOPSIS method based on correlation coefficient and entropy measure for linguistic Pythagorean fuzzy sets and its application to multiple attribute decision making. Complexity, 6967390 (2019). https://doi.org/10.1155/2019/6967390
24. Liu, S.T., Kao, C.: Fuzzy measures for correlation coefficient of fuzzy numbers. Fuzzy Sets Syst. **128**(2), 267–275 (2002)
25. Mitchell, H.B.: A correlation coefficient for intuitionistic fuzzy sets. Int. J. Intell. Syst. **19**, 483–490 (2004)
26. Senapati, T., Yager, R.R.: Some new operations over Fermatean fuzzy numbers and application of Fermatean fuzzy WPM in multiple criteria decision making. Informatica **30**(2), 391–412 (2019). https://doi.org/10.15388/Informatica.2019.211
27. Senapati, T., Yager, R.R.: Fermatean fuzzy weighted averaging/geometric operators and its application in multi-criteria decision-making methods. Eng. Appl. Artif. Intell. **85**, 112–121 (2019). https://doi.org/10.1016/j.engappai.2019.05.012
28. Senapati, T., Yager, R.R.: Fermatean fuzzy sets. J. Ambient Intell. Human. Comput. **11**, 663–674 (2020). https://doi.org/10.1007/s12652-019-01377-0
29. Shahzadi, G., Akram, M.: Group decision-making for the selection of an antivirus mask under Fermatean fuzzy soft information. J. Intell. Fuzzy Syst. **40**(1), 1401–1416 (2021). https://doi.org/10.3233/JIFS-201760
30. Szmidt, E., Kacprzyk, J.: Distances between intuitionistic fuzzy sets. Fuzzy Sets Syst. **114**, 505–518 (2000)
31. Wei, G.W., Wang, H.J., Lin, R.: Application of correlation coefficient to interval-valued intuitionistic fuzzy multiple attribute decision-making with incomplete weight information. Knowl. Inf. Syst. **26**(2), 337–349 (2011). https://doi.org/10.1007/s10115-009-0276-1
32. Yager, R.R.: Pythagorean fuzzy subsets. In: Proc. Joint IFSA World Congress and NAFIPS Annual Meeting, Edmonton, Canada (2013)
33. Yager, R.R.: Pythagorean membership grade in multicriteria decision making. IEEE Fuzzy Syst. **22**, 958–965 (2014). https://doi.org/10.1109/TFUZZ.2013.2278989
34. Yager, R.R., Abbasov, A.M.: Pythagorean membership grades, complex numbers, and decision-making. Int. J. Intell. Syst. **28**, 436–452 (2013). https://doi.org/10.1002/int.21584
35. Zadeh, L.A.: Fuzzy sets. Inf. Comput. **8**, 338–353 (1965)
36. Zeng, W., Li, H.: Correlation coefficient of intuitionistic fuzzy sets. J. Ind. Eng. Int. **3**, 33–40 (2007)
37. Zhang, X.L., Xu, Z.S.: Extension of TOPSIS to multi-criteria decision-making with Pythagorean fuzzy sets. Int. J. Intell. Syst. **29**, 1061–1078 (2014). https://doi.org/10.1002/int.21676

On Some Gaussian Oresme Numbers

Serpil Halici and Elifcan Sayin

Abstract Sequences with recurrence relations are used in many branches of science such as mathematics, physics and engineering. The most well-known and studied sequence among these is Horadam sequence. This sequence is a generalization of many sequences and has an important place in the literature. New sequences can also be obtained from this sequence by changing the initial conditions. The most common sequences obtained by this method in the literature are Fibonacci sequence, Lucas sequence and Jacobstal sequence. The Fibonacci sequence, which has an important place among these sequences, has been studied by many authors. Horadam, who also worked on Fibonacci sequences, defined these sequences in complex space and discussed Gaussian Fibonacci sequences. A new sequence, which was defined by Nicole Oresme in the fourteenth century and called the Oresme sequence, is a special case of the Horadam sequence whose initial conditions are rational numbers. Subsequently, this was reviewed by many authors. Based on these studies, Gaussian Oresme number sequences were studied.

In this study, we obtained some identities of Gaussian Oresme sequences. We have obtained some sum formulas of Gaussian Oresme numbers. We obtained the matrix for the Gaussian Oresme sequences and calculated the nth power.

1 Introduction

The Horadam sequence is actually a generalization of many integer sequences. For $n \geq 0$ and the initial values $a = w_0$, $b = w_1$ this sequence is determined by the following equation [4].

$$w_n = pw_{n-1} - qw_{n-2}. \tag{1}$$

S. Halici · E. Sayin (✉)
Department of Mathematics, Pamukkale University, Denizli, Turkey
e-mail: shalici@pau.edu.tr

V. Gayoso Martínez et al. (eds.), *Mathematical Methods for Engineering Applications*, Springer Proceedings in Mathematics & Statistics 439,
https://doi.org/10.1007/978-3-031-49218-1_25

Two of the sequences that have been studied frequently recently are the Lucas sequence $L_n = w_n(2, 1; 1, -1)$ and the Jacobstal sequence $J_n = w_n(0, 1; 1, -2)$. The Horadam sequence, which was defined and studied extensively by Horadam, has also been discussed and studied in different spaces. One of these studies was handled in complex space and it was called Gaussian number sequences. The following iterative relation using Fibonacci numbers was given by Pethe and Horadam in 1986 [8].

$$GF_{n+1} = F_{n+1} + iF_n, GF_0 = 0, GF_1 = 1. \tag{2}$$

For some basic and important studies on Gaussian sequences the references [6, 8] can be consulted. Nicole Oresme was the first author to define and work on the Horadam sequence on rational numbers. The elements of this sequence, which was later called the Oresme sequence, is given as follows [7].

$$\left\{0, \frac{1}{2}, \frac{1}{2}, \frac{3}{8}, \frac{1}{4}, \dots, \frac{n}{2^n}\right\}$$

Horadam [2] made a study on the generalization of Oresme sequences in 1974. For $n \geq 0$, the author defined $O_n(0, \frac{1}{2}; 1, \frac{1}{4})$. Also this sequence can be written as

$$O_{n+2} = O_{n+1} - \frac{1}{4}O_n, O_n = \frac{n}{2^n}. \tag{3}$$

The relationship between the Oresme sequence and the Horadam sequence is given by the equation below [4].

$$O_n = \frac{1}{2}U_{n-1}. \tag{4}$$

Two important sequences studied in detail by Lucas are the generalized Fibonacci sequences $U_n = w_n(1, p; p, q)$ and the generalized Lucas sequences $V_n = w_n(2, p; p, q)$. Some properties have been defined for Oresme numbers. We see them in [5].

$$O_{n+3} = \frac{3}{4}O_{n+1} - \frac{1}{4}O_n, \tag{5}$$

$$O_{n+3} = \frac{3}{4}O_{n+2} - \frac{1}{16}O_n, \tag{6}$$

$$\sum_{j=0}^{n} O_j = 4\left(\frac{1}{2} - O_{n+2}\right). \tag{7}$$

Table 1 Gaussian Oresme numbers

n	0	1	2	3	4	...
GO_n	$-2i$	$\frac{1}{2^1}$	$\frac{1}{2^2}(2+2i)$	$\frac{1}{2^3}(3+4i)$	$\frac{1}{2^4}(4+6i)$	...

The generalization of these numbers has been studied by Cook. He defined these numbers for $k \geq 2$ and $n \geq 0$ as follows [2].

$$O_{n+2}^{(k)} = O_{n+1}^{(k)} - \frac{1}{k^2} O_n^{(k)},\ O_0^{(k)} = 0,\ O_1^{(k)} = \frac{1}{k}. \tag{8}$$

Here, when we take $k = 2$, it turns into classical Oresme numbers. The below are some important properties for k-Oresme numbers (see [2, 5]).

$$O_{n+2}^{(k)} = \left(\frac{k^2-1}{k^2}\right) O_{n+1}^{(k)} - \frac{1}{k^4} O_{n-1}^{(k)},\ n \geq 1, \tag{9}$$

$$\sum_{j=0}^{n} O_j^{(k)} = k^2 \left(\frac{1}{k} - O_{n+2}^{(k)}\right). \tag{10}$$

The Oresme numbers have been interesting and studied by many authors (see [1, 3, 9]). The authors defined Gaussian Oresme numbers using the definition of Oresme numbers [10]. The Gaussian Oresme numbers are as follows.

$$GO_n = O_n + i\,O_{n-1},\ n \geq 1. \tag{11}$$

where $GO_0 = -2i, GO_1 = \frac{1}{2}$. The recurrence relation of these numbers, the Binet formula and generating function are as follows.

$$GO_{n+2} = GO_{n+1} - \frac{1}{4} GO_n, \tag{12}$$

$$GO_n = \frac{1}{2^n}\left[n + 2i\,(n-1)\right], \tag{13}$$

$$\sum_{k=0}^{\infty} GO_n x^n = \frac{2x}{(x-2)^2}. \tag{14}$$

Let us give the elements of the sequence $\{GO_n\}$ with Table 1 for later use [10].

2 On Some Gaussian Oresme Numbers

Corollary 1 *Let's denote the Gaussian Oresme numbers by* $\{GO_n\}$ *and the arithmetic operations of these numbers are as follows. Here* GO_x *and* GO_y.

$$(i) GO_x + GO_y = \frac{1}{2^{x+y}}\left[2^y\left[x + 2i\left(x-1\right)\right] + 2^x\left[y + 2i\left(y-1\right)\right]\right], \tag{15}$$

$$(ii) GO_x - GO_y = \frac{1}{2^{x+y}}\left[2^y\left[x + 2i\left(x-1\right)\right] - 2^x\left[y + 2i\left(y-1\right)\right]\right], \tag{16}$$

$$(iii) GO_x GO_y = \frac{1}{2^{x+y}}\left[\left(4x + 4y - 3xy - 4\right) + 2i\left(2xy - x - y\right)\right]. \tag{17}$$

If we substitute Eq. (4) for Eq. (11), the following definition is obtained.

Definition 1 For $n \geq 2$, the equation,

$$GO_n = \frac{1}{2}\left(U_{n-1} + iU_{n-2}\right) \tag{18}$$

is obtained.

In the next theorem, we give the Vajda identity.

Theorem 1 *The Vajda identity is given for* $n, r, s \in Z^+$.

$$GO_{n+r}GO_{n+s} - GO_nGO_{n+r+s} = -\frac{1}{2^{2n+r+s}}rs\left(3 - 4i\right). \tag{19}$$

Proof Let us substitute the Binet formula of the Oresme numbers in Eq. (13) in the equations GO_{n+r}, GO_{n+s}, GO_n and GO_{n+r+s} . Here,

$$GO_{n+r} = \frac{1}{2^{n+r}}\left[\left(n+r\right) + 2i\left(n+r-1\right)\right],$$

$$GO_{n+s} = \frac{1}{2^{n+s}}\left[\left(n+s\right) + 2i\left(n+s-1\right)\right],$$

$$GO_{n+r+s} = \frac{1}{2^{n+r+s}}\left[\left(n+r+s\right) + 2i\left(n+r+s-1\right)\right].$$

If we substitute it in the equation $GO_{n+r}GO_{n+s} - GO_nGO_{n+r+s}$.

$$LHS = \frac{1}{2^{2n+r+s}}rs(4i - 3).$$

Example 1 Let us see. For $n = 1, r = 2$ and $s = 3$, then we have

$$RHS = GO_3GO_4 - GO_1GO_6 = \frac{24i - 18}{128}$$

and

$$LHS = \frac{1}{2^{2+2+3}}6(4i - 3) = \frac{24i - 18}{128}.$$

In the next theorem, we give the Honsberger's identity.

Theorem 2 *For $n, m > 1$, Honsberger's identity is*

$$GO_{m-1}GO_n + GO_mGO_{n+1} = \frac{1}{2^{n+m+1}}\left[A + iB\right]. \tag{20}$$

Where

$$A = -4n^2 + 32n + 17m - 11nm - 32$$

and

$$B = 2n^2 - 24n - 8m + 18nm + 6.$$

Proof Let us substitute the Binet formula of the $\{GO_n\}$ sequence in the equations GO_{m-1}, GO_n, GO_m and GO_{n+1}. Here from,

$$GO_n = \frac{1}{2^n}\left[n + 2i(n-1)\right],$$

$$GO_{n+1} = \frac{1}{2^{n+1}}\left[(n+1) + 2in\right],$$

$$GO_{m-1} = \frac{1}{2^{m-1}}\left[(m-1) + 2i(m-2)\right],$$

and

$$GO_m = \frac{1}{2^m}\left[m + 2i(m-1)\right]$$

it is obtained. If we substitute these equations in (20), we get

$$LHS = \frac{1}{2^{n+m-1}} [(m-1) + 2i(m-2)] [n + 2i(n-1)] + \frac{1}{2^{n+m+1}} [m + 2i(m-1)] [(n+1) + 2in],$$

$$LHS = \frac{1}{2^{n+m-1}} [4inm - 3nm + 7n - 6in - 2im + 2i + 4m - 8] + \frac{1}{2^{n+m+1}} \left[nm + m + 2inm + 2in^2 - 2i - 4n^2 + 4n\right].$$

If necessary operations are performed,

$$LHS = \frac{1}{2^{n+m+1}} [18inm - 11nm + 32n - 24in - 8im + 6i + 17m - 32 + 2in^2 - 4n^2],$$

$$LHS = \frac{1}{2^{n+m+1}} \left[(-4n^2 + 32n + 17m - 11nm - 32) + i(2n^2 - 24n - 8m + 18nm + 6)\right]$$

is obtained. Thus, the proof is completed.

Example 2 Let us see. For $n = 2$, $m = 2$, then we have

$$LHS = \frac{1}{2^5} [(64 + 34 - 44 - 32 - 16) + i(27 - 48 - 16 + 6 + 8)] = \frac{1}{32}(6 + 22i)$$

and

$$RHS = GO_1GO_2 + GO_2GO_3 = \frac{1}{8}(2 + 2i) + \frac{1}{32}(14i - 2) = \frac{1}{32}(6 + 22i).$$

In the following theorem, we have given the finite sum formula of the Gaussian Oresme sequence.

Theorem 3 *The finite sum formula of the sequence $\{GO_n\}$ for $k \geq 1$ is as follows.*

$$\sum_{n=0}^{k} GO_n = \frac{1}{2^k} \left[(2^{k+1} - (k+2)) - 2i(k+1)\right]. \tag{21}$$

Proof For $k = 1$, then we give

$$GO_0 + GO_1 = \frac{1}{2} \left[(2^2 - 3) - 2i(1+1)\right] = \frac{-4i + 1}{2}.$$

For $k+1$, let's see that it's true.

$$\sum_{n=0}^{k+1} GO_n = GO_0 + GO_1 + GO_2 + GO_3 + \cdots + GO_k + GO_{k+1},$$

$$\sum_{n=0}^{k+1} GO_n = \sum_{n=0}^{k} GO_n + GO_{k+1}.$$

Here, let's substitute the Binet formula for GO_{k+1}.

$$LHS = \frac{1}{2^k}\left[(2^{k+1} - (k+2)) + -2i(k+1)\right] + \frac{1}{2^{k+1}}[(k+1) + 2ik],$$

$$LHS = \frac{1}{2^{k+1}}[(2^{(k+2)} - (k+3)) - 2i(k+2))].$$

Thus, the proof is completed.

In the following theorem, we have given the sum formula of the even terms of the Gaussian Oresme sequence.

Theorem 4 *The even sum formula of the sequence $\{GO_{2n}\}$ for $k \geq 3$ is as follows.*

$$\sum_{n=0}^{k} GO_{2n} = \frac{1}{2^{2k}}\left[(2^{2k} - 2k - 4(k-2)3^{k-3}) - i(2^{2k} - 4(k-2)3^{k-3} + 2)\right]. \tag{22}$$

Proof For $k = 3$, then we give

$$GO_0 + GO_2 + GO_4 + GO_6 = \frac{1}{64}\left[(2^6 - 6 - 4) - i(2^6 - -4 + 2)\right] = \frac{54i - 62}{64}.$$

For $k+1$, let's see that it's true.

$$\sum_{n=0}^{k+1} GO_{2n} = GO_0 + GO_2 + GO_4 + GO_6 + \cdots + GO_{2k} + GO_{2k+2},$$

$$\sum_{n=0}^{k+1} GO_{2n} = \sum_{n=0}^{k} GO_{2n} + GO_{2k+2}.$$

Here, let's substitute the Binet formula for GO_{2k+2}.

$$LHS = \frac{1}{2^{2k}}\left[(2^{2k} - 2k - 4(k-2)3^{k-3}) - i(2^{2k} - 4(k-2)3^{k-3} + 2)\right] + \frac{1}{2^{2k+2}}\left[(2k+2) + 2i\,(2k+1)\right].$$

Thus, the proof is completed.

Example 3

$$\sum_{n=0}^{4} GO_{2n} = GO_0 + GO_2 + GO_4 + GO_6 + GO_8 = \frac{224 - 234i}{256},$$

$$LHS = \frac{1}{2^8}\left[(2^8 - 8 - 4(4-2)3^{4-3}) - i(2^8 - 4(4-2)3^{4-3} + 2)\right] = \frac{224 - 234i}{256}.$$

Now, we have given the matrix of Gaussian Oresme numbers.

Definition 2 Let be

$$O = \begin{bmatrix} 1 & -\frac{1}{4} & 0 \\ 1 & 0 & 0 \\ 0 & 0 & 1 \end{bmatrix} \tag{23}$$

the matrix of the Oresme numbers and

$$G = \begin{bmatrix} GO_2 \\ GO_1 \\ GO_0 \end{bmatrix} \tag{24}$$

the matrix of the Gaussian Oresme numbers. In this case, the product of the matrices is as follows.

$$\begin{bmatrix} 1 & -\frac{1}{4} & 0 \\ 1 & 0 & 0 \\ 0 & 0 & 1 \end{bmatrix}\begin{bmatrix} GO_2 \\ GO_1 \\ GO_0 \end{bmatrix} = \begin{bmatrix} GO_2 - \frac{1}{4}GO_1 \\ GO_2 \\ GO_1 \end{bmatrix} = \begin{bmatrix} GO_3 \\ GO_2 \\ GO_1 \end{bmatrix}.$$

For $n > 1$, it is seen by induction method for the following equation.

$$\begin{bmatrix} GO_{n+2} \\ GO_{n+1} \\ GO_n \end{bmatrix} = \begin{bmatrix} 1 & -\frac{1}{4} & 0 \\ 1 & 0 & 0 \\ 0 & 0 & 1 \end{bmatrix}^n \begin{bmatrix} GO_2 \\ GO_1 \\ GO_0 \end{bmatrix}. \tag{25}$$

Example 4 Let us see for $n = 2$. It is equal to

$$\begin{bmatrix} GO_4 \\ GO_3 \\ GO_2 \end{bmatrix} = \begin{bmatrix} 1 & -\frac{1}{4} & 0 \\ 1 & 0 & 0 \\ 0 & 0 & 1 \end{bmatrix}^2 \begin{bmatrix} GO_2 \\ GO_1 \\ GO_0 \end{bmatrix},$$

$$\begin{bmatrix} GO_4 \\ GO_3 \\ GO_2 \end{bmatrix} = \begin{bmatrix} \frac{3}{4} & -\frac{1}{4} & 0 \\ 1 & \frac{1}{4} & 0 \\ 1 & 0 & 0 \end{bmatrix} \begin{bmatrix} GO_2 \\ GO_1 \\ GO_0 \end{bmatrix},$$

$$GO_4 = \frac{3}{4}GO_2 - \frac{1}{4}GO_1 = \frac{4+6i}{16}.$$

In the following theorem, we have given the generalization of these matrices.

Theorem 5 *The following matrix equation is written for n* ≥ 0.

$$\begin{bmatrix} GO_{2n+2} \\ GO_{2n+1} \\ GO_{2n} \end{bmatrix} = \begin{bmatrix} 1 & -\frac{1}{4} & 0 \\ 1 & 0 & 0 \\ 0 & 0 & 1 \end{bmatrix}^n \begin{bmatrix} GO_{n+2} \\ GO_{n+1} \\ GO_n \end{bmatrix}. \tag{26}$$

Proof It is true for $n = 1$. Let us see that it is true for $n + 1$.

$$\begin{bmatrix} GO_{2n+4} \\ GO_{2n+3} \\ GO_{2n+2} \end{bmatrix} = \begin{bmatrix} 1 & -\frac{1}{4} & 0 \\ 1 & 0 & 0 \\ 0 & 0 & 1 \end{bmatrix}^{n+1} \begin{bmatrix} GO_{n+3} \\ GO_{n+2} \\ GO_{n+1} \end{bmatrix},$$

it is easily seen that it is.

3 Conclusion

In this study, we performed arithmetic operations for Gaussian Oresme sequences. We have obtained some identities of the Gaussian Oresme sequences. We have given the finite sum formula and even sums. We also defined the matrix of the Gaussian Oresme sequence and calculated its nth power. We got a different equation for this matrix.

References

1. Cerda-Morales, G.: Oresme polynomials and their derivatives. arXiv preprint arXiv:1904.01165 (2019)
2. Cook, C.K.: Some sums related to sums of Oresme numbers. In: Applications of Fibonacci Numbers: Volume 9: Proceedings of The Tenth International Research Conference on Fibonacci Numbers and Their Applications, pp. 87–99. Springer Netherlands, Dordrecht (2004)
3. Dos Santos Mangueira, M.C., Vieira, R.P.M., Alves, F.R.V., Catarino, P.M.M.C.: Corrigendum to The Oresme sequence: The generalization of its matrix form and its. Discrete Math. **27**(1), 101–111 (2021)
4. Horadam, A.F.: Basic properties of a certain generalized sequence of numbers. Fibonacci Quart. **3**(3), 161–176 (1965)
5. Horadam, A.F.: Oresme numbers. Fibonacci Quart. **12**(3), 267–271 (1974)
6. Kara, N., Yilmaz, F.: On hybrid numbers with Gaussian Leonardo coefficients. Mathematics **11**(6), 1551 (2023)
7. Oresme, N.: In: Busard, H.L.L. (ed.) Quaestiones super geometriam Euclidis, 2 vols. Brill, Leiden (1961)
8. Pethe, S., Horadam, A.F.: Generalised Gaussian Fibonacci numbers. Bull. Aust. Math. Soc. **33**(1), 37–48 (1986)
9. Sentürk, G.Y., Gurses, N., Yuce, S.: A New look on Oresme numbers: dual-generalized complex component extension. In: Conference Proceeding Science and Technology, vol. 1, no. 1, pp. 254–265 (2018)
10. Yilmaz, F., Ertas, A.: On quaternions with Gaussian Oresme coefficients. Turk. J. Math. Comput. Sci. **15**(1), 192–202 (2023)